Mathematical Modeling
for Business Analytics

TEXTBOOKS in MATHEMATICS

Series Editors: Al Boggess and Ken Rosen

PUBLISHED TITLES

ABSTRACT ALGEBRA: A GENTLE INTRODUCTION
Gary L. Mullen and James A. Sellers

ABSTRACT ALGEBRA: AN INTERACTIVE APPROACH, SECOND EDITION
William Paulsen

ABSTRACT ALGEBRA: AN INQUIRY-BASED APPROACH
Jonathan K. Hodge, Steven Schlicker, and Ted Sundstrom

ADVANCED LINEAR ALGEBRA
Hugo Woerdeman

ADVANCED LINEAR ALGEBRA
Nicholas Loehr

ADVANCED LINEAR ALGEBRA, SECOND EDITION
Bruce Cooperstein

APPLIED ABSTRACT ALGEBRA WITH MAPLE™ AND MATLAB®, THIRD EDITION
Richard Klima, Neil Sigmon, and Ernest Stitzinger

APPLIED DIFFERENTIAL EQUATIONS: THE PRIMARY COURSE
Vladimir Dobrushkin

APPLIED DIFFERENTIAL EQUATIONS WITH BOUNDARY VALUE PROBLEMS
Vladimir Dobrushkin

APPLIED FUNCTIONAL ANALYSIS, THIRD EDITION
J. Tinsley Oden and Leszek Demkowicz

A BRIDGE TO HIGHER MATHEMATICS
Valentin Deaconu and Donald C. Pfaff

COMPUTATIONAL MATHEMATICS: MODELS, METHODS, AND ANALYSIS WITH MATLAB® AND MPI,
SECOND EDITION
Robert E. White

A CONCRETE INTRODUCTION TO REAL ANALYSIS, SECOND EDITION
Robert Carlson

A COURSE IN DIFFERENTIAL EQUATIONS WITH BOUNDARY VALUE PROBLEMS, SECOND EDITION
Stephen A. Wirkus, Randall J. Swift, and Ryan Szypowski

A COURSE IN ORDINARY DIFFERENTIAL EQUATIONS, SECOND EDITION
Stephen A. Wirkus and Randall J. Swift

PUBLISHED TITLES CONTINUED

A TOUR THROUGH GRAPH THEORY
Karin R. Saoub

TRANSITION TO ANALYSIS WITH PROOF
Steven G. Krantz

TRANSFORMATIONAL PLANE GEOMETRY
Ronald N. Umble and Zhigang Han

Mathematical Modeling
for Business Analytics

by
William P. Fox

CRC Press
Taylor & Francis Group
Boca Raton London New York

CRC Press is an imprint of the
Taylor & Francis Group, an **informa** business

CRC Press
Taylor & Francis Group
6000 Broken Sound Parkway NW, Suite 300
Boca Raton, FL 33487-2742

First issued in paperback 2022

ISBN 13: 978-1-03-247640-7 (pbk)
ISBN 13: 978-1-138-55661-4 (hbk)

DOI: 10.1201/9781315150208

Library of Congress Cataloging-in-Publication Data

Names: Fox, William P., 1949- author.
Title: Mathematical modeling for business analytics / William P. Fox.
Description: Boca Raton, FL : CRC Press, 2018.
Identifiers: LCCN 2017034022 | ISBN 9781138556614
Subjects: LCSH: Decision making--Mathematical models. |
Management--Mathematical models.
Classification: LCC HD30.23 .F7325 2018 | DDC 658.4/033--dc23
LC record available at https://lccn.loc.gov/2017034022

**Visit the Taylor & Francis Web site at
http://www.taylorandfrancis.com**

**and the CRC Press Web site at
http://www.crcpress.com**

This book is dedicated to my wife, Hamilton Dix-Fox,

who encouraged me to write this book.

Contents

Preface

Addressing the Current Needs

In recent years of teaching mathematical modeling for decision-making coupled with conducting applied mathematical modeling research, I have found that (a) decision-makers at all levels must be exposed to the tools and techniques that are available to help them in the decision-making process, (b) decision-makers and analysts need to have and use technology to assist in the analysis process, and (c) the interpretation and explanation of the results are crucial to understanding the strengths and limitations of modeling. With this in mind this book emphasizes and focuses on the model formulation and modeling building skills that are required for decision analysis and the technology to support the analysis.

Audience

This book will be best used for a senior-level discrete modeling course in mathematics, operations research, or industrial engineering departments or graduate-level discrete choice modeling courses, or decision models courses offered in business schools offering business analytics. The book *would be* of interest to mathematics departments that offer mathematical modeling courses focused on discrete modeling.

The following groups would benefit from using this book:

- Undergraduate students who are involved in quantitative methods courses in business, operations research, industrial engineering, management sciences, industrial engineering, or applied mathematics.
- Graduate students in discrete mathematical modeling courses covering topics from business, operations research, industrial engineering, management sciences, industrial engineering, or applied mathematics.
- Junior analysts who want a comprehensive review of decision-making topics.
- Practitioners desiring a reference book.

Objective

The primary objective of this book is illustrative in nature. It seeks the tone in Chapter 1 through the introduction to mathematical modeling. In this chapter, we provide a process for formally thinking about the problem. We illustrate many scenarios and illustrative examples. In these examples, we begin the setup of the solution process. We also provide solutions in this chapter that we will present more in depth in later chapters.

We thought for years about which techniques should be included or excluded in this book. Finally, we decided on the main techniques that we cover in our three-course sequence in mathematical modeling for decision-making in the Department of Defense Analysis. We feel these subjects have served our students as well as they have gone on as leaders and decision-makers for our nation.

Organization

This book contains information that could easily be covered in a two-semester course or a one-semester overview of topics. This allows instructors the flexibility to pick and choose topics that are consistent with their course and consistent with the current needs.

In Chapters 2 through 8, we present materials to solve the typical problems introduced in Chapter 1. The contexts of these problems are in business, industry, and government (BIG). Thus, the problems visited are BIG problems.

In Chapter 2, we present a decision theory. We discuss decision under uncertainty and risk.

We present the process of using decision trees to draw out the problem and then use the expected value to solve for the *best* decision.

Chapter 3 has developed into a mathematical programming covering techniques and applications in technology to solve linear, integer, and nonlinear optimization problems. Among the problems illustrated are a supply chain operation, a recruiting office analysis, emergency service planning, optimal path to transport hazardous materials, a minimum variance investments, and cable installation.

Chapter 4 covers the techniques to rank entities or alternatives when the decision-maker has criteria to be considered that impact the decision-making process. Discussions and illustrative problems are presented in data envelopment analysis using linear programming (LP), sum of additive weights, analytical hierarchy process (AHP), and technique for order of preference by similarity to ideal solution (TOPSIS). In our current modeling courses, we present these techniques to our students who are finding more and more applications in BIG.

Chapter 5 covers game theory when decisions are made versus an opponent. We assume rational players, simultaneous games with perfect knowledge in our presentation. We do cover both total conflict and partial conflict applications. Applications are presented that covers the following:

Battle of the Bismarck Sea, Penalty kicks in soccer, Batter–Pitcher Duel in baseball, Operation Overlord, Choosing the right Course of Action, Cuban Missile Crisis, 2007–2008 Writer's Guild Strike, Dark Money Network (DMN) Game, and Course of Actions Revisited.

In Chapter 6, we present an overview of regression techniques. We have found in our advising that students (future decision-makers) often use the wrong technique in the work. Here, we present simple linear regression, multiple linear regression, nonlinear regression, logistics regression, and Poisson regression. The approach here is when to use each type of regression based on the data available.

Chapter 7 presents discrete dynamical systems (DDS). We cover not only simple linear models but nonlinear as well as system of systems. We discuss stability and equilibrium values.

Illustrative example include the following: drug dosage, time value of money, simple mortgage, population growth, spread of a contagious disease, inventory systems analysis, competitive hunter model, Lanchester's combat models, discrete predator–prey model, and a systems model for disease known as the susceptible (S), infected (I), and resistant (R) (SIR) model.

Chapter 8 is a brief presentation of discrete simulation models. Often models cannot be constructed to adequately reflect the fidelity of the system of interest. In these cases, simulation models, especially Monte Carlo simulations, provide a look into the system that provides information to the decision-makers.

Chapter 9 presents an introduction into financial mathematics. The topics from engineering economic analysis such as rates of interest, depreciation, discounting, annual percentage rate (APR), compounding (discrete and continuous), net present value (NPV), bonds, annuities, and shrinking funds are presented and explained so that they can be easily understood and used. We also discuss examples using previously covered techniques such as multiattribute decision-making, dynamical systems, and mathematical programming that are applied directly to financial mathematics. In our research on institutions that provided interest, we found none that gave a continuous interest. They all gave interest at discrete intervals. At our local credit union, their sign says money deposited after 10 a.m. will be credited until after 10 a.m. the following day. So many examples from engineering economic analysis are derived in the discrete model using DDS. Examples for pensions with stock portfolios, financial planning, optimization of interest, and cash flow are discussed using the appropriate mathematical technique.

This book shows the power and limitations for mathematical modeling to solve real problems. The solutions shown might not be the best solution but they are certainly solutions that are or could be considered in the analysis. As evidenced by previous textbooks in mathematical modeling, such as a *First Course in Mathematical Modeling*, scenarios are revisited to illustrate alternative techniques in solving problems. As we have seen from many years of the Mathematical Contest in Modeling, student ingenuity and creativity in modeling methods and solution techniques are always present.

In this book, we cannot address every nuance in modeling real-world problems. What we can do is provide a sample of models and possible appropriate techniques to obtain useful results. We can establish a process to *do modeling*. We can illustrate many examples of modeling and illustrate a technique in order to solve the problem. In the techniques chapters, we must assume no background and spend a little time establishing the procedure before we return to providing examples.

The data used in this book are unclassified and often the real data are not displayed. Data similar to nature and design are used in the examples.

This book can apply to analysts to allow them to see the range and type of problems that fit into specific mathematical techniques understanding that we did address all the possible mathematics techniques. Some important techniques that we left out include differential equations.

This book is also applicable to decision-makers. It shows the decision-maker the wide range of applications of quantitative approaches to aid in the decision-making process. As we say in class every day, mathematics does not tell what to do but it does provide insights and allows critical thinking into the decision-making process. In our discussion, we consider the mathematical modeling process as a framework for decision-makers. For a decision-maker there are four key elements: (1) the formulation process, (2) the solution process, (3) interpretation of the mathematical answer in the context of the actual problem, and (4) sensitivity analysis. At every step along the way in the process the decision-maker should question procedures and techniques and ask for further explanations as well as assumptions used in the process. One major question could be, "Did you use an appropriate technique?" to obtain a solution and "Why were other techniques not considered or used?" Another question could be "Did you over simplify the process?" so much that the solution does not really apply in order or were the assumptions make fundamental to even being able to solve the problem?

We thank all the mathematical modeling students that we have had over this time as well as all the colleagues who have taught mathematical modeling with us during this adventure. I am especially appreciative of the mentorship of Frank R. Giordano over the past thirty-plus years.

William P. Fox
Naval Postgraduate School

Author

Dr. William P. Fox is currently a professor in the Department of Defense Analysis at the Naval Postgraduate School, Monterey, California and teaches a three-course sequence in mathematical modeling for decision-making. He earned his BS degree from the United States Military Academy at West Point, New York, MS in operations research from the Naval Postgraduate School, and PhD in industrial engineering from Clemson University, Clemson, South Carolina. He has a teaching experience of 12 years at the United States Military Academy until retiring for his active military service and he was the chair of mathematics for 8 years at the Francis Marion University, Florence, South Carolina. He has many publications and scholarly activities, including 16 books, 21 book chapters and technical reports, 150 journal articles, and more than 150 conference presentations and mathematical modeling workshops. He has directed several international mathematical modeling contests through the Consortium of Mathematics and Its Applications (COMAP): the HiMCM and the MCM. His interests include applied mathematics, optimization (linear and nonlinear), mathematical modeling, statistical models, models for decision-making in business, industry, medical, and government, and computer simulations. He is a member of the Institute for Operations Research and the Management Sciences (INFORMS), the Military Application Society of INFORMS, the Mathematical Association of America, and the Society for Industrial and Applied mathematics where he has held numerous positions.

1

Introduction to Mathematical Modeling for Business Analytics

OBJECTIVES

1. Understand the modeling process.
2. Know and use the steps in modeling.
3. Experience a wide variety of examples.

1.1 Introduction

Consider the importance of modeling for decision-making in business (B), industry (I), and government (G), BIG. BIG decision-making is essential for success at all levels. We do not encourage *shooting from the hip* or simply flipping a coin to make a decision. We recommend good analysis that enables the decision-maker to examine and question results to find the best alternative to choose or decision to make. This book presents, explains, and illustrates a modeling process and provides examples of decision-making analysis throughout.

Let us describe a mathematical model as a mathematical description of a system by using the language of mathematics. Why mathematical modeling? Mathematical modeling, business analytics, and operations research are all similar descriptions that represent the use of quantitative analysis to solve real problems. This process of developing such a mathematical model is termed as mathematical modeling. Mathematical models are used in the natural sciences (such as physics, biology, earth science, and meteorology), engineering disciplines (e.g., computer science, systems engineering, operations research, and industrial engineering), and in the social sciences (such as business, economics, psychology, sociology, political science, and social networks). The professionals in these areas use mathematical models all the time.

A mathematical model may be used to help explain a system, to study the effects of different components, and to make *predictions* about behavior (Giordano et al., 2014, pp. 58–60). So let us make a more formal definition of a mathematical model: a mathematical model is the application of mathematics to a real-world problem.

Mathematical models can take many forms, including but not limited to dynamical systems: statistical models, differential equations, optimization models, or game theoretic models. These and other types of models can overlap, of which one output becomes the input for another similar or different model forms. In many cases, the quality of a scientific field depends on how well the mathematical models developed on the theoretical side agree with the results of repeatable experiments (Giordano et al., 2014, pp. 58–60). Any lack of agreement between theoretical mathematical models and experimental measurements leads to model refinements and better models. We do not plan to cover all the mathematical modeling processes here. We only provide an overview to the decision-makers. Our goal is to offer *competent, confident problem solvers* for the twenty-first century. We suggest the books listed in the reference section to become familiar with many more modeling forms.

1.2 Background

1.2.1 Overview and Process of Mathematical Modeling

Bender (2000, pp. 1–8) first introduced a process for modeling. He highlighted the following: formulate the model, outline the model, ask if it is useful, and test the model. Others have expanded this simple outlined process. Giordano et al. (2014, p. 64) presented a six-step process: identify the problem to be solved, make assumptions, solve the model, verify the model, implement the model, and maintain the model. Myer (2004, pp. 13–15) suggested some guidelines for modeling, including formulation, mathematical manipulation, and evaluation. Meerschaert (1999) developed a five-step process: ask the question, select the modeling approach, formulate the model, solve the model, and answer the question. Albright (2010) subscribed mostly to concepts and process described in previous editions of Giordano et al. (2014). Fox (2012, pp. 21–22) suggested an eight-step approach: understand the problem or question, make simplifying assumptions, define all variables, construct the model, solve and interpret the model, verify the model, consider the model's strengths and weaknesses, and implement the model.

Most of these pioneers in modeling have suggested similar starts in understanding the problem or question to be answered and in making key assumptions to help enable the model to be built. We add the need for sensitivity analysis and model testing in this process to help ensure that we have a model that is performing correctly to answer the appropriate questions.

For example, student teams in the Mathematical Contest in Modeling were building models to determine the all-time best college sports coach. One team picked a coach who coached less than a year, went undefeated for the remaining part of the year, and won their bowl game. Thus, his season was a perfect season. Their algorithm picked this person as the all-time best coach. Sensitivity analysis and model testing could have shown the fallacy to their model.

Someplace between the defining of the variables and the assumptions, we begin to consider the model's form and technique that might be used to solve the model. The list of techniques is boundless in mathematics, and we will not list them here. Suffice it to say that it might be good to initially decide among the forms: deterministic or stochastic for the model, linear or nonlinear for the relationship of the variables, and continuous or discrete.

For example, consider the following scenarios:

Two observation posts that are 5.43 miles apart pick up a brief radio signal. The sensing devices were oriented at 110° and 119°, respectively, when a signal was detected. The devices are accurate to within 2° (that is ±2° of their respective angle of orientation). According to intelligence, the reading of the signal came from a region of active terrorist exchange, and it is inferred that there is a boat waiting for someone to pick up the terrorists. It is dusk, the weather is calm, and there are no currents. A small helicopter leaves a pad from Post 1 and is able to fly accurately along the 110° angle direction. This helicopter has only one detection device, a searchlight. At 200 ft, it can just illuminate a circular region with a radius of 25 ft. The helicopter can fly 225 miles in support of this mission due to its fuel capacity. Where do we search for the boat? How many search helicopters should you use to have a *good* chance of finding the target (Fox and Jaye, 2011, pp. 82–93)?

The writers of TV and movies decide that they are not receiving fair compensation for their work as their shows continue to be played on cable and DVDs. The writers decide to strike. Management refuses to budge. Can we analyze this to prevent this from reoccurring? Can we build a model to examine this?

Consider locating emergency response teams within a county or region. Can we model the location of ambulances to ensure that the maximum number of potential patients is covered by the emergency response teams? Can we find the minimum number of ambulances required?

You are a new manager of a bank. You set new goals for your tenure as manager. You analyze the current status of service to measure against your goals. Are you meeting demand? If not what can be done to improve service? You want to prevent catastrophic failure at your bank.

You have many alternatives to choose from for your venture. You have
certain decision criteria that you consider to sue to help in making
this future. Can we build a mathematical model to assist us in this
decision?

These are all events that we can model using mathematics. This chapter will
help a decision-maker understand what a mathematical modeler might do
for them as a confident problem solver using the techniques of mathematical
modeling. As a decision-maker, understanding the possibilities and asking
the key questions will enable better decisions to be made and will lower the
risks.

1.2.2 The Modeling Process

We introduce the process of modeling and examine many different scenarios
in which mathematical modeling can play a role.

The art of mathematical modeling is learned through experience of
building and solving models. Modelers must be creative, innovative,
inquisitive, and willing to try new techniques as well as being able to
refine their models, if necessary. A major step in the process is *passing the
common sense* test for use of the model.

In its basic form, modeling consists of three steps:

1. Make assumptions
2. Do some *math*
3. Derive and interpret conclusions

To that end, one cannot question the mathematics and its solution, but one
can always question the assumptions used.

To gain insight, we will consider one framework that will enable the
modeler to address the largest number of problems. The key is that there
is something *changing for which we want to know the effects and the results of
the effects*. The problem might involve any system under analysis. The real-
world system can be very simplistic or very complicated. This requires
both types of real-world systems to be modeled with the same logical step-
wise process.

Consider modeling an investment. Our first inclination is to use the equations about compound interest rates that we used in high school or college
algebra. The compound interest formula calculates the value of a compound
interest investment after "n" interest periods.

$$A = P(1-i)^n$$

where:

    A is the amount after n interest periods
    P is the principal, the amount invested at the start
    i is the interest rate applying to each period
    n is the number of interest periods

This is a continuous formula. Have you seen any banking institutions that give continuous interest? In our research, we have not. As a matter of fact at our local credit union, they have a sign that says, money deposited after 10 a.m. do not get credited until the night after the deposit. This makes discrete compound interest on the balance in a more compelling assumption.

A powerful paradigm that we use to model with discrete dynamical systems is as follows:

$$\text{Future value} = \text{present value} + \text{change}$$

The dynamical systems that we will study with this paradigm may differ in appearance and composition, but we will be able to solve a large class of these *seemingly* different dynamical systems with similar methods. In this chapter, we will use iteration and graphical methods to answer questions about the discrete dynamical systems.

We could use flow diagrams to help us see how the dependent variable changes. These flow diagrams help to see the paradigm and put it into mathematical terms. Let us consider financing a new Ford Mustang. The cost is $25,000 and you can put down $2,000, so you need to finance $23,000. The dealership offers you 2% financing over 72 months. Consider the change diagram, Figure 1.1, for financing the car that depicts this situation.

We use this change diagram to help build the discrete dynamical model. Let $A(n)$ be the amount owed after n months. Notice that the arrow pointing into the circle is the interest to the unpaid balance that increases your debt. The arrow pointing out of the circle is your monthly payment that decreases your debt. We define the following variables:

    $A(n + 1)$ is the amount owed in the future
    $A(n)$ is the amount currently owed

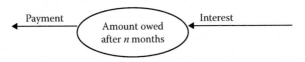

FIGURE 1.1
Flow diagram for buying a new car.

Change as depicted in the change diagram is $i\,A(n) - P$, so the model is

$$A(n+1) = A(n) + i\,A(n) - P$$

where:
  i is the monthly interest rate
  P is the monthly payment

We solve by iterating this equation in a spreadsheet. First, we assume that we deposit $1000 at 2.5% annual interest.

We will use the notation $A(n)$ to be the amount of money that we have in our account after n periods. Using the paradigm we have the initial model as

$$A(n+1) = A(n) + \text{change}$$

We model change with the interest rate information and how it is compounded. A change diagram is suggested for more complicated dynamical system models (Figure 1.2).

$$A(0) = \$1000$$

After 1 month, we have

$$1000 + (1000 * 0.025 / 12) = \$1002.08$$

If we want to see after 10 years or 120 months, we are better off using technology such as Excel. We will have $1281.02. We did not fare well, so perhaps we would seek other ways to invest and make this money grow.

Consider installing cable to link computers together in a new computer room. Our first inclination is to use the equations about distance that we used in high school mathematics class. These equations are very simplistic and ignore many factors that could impact the installation of wire lines such as location of terminals, ground or ceiling, and other factors. As we add more factors, we can improve the precision of the model. Adding these additional factors makes the model more realistic but possibly more complicated to produce and solve.

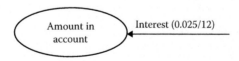

FIGURE 1.2
Change diagram for a savings account.

Figure 1.1 provides a closed loop process for modeling. Given a real-world situation similar to the one that is mentioned earlier, we collect data to formulate a mathematical model. This mathematical model can be one that we derive or select from a collection of already existing mathematical models. Then, we analyze the model that we used and reach mathematical conclusions about it. Next, we interpret the model and either makes predict about what has occurred or offer explanation as to why something has occurred. Finally, we test our conclusion about the real-world system with new data. We use sensitivity analysis of the parameters or inputs to see how they affect the model. We may refine or improve the model to improve its ability to predict or explain the phenomena. We might even go back and reformulate a new mathematical model.

1.2.3 Mathematical Modeling for Business Analytics as a Process

We will illustrate some mathematical models that describe change in the real world. We will solve some of these models and will analyze how good our resulting mathematical explanations and predictions are in context of the problem. The solution techniques that we employ take advantage of certain characteristics that the various models enjoy as realized through the formulation of the model.

When we observe change, we are often interested in understanding or explaining why or how a particular change occurs. Maybe we need or want to analyze the effects under different conditions or perhaps to predict what could happen in the future. Consider the firing of a weapon system or the shooting of a ball from a catapult as shown in Figure 1.3. Understanding how the system behaves in different environments under differing weather or operators, or predicting how well it hits the targets are all of interest. For the catapult, the critical elements of the ball, the tension, and angle of the firing arm are found as important elements (Fox, 2013b). For our purposes, we will consider a mathematical model to be a mathematical construct that

FIGURE 1.3
The catapult and balls.

is designed to study a particular real-world system or behavior (Giordano et al., 2014). The model allows us to use mathematical operations to reach mathematical conclusions about the model as illustrated in Figure 1.4. It is the arrow going from real-world system and observations to the mathematical model using the assumptions, variables, and formulations that are critical in the process.

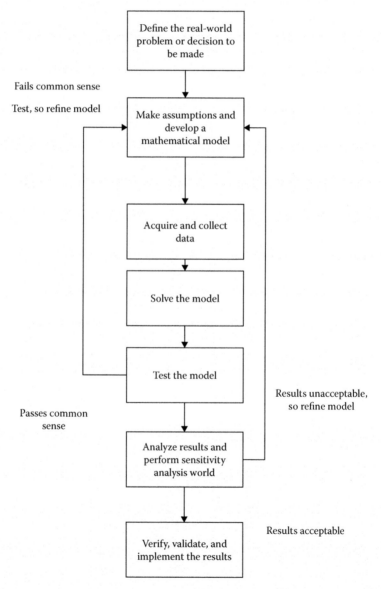

FIGURE 1.4
Modeling real-world systems with mathematics.

We define a system as a set of objects joined by some regular interaction or interdependence in order for the complete system to work together. Think of a larger business with many companies that work independently and interact together to make the business prosper. Other examples might include a bass and trout population living in a lake, a communication, cable TV, or weather satellite orbiting the earth, delivering Amazon Prime packages, U.S. postal service mail or packages, locations of emergency services or computer terminals, or large companies' online customer buying systems. The person modeling is interested in understanding how a system works, what causes change in a system, and the sensitivity of the system to change. Understanding all these elements will help in building an adequate model to replicate reality. The person modeling is also interested in predicting what changes might occur and when these changes might occur.

Figure 1.5 suggests how we can obtain real-world conclusions from a mathematical model. First, observations identify the factors that seem to be involved in the behavior of interest. Often we cannot consider, or even identify, all the relevant factors, so we make simplifying assumptions excluding some of those factors (Giordano et al., 2014). Next, we determine what data are available and what variables they represent. We might build or test tentative relationships among the remaining identified factors. For example, if we were modeling the weight of a fish in a fish contest and if we were collecting data on the length and girth of a fish, perhaps we would want to test the relationship of length and girth before we used both variables. This might give us a reasonable first cut at a model. We then solve the model and determine the reasonableness of the model's conclusions. Passing the *common sense* test is important. However, as these results apply only to the model, they may or may not apply to the actual real-world system in question. Simplifications

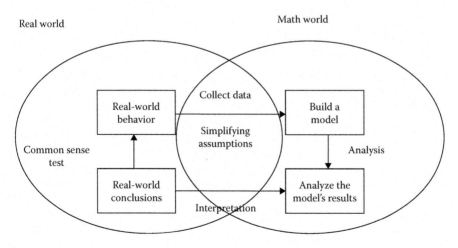

FIGURE 1.5
A closed system for modeling.

were made in constructing the model, and the observations on which the model is based invariably contain errors and limitations. Thus, we must carefully account for these anomalies and must test the conclusions of the model against real-world observations. If the model is reasonably valid, we can then draw inferences about the real-world behavior from the conclusions drawn from the model. In summary, the critical elements are the modeling assumptions and variables. The mathematical model used is usually not questionable but why we used it might be questionable. Therefore, we have the following procedure for investigating real-world behavior and for building a mathematical representation:

1. Observe the system and identify the factors and variables involved in the real-world behavior, possibly making simplifying assumptions as necessary.
2. Build initial relationships among the factors and variables.
3. Build the model and analyze the model's results.
4. Interpret the mathematical results both mathematically and in terms of the real-world system.
5. Test the model results and conclusions against real-world observations. Do the results and use of the model pass the common sense test? If not go back and remodel the system.

There are various kinds of models. A good mathematical modeler will build a library of models to recognize various real-world situations to which they apply. Most models simplify reality. Generally, models can only approximate real-world behavior. Next, let us summarize a multistep *process* for formulating a mathematical model.

1.2.4 Steps in Model Construction

An outline is presented as a procedure to help construct mathematical models. In the next section, we will illustrate this procedure with a few examples. We suggest a nine-step process.

These nine steps are summarized in Figure 1.6. These steps act as a guide for thinking about the problem and getting started in the modeling process. We choose these steps from the compilation of steps by other authors listed in additional readings and put them together in these nine steps.

We illustrate the process through an example. Consider building a model where we want to identify the spread of a contagious disease.

Step 1: Understand the decision to be made, the question to be asked, or the problem to be solved.

Understanding the decision is the same as identifying the problem to be solved. Identifying the problem to study is usually difficult.

Step 1: Understand the decision to be made or the question asked.
Step 2: Make simplifying assumptions.
Step 3: Define all variables.
Step 4: Construct a model.
Step 5: Solve and interpret the model. Test the model. Do the results pass the *common sense test?*
Step 6: Verify the model. Validate the model, if possible
Step 7: Identify the strenghts and weaknesses as a reflection of your model.
Step 8: Sensitivity analysis and/or model testing
Step 9: Implement and maintain the model for future use if it passes the common sense test

FIGURE 1.6
Mathematical modeling process.

In real life, no one walks up to you and hands you an equation to be solved. Usually, it is a comment like "we need to make more money" or "we need to improve our efficiency." Perhaps, we need to make better decisions or we need all our units that are not 100% efficient to become more efficient. We need to be precise in our formulation of the mathematics to actually describe the situation that we need to solve. In our example, we want to identify the spread of a contagious disease to determine how fast it will spread within our region. Perhaps, we will want to use the model to answer the following questions:

1. How long will it take until one thousand people get the disease?

2. What actions may be taken to slow or eradicate the disease?

Step 2: Make simplifying assumptions.

Giordano et al. (2014, pp. 62–65) described this well. Again, we suggest starting by brain storming the situation. Make a list of as many factors, or variables, as you can. Now, we realize that we usually cannot capture all these factors influencing a problem in our initial model. The task now is simplified by reducing the number of factors under consideration. We do this by making simplifying assumptions about the factors, such as holding certain factors as constants or ignoring some in the initial modeling phase. We might then examine to see if relationships exist between the remaining factors (or variables). Assuming simple relationships might reduce the complexity of the problem. Once you have a shorter list of variables, classify them as independent variables, dependent variables, or neither.

In our example, we assume we know the type of disease, how it is spread, the number of susceptible people within our region, and what type of medicine is needed to combat the disease. Perhaps, we assume that we know the size of population and the approximate number susceptible to getting the disease.

Step 3: Define all variables.

It is critical to define all your variables and provide the mathematical notation for each. In addition, if your variables have units, include them as well. Include all variables, even those you might think that you will not initially use. Often we find that we need these variables later in the refinement process.

In our example, the variables of interest are the number of people currently infected, the number of people who are susceptible to the disease, and the number of people who recently recovered from the disease.

Step 4: Construct a model.

Using tools at your disposal or after learning new mathematical tools, you use your own creativity to build a model that describes the situation and the solution of which helps to answer important questions that have been asked. Generally, three methods might be applied here. From first principles, your assumptions, and your variable list, construct a useable mathematical model. From a data set, perform data analysis to examine useful patterns that might suggest a useful model form such as a regression model. From research, take a model off the shelf and either use it directly or modify it appropriately for your use (a good discussion is found in Giordano et al. 2014).

In our example, we find that we might be able to initially use the susceptible (S), infected (I), and resistant (R) (SIR) model off the shelf.

Step 5: Solve and interpret the model.

We take the model that we have constructed in Steps 1–4 and we solve it using mathematical tools. Often this model might be too complex or unwieldy, so we cannot solve it or interpret it. If this happens, we return to Steps 2–4 and simplify the model further. We can always try to enhance the model later. We must also ensure that the model yields useable results for which the model was proposed. We will call this *passing the common sense test*.

In our example, the system of discrete dynamical system SIR equations has no closed form analytical solution. It does have graphical and numerical solutions that can be analyzed.

Step 6: Verify the model (Giordano et al., 2014, pp. 63–64).

Before we use the model, we should test it out. There are several questions we must ask. Does the model directly answer the question or does the model allow for the answer to the questions to be answered? Is the model useable in a practical sense (can we obtain data to use the model)? Does the model pass the common sense test?

We provide an example to show this. We used a data set whose plot was reasonably a decreasing linear model. The correlation value was 0.94, which meant that the data were strongly linear. We built the linear model and used it to predict the value of a future y, where y had to be a positive value. The value of y, for our input x, was −23.5. So although many of the diagnostics were telling us that a linear model was adequate, the common sense test for using the model caused major refinements until we got a nonlinear model that has good diagnostics and passed the common sense tests (Fox, 2011; 2012).

Step 7: Strengths and weaknesses.

No model is complete without self-reflection of the modeling process. We need to consider not only what we did right but also what we did that might be suspected as well as what we could do better. This reflection also helps in refining models in the future.

Step 8: Sensitivity analysis and model testing.

Sensitivity analysis is used to determine how *sensitive* a model is to changes in the value of the parameters of the model and to changes in the structure of the model.

Parameter sensitivity is usually performed as a series of tests in which the modeler sets different parameter values to see how a change in the parameter causes a change in the dynamic behavior of the stocks. By showing how the model behavior responds to changes in parameter values, sensitivity analysis is a useful tool in model building as well as in model evaluation.

Sensitivity analysis helps to build confidence in the model by studying the uncertainties that are often associated with parameters in models. Many parameters in system dynamics models represent quantities that are very difficult, or even impossible to measure to a great deal of accuracy in the real world. In addition, some parameter values change in the real world. Therefore, when building a system dynamics model, the modeler is usually at least somewhat uncertain about the parameter values he chooses and must use estimates. Sensitivity analysis allows him to determine what level of accuracy is necessary for a parameter to make the model sufficiently useful and valid. If the tests reveal that the model is insensitive, then it may be possible to use an estimate rather than a value with greater precision. Sensitivity analysis can also indicate which parameter values are reasonable to use in the model. If the model behaves as expected from real-world observations, it gives some indication that the parameter values reflect, at least in part, the *real world*. Sensitivity tests help the modeler to understand the dynamics of a system.

Experimenting with a wide range of values can offer insights into the behavior of a system in extreme situations. Discovering that the system behavior greatly changes for a change in a parameter value can identify a leverage point in the model—a parameter whose specific value can significantly influence the behavior mode of the system.

In our SIR example, we would test how the changes in the parameters affect the solution.

Step 9: Implement and maintain the model if it passes the common sense test.

A model is pointless if we do not use it. The more user-friendly the model, the more it will be used. Sometimes, the ease of obtaining data for the model can dictate the model's success or failure. The model must also remain current. Often this entails updating the data used for the model and updating the parameters used in the model.

1.2.5 Illustrative Examples: Starting the Modeling Process

We now demonstrate the modeling process that was presented in Section 1.2.4. Emphasis is placed on identifying the problem and on choosing appropriate (useable) variables in this section.

Example 1.1: Prescribed Drug Dosage

Scenario: Consider a patient who needs to take a newly marketed prescribed drug. To prescribe a safe and effective regimen for treating the disease, one must maintain a blood concentration above some effective level and below any unsafe level. How is this determined?

Understanding the decision and problem: Our goal is a mathematical model that relates dosage and time between dosages to the level of the drug in the bloodstream. What is the relationship between the amount of drug taken and the amount of drug in the blood after time, *t*? By answering this question, we are empowered to examine other facets of the problem of taking a prescribed drug.

Assumptions: We should choose or know the disease in question and the type (name) of the drug that is to be taken. We will assume in this example that the drug is rythmol, a drug taken for the regulation of the heartbeat. We need to know or find decaying rate of rythmol in the bloodstream. This might be found from data that have been previously collected. We need to find the safe and unsafe levels of rythmol based on the drug's *effects* within the body. This will serve as bounds for our model. Initially, we might assume that the patient size and weight have no effect on the drug's decay rate. We might assume that all patients are about the same size and weight. All are in good health and no one takes other drugs that affect the prescribed drug. We assume that all internal organs are functioning properly. We might assume that we can model this using a discrete time period even though the absorption rate is a continuous function. These assumptions help simplify the model.

Example 1.2: Emergency Medical Response

The Emergency Service Coordinator (ESC) for a county is interested in locating the county's three ambulances to maximize the residents that can be reached within 8 minutes in emergency situations. The county is divided into six zones, and the average time required to travel from one region to the next under semiperfect conditions is summarized in Table 1.1.
The population in zones 1, 2, 3, 4, 5, and 6 is given in Table 1.2.

Understanding the decision and problem: We want better coverage to improve the ability to take care of patients who require an ambulance to go to a hospital. Determine the location for placement of the ambulances to maximize coverage within the predetermined allotted time.

Assumptions: We initially assume that time travel between zones is negligible. We further assume that the times in the data are averages under ideal circumstances.

TABLE 1.1

Average Travel Times from Zone i to Zone j in Perfect Conditions

	1	2	3	4	5	6
1	1	8	12	14	10	16
2	8	1	6	18	16	16
3	12	18	1.5	12	6	4
4	16	14	4	1	16	12
5	18	16	10	4	2	2
6	16	18	4	12	2	2

TABLE 1.2

Populations in Each Zone

1	50,000
2	80,000
3	30,000
4	55,000
5	35,000
6	20,000
Total	270,000

Example 1.3: Bank Service Problem

The bank manager is trying to improve customer satisfaction by offering better service. The management wants the average waiting time for the customer to be less than 2 minutes and the average length of the queue (length of the line waiting) to be two persons or fewer. The bank estimates about 150 customers per day. The existing arrival and service times are given in Tables 1.3 and 1.4.

> *Understand the decision and problem*: The bank wants to improve customer satisfaction. First, we must determine if we are meeting the goal or not. Build a mathematical model to determine if the bank is meeting its goals and if not, come up with some recommendations to improve customer satisfaction.
>
> *Assumptions*: Determine if the current customer service is satisfactory according to the manager guidelines. If not, determine the minimal changes for servers that are required to accomplish the manager's goal through modeling. We might begin by selecting a queuing model off the shelf to obtain some benchmark values.

TABLE 1.3

Arrival Times

Time between Arrivals in Minutes	Probability
0	0.10
1	0.15
2	0.10
3	0.35
4	0.25
5	0.05

TABLE 1.4

Service Times

Service Time in Minutes	Probability
1	0.25
2	0.20
3	0.40
4	0.15

Determine if the current customer service is satisfactory according to the manager guidelines. If not, determine the minimal changes for servers that are required to accomplish the manager's goal through modeling. We might begin by selecting a queuing model off the shelf to obtain some bench mark values.

Example 1.4: Measuring Efficiency of Units

We have three major units where each unit has two inputs and three outputs as shown in Table 1.5.

> *Understand the decision and problem*: We want to improve efficiency of our operation. We want to be able to find *best practices* to share. First, we have to measure efficiency. We need to build a mathematic model to examine efficiency of a unit based on their inputs and outputs and be able to compare efficiency to other units.
>
> *Assumptions and variable definitions*: We define the following decision variables:
>
> t_i is the value of a single unit of output of decision-making unit (DMUi), for $i = 1,2,3$
>
> w_i is the cost or weights for one unit of inputs of DMUi, for $i = 1,2$
>
> Efficiency$_i$ = (total value of i's outputs)/(total cost of i's inputs), for $i = 1,2,3$

The following modeling initial assumptions are made:

1. No unit will have an efficiency more than 100%.
2. If any efficiency is less than 1, then it is inefficient.

Example 1.5: World War II Battle of the Bismarck Sea

In February 1943 at a critical stage of the struggle for New Guinea, the Japanese decided to bring reinforcements from the nearby island of New Britain. In moving their troops, the Japanese could either route north where rain and poor visibility were expected or south where clear weather was expected. In either case, the trip would be 3 days. Which route should they take? If the Japanese were only interested in time, they would be indifferent to the two routes. Perhaps, they wanted to minimize their convoy to expose by U.S. bombers. For the United States, General Kenney also faced a difficult choice. Allied intelligence had detected evidence of the Japanese convoy that assemble at the far side of New Britain. Kenney, of course, wanted to maximize the days that the bombers could attack the convoy, but he did not have enough reconnaissance planes to saturate both routes. What should he do?

TABLE 1.5

Input and Outputs

Unit	Input #1	Input #2	Output #1	Output #2	Output #3
1	5	14	9	4	16
2	8	15	5	7	10
3	7	12	4	9	13

Understand the decision and problem: We want to build and use a mathematical model of conflict between players to determine the *best* strategy option for each player.

Assumptions: Let us assume that General Kenney can search only south or north. We will put these into rows. Let us further assume that the Japanese can actually sail north or south and let us put these in columns. Assume that we get additional information from the intelligence community of the U.S. Armed Forces, and that this information is accurate. This information states that if there is a clear exposure, then we bomb all 3 days. If we search south and do not find the enemy (then have to search north in the poorer weather that will waste 2 days of searching), then we have only 1 day to bomb. If we search north and Japanese sail north, the enemy will be exposed in 2 days. If we search north and the Japanese sail south, the enemy will be exposed in 2 days.

Example 1.6: Risk Analysis for Homeland Security

Consider providing support to the Department of Homeland Security. The department only has so many assets and a finite amount of time to conduct investigations, thus priorities might be established. The risk assessment office has collected the data for the morning meeting as shown in Table 1.6. Your operations research team must analyze the information and must provide a priority list to the risk assessment team for that meeting.

Understand the decision and problem: There are more risks than we can possibly investigate. Perhaps if we rank these based on useful criteria, we can determine a priority for investigating these risks. We need to construct a useful mathematical model that ranks the incidents or risks in a priority order.

Assumptions: We have past decision that will give us insights into the decision maker's process. We have data only on reliability, approximate number of deaths, approximate costs to fix or rebuild, location, destructive influence, and on number of intelligence gathering tips. These will be the criteria for our analysis. The data are accurate and precise. We can convert word data into ordinal numbers.

Model: We could use multiattribute decision-making techniques for our model. We decide on a hybrid approach of analytical hierarchy process (AHP) and technique for order of preference by similarity to ideal solution (TOPSIS), that is, AHP-TOPSIS. We will use AHP with Saaty's (1980) pairwise comparison to obtain the decision-maker weights. We will also use the pairwise comparison to obtain numerical values for the criteria: location and destructive influence. Then we will use TOPSIS.

TABLE 1.6

Risk Assessment Priority

Threat Alternatives/ Criterion	Reliability of Threat Assessment	Approximate Associated Deaths (000)	Cost to Fix Damages in (Millions)	Location	Destructive Psychological Influence	Number of Intelligence-Related Tips
1. Dirty bomb threat	0.40	10	150	Urban dense	Extremely intense	3
2. Anthrax-bioterror threat	0.45	.8	10	Urban dense	Intense	12
3. DC-road and bridge network threat	0.35	0.005	300	Urban & rural	Strong	8
4. NY subway threat	0.73	12	200	Urban dense	Very strong	5
5. DC metro threat	0.69	11	200	Both Urban dense and rural	Very strong	5
6. Major bank robbery	0.81	0.0002	10	Urban dense	Weak	16
7. FAA threat	0.70	0.001	5	Rural dense	Moderate	15

Example 1.7: Discrete SIR Models of Epidemics

Consider a disease that is spreading throughout the United States such as the new deadly flu. The Centers for Disease Control and Prevention is interested in knowing and experimenting with a model for this new disease before it actually becomes a *real* epidemic. Let us consider the population that is beingdivided into three categories: susceptible, infected, and removed. We make the following assumptions for our model:

- No one enters or leaves the community, and there is no contact outside the community.
- Each person is either susceptible, S (able to catch this new flu); infected, I (currently has the flu and can spread the flu); or removed, R (already had the flu and will not get it again that includes death).
- Initially every person is either S or I.
- Once someone gets the flu this year, they will not get again.
- The average length of the disease is 2 weeks over which the person is deemed to be infected and can spread the disease.
- Our time period for the model will be per week.

The model we will consider is an off-the-shelf model, the SIR model (Allman and Rhodes, 2004).

Let us assume the following definition for our variables:

$S(n)$ is the number in the population susceptible after period n.

$I(n)$ is the number infected after period n.

$R(n)$ is the number removed after period n.

Let us start our modeling process with $R(n)$. Our assumption for the length of time someone has the flu is 2 weeks. Thus, half of the infected people will be removed each week:

$$R(n+1) = R(n) + 0.5I(n)$$

The value, 0.5, is called the removal rate per week. It represents the proportion of the infected persons who are removed from infection each week. If real data are available, then we could do *data analysis* to obtain the removal rate.

$I(n)$ will have terms that both increase and decrease its amount over time. It is decreased by the number that is removed each week, $0.5 * I(n)$. It is increased by the number of susceptible people who come into contact with an infected person and catch the disease, $aS(n)I(n)$. We define a as the rate at which the disease is spread or as the transmission coefficient. We realize that this is a probabilistic coefficient. We will assume, initially, that this rate is a constant value that can be found from initial conditions.

Let us illustrate as follows: Assume that we have a population of 1000 students in the dorms. Our nurse found that three students were reporting to the infirmary initially. The next week, five students came to the infirmary with flu-like symptoms. $I(0) = 3$, $S(0) = 997$. In week 1, the number of newly infected students is 30.

$$5 = a\,I(n)S(n) = a(3)*(995)$$

$$a = 0.00167$$

Let us consider $S(n)$. This number is decreased only by the number that becomes infected. We may use the same rate, a, as before to obtain the model:

$$S(n+1) = S(n) - aS(n)I(n)$$

Our coupled SIR model is

$$R(n+1) = R(n) + 0.5I(n)$$

$$I(n+1) = I(n) - 0.5I(n) + 0.00167I(n)S(n)$$

$$S(n+1) = S(n) - 0.00167S(n)I(n)$$

$$I(0) = 3, S(0) = 997, R(0) = 0$$

The SIR model can be solved iteratively and can be viewed graphically. We will revisit this again in Chapter 7. In Chapter 7, we determine that the worse of the flu epidemic occurs around week 8, at the maximum of the infected graph. The maximum number is slightly larger than 400; from Figures 7.30–7.33 in Chapter 7, it is approximated as 427. After 25 weeks, slightly more than 9 people never got the flu.

These examples will be solved in subsequent chapters.

1.3 Technology

Most real-world problems that we have been involved in solving model require technology to assist the analyst, the modeler, and the decision-maker. Microsoft Excel is available on most computers and represents a fairly good technological support for analysis of the average problems, especially with *Analysis ToolPak* and the *Solver* installed. Other specialized software to assist analysts include MATLAB®, Maple, Mathematica, LINDO, LINGO, GAMS, as well as some additional add-ins for Excel such as the simulation package and Crystal Ball. Analysts should avail themselves to have access to as many of these packages as necessary. In this book, we illustrate Excel and Maple although other software may be easily substituted.

1.4 Conclusion

We have provided a clear and simple process to begin mathematical modeling in applied situation that requires the stewardship of applied mathematics, operations research analysis, or risk assessment. We did not cover all the possible models but highlighted a few through illustrative examples. We emphasize that sensitivity analysis is extremely important in all models and should be accomplished before making any decision. We show this in more detail in the following chapters that cover the techniques.

Exercises

Using Steps 1–3 of the modeling process, identify a problem from scenario 1–11 that you could study. There are no *right* or *wrong* answers, just measures of difficulty.

1. The population of deer in your community.
2. A new outdoor shopping mall is being constructed. How should you design the illumination of the parking lot?
3. A farmer wants a successful season with his crops. He thinks that this is accomplished by growing a lot of anything as long as he uses all his land.
4. Ford Motor Company bought Volvo. Are Volvos still *top quality* cars?
5. A minor Forbes 500 company wants to go mobile with internet access and computer upgrades, but cost might be a problem.
6. Starbucks has many varieties of coffee available. How can Starbucks make more money?
7. A student does not like math or math-related courses. How can a student maximize their chances for a good grade in a math class to improve their overall GPA?
8. Freshmen think that their first semester courses should be pass–fail for credit or no credit only.
9. Alumni are clamoring to fire the college's head football coach.
10. Stocking a fish pond with bass and trout.
11. Safety airbags in millions of cars are to be replaced. Can this be done in a timely manner?

Projects

1. Is Michael Jordan or Stephan Curry the greatest basketball player of the century? What variables and factors need to be considered?

2. What kind of car should you buy when you graduate from college? What factors should be in your decision? Are car companies modeling your needs?

3. Consider domestic decaffeinated coffee brewing. Suggest some objectives that could be used if you wanted to market your new brew. What variables and data would be useful?

4. Replacing a coaching legend at a school is a difficult task. How would you model this? What factors and data would you consider? Would you equally weigh all factors?

5. How would you go about building a model for the *best pro football player of all time*?

6. Rumors abound in major league baseball about steroid use. How would you go about creating a model that could imply the use of steroids?

References and Suggested Readings

Albright, B. 2010. *Mathematical Modeling with Excel.* Sudbury, MA: Jones and Bartlett Publishers.

Allman, E. and J. Rhodes. 2004. *Mathematical Modeling with Biology: An Introduction.* Cambridge, UK: Cambridge University Press.

Bender, E. 2000. Chapter 1, *Mathematical Modeling.* Mineola, NY: Dover Press, pp. 1–8.

Fox, W. P. 2011. Using the EXCEL solver for nonlinear regression. *Computers in Education Journal,* 2(4), 77–86.

Fox, W. P. (2012). Issues and importance of "good" starting points for nonlinear regression for mathematical modeling with maple: Basic model fitting to make predictions with oscillating data. *Journal of Computers in Mathematics and Science Teaching,* 31(1), 1–16.

Fox, W. P. 2013a. Modeling engineering management decisions with game theory. In F. P. García Márquez and B. Lev (Eds.), *Engineering Management.* Rijeka, Croatia: InTech.

Fox, W. 2013b. Mathematical modeling and analysis: An example using a Catapult. *Computer in Education Journal,* 4(3): 69–77.

Fox, W. 2014a. Chapter 17, Game theory in business and industry. In *Encyclopedia of Business Analytics and Optimization.* Hershey, PA: IGI Global and Sage Publications, V(1), pp. 162–173.

Fox, W. 2014b. Chapter 221, TOPSIS in business analytics. In *Encyclopedia of Business Analytics and Optimization.* Hershey, PA: IGI Global and Sage Publications, V(5), pp. 281–291.

Fox, W. P. 2016. Applications and modeling using multi-attribute decision making to rank terrorist threats. *Journal of Socialomics,* 5(2), 1–12.

Fox, W. P. and M. J. Jaye. 2011. Search and rescue carry through mathematical modeling problem. *Computers in Education Journal,* 2(3), 82–93.

Giordano, F. R., W. Fox, and S. Horton. 2014. *A First Course in Mathematical Modeling* (5th ed.). Boston, MA: Brooks-Cole Publishers.

Meerschaert, M. M. 1999. *Mathematical Modeling* (2nd ed.). San Diego, CA: Academic Press.

Myer, W. 2004. *Concepts of Mathematical Modeling*. New York: Dover Press, pp. 13–15.

Saaty, T. 1980. *The Analytical Hierarchy Process*. New York: McGraw Hill.

Winston, W. 1995. *Introduction to Mathematical Programming*. Belmont, CA: Duxbury Press, pp. 323–325.

2

Introduction to Stochastic Decision-Making Models for Business Analytics

OBJECTIVES

1. Know the basic concepts of probability and expected value.
2. Understand and use expected value in decision-making under risk and uncertainty.
3. Apply Bayesian probability to decision-making.
4. Develop a decision tree to compare alternatives.
5. Be able to use appropriate technology.

Consider a company that needs to decide whether they should build and market large or small outdoor play sets. The three alternatives under consideration with their respective demand revenues and losses under the estimated demand probabilities are as follows:

	Outcomes		
	High Demand	**Moderate Demand**	**Low Demand**
Alternatives	($p_{hd} = 0.35$)	($p_{md} = 0.40$)	($p_{ld} = 0.25$)
Large play sets	$200,000	$120,000	−$125,000
Small play sets	$100,000	$55,000	−$25,000
No play sets	$0	$0	$0

Can we determine the best course of action for this company? We will analyze and solve this problem later in this chapter.

2.1 Introduction

What is stochastic modeling? Stochastic modeling concerns the use of probability to model real-world situations in which uncertainty is present. As uncertainty is pervasive, this means that the tools that we present can potentially prove useful in almost all facets of one's professional life and in one's personal life. Such topics are as follows:

- Gambling
- Personal or business finances
- Disease treatment options
- Economic forecasting
- Product demand
- Call center provisioning
- Product reliability and warranty analysis, and so on

The use of a stochastic model does not imply that the modeler fundamentally believes that the system under consideration behaves *randomly*. For example, the behavior of an individual may appear *random*. But an interview with that individual may reveal a set of preferences under which that person's behavior is then revealed as totally predictable. The use of a stochastic model reflects only a logical and reasonable decision on the part of the modeler that such a model represents the best currently available description of the phenomenon under consideration, given the data that are available and models known to the modeler.

In this chapter, we discuss the role of using probability in decision-making. Let us begin with simple but useful definition of decision theory.

Decision theory, or decision analysis, is a collection of mathematical models and tools that are developed to assist decision-makers, at any level, in choosing among alternatives in complex situations.

In many cases, we might know the possible outcomes that are affected by some states of nature that occur with some probability or likelihood. We are uncertain exactly which of these outcomes will happen. Between the time that we make a decision to act and when the process concludes, there might be other factors affecting the action, factors over which we have no control. We might have information about the chance of each outcome from each action. These likelihoods or chances of occurrence will be probabilistic.

There are four states where decisions are made. These are initially categorized as *decisions under certainty, decisions under chance,* and *decision under conflict (game theory)*. In management science, decision science, and operations research, *decision under chance* is broken down into two separate types: *Decision under risk* where probabilities are known or estimated and *decision under uncertainty* where probabilities are neither known nor estimated (Lee et al., 1985).

This chapter discusses only *decisions under uncertainty* and *decisions under risk*. Decision under conflict, known as game theory, will be discussed in Chapter 5. Decisions under certainty will be discussed in the optimization chapter on linear programming where the coefficients are known with certainty.

In addition, we consider whether a decision is a repetitive action or a one-time decision. Repetitive action might allow us to use probability, whereas in a one-time decision, we may or may not consider any knowledge of estimated or assumed probability. If we do not consider probability, then other factors or assumptions might take precedence in our decision process.

The decisions under risk exist when perfect information (certainty) is not available but the probabilities that outcomes will occur can be estimated or are available. We might use probability theory to find or estimate the chance of each state of nature. A state of chance might also occur where the probabilities of occurrences in a decision situation are neither known nor could be adequately estimated.

For example, consider whether or not to put your convertible top down while going to work. It is a warm, cloudy, and humid morning and you are going to work. You neither watched the weather report nor got a morning newspaper. You must decide whether or not to put the top down on your convertible. You have two alternatives: put the top down or do not put the top down. The state of nature is that it can rain or not rain. There are four possible outcomes:

A: Do not put the top down, it rains

B: Put the top down, it does not rain

C: Put the top down, it rains

D: Do not put the top down, it does not rain

You will be the happiest if you put the top down and it does not rain, and your worst case is to put the top down and it rains. The other two outcomes are somewhere in between. We will examine the criterion later in the chapter to help with our decision under the uncertainty of rain.

But what if you watched the evening weather or read the morning paper, where weather forecasters said there is 75% chance of rain. With this estimated probability, we can examine decisions under risk. What we may consider is whether the long-term probability of an event affects our short-term decision-making (Giordano et al., 2014).

A large skiing group in Lake Tahoe runs a very profitable business renting skis and ski wear. The management is planning on next year's skiing season and wants to know how many sets of skis to order? They can order a small amount, a medium amount, and a large amount of skis for next season. The accounting staff has estimated the demand and the estimated profits/losses based on the amount of predicted snowfall. We summarize the outcomes in the following table:

	Outcomes		
Alternatives	**Light Snow** ($p = 0.3$)	**Moderate Snow** ($p = 0.6$)	**Heavy Snow** ($p = 0.1$)
Small order	$4,000	$5,700	$7,200
Medium order	$2,400	$8,000	$10,000
Large order	−$1,600	$4,200	$12,000

Let us begin with a discussion of probability and expected value and then return to this skiing group decision problem.

2.2 Probability and Expected Value

First, let us discuss games of chance. Games of chance include, but are not limited to, games involving flipping coin, rolling dice, drawing cards from a deck, spinning a wheel, and such. Suppose you are flipping a coin with a friend. If you both flip the same, both heads or both tails, you win $1. If you flip a head and a tail each, you lose $1. If you play this game with this bet 100 times over the course of the evening, how much do you expect to win? Or lose? To answer such questions, we need two concepts: the probability of an event and the expected value of a random variable.

In situations like this, it makes sense to count the number of times something occurs. An efficient way to do this is to use the frequency definition of the probability of an event. The probability of the event two heads or two tails is the number of ways we can achieve these results divided by the total number of possible outcomes. That is, we define both coins being flipped that land on both heads or both tails as a favorable event A. We can define event A as the set of outcomes that include {HH, TT}.

Favorable outcomes are those that consist of {HH, TT}.

$$\text{Probability of an event} = \frac{\text{Favorable outcomes}}{\text{Total outcomes}}$$

$$\text{Probability of an event } \{A\} = \frac{\text{Number of outcomes of } \{A\}}{\text{Total outcomes}}$$

Of course, the probability of an event (flip of a fair coin) must be equal or greater than zero, and equal to or less than 1. And the sum of the probabilities of all possible events must equal 1. That is,

$$0 \le p_i \le 1$$

$$\sum_{i=1}^{n} p_i = 1, i = 1, 2, \dots, n$$

We need to compute all the possible outcomes of flipping two coins, and then determine how many result in the same results defined by event A. A tree is useful for visualizing the outcomes. These outcomes constitute the sample space. On the first flip, the possible results are H or T. And on the second flip, the same outcomes are still available. We assume that these events are equally likely to occur based on flipping and obtaining either a head or tail of each flip.

We can summarize the possible outcomes. There are four outcomes {HH, HT, TH, TT} and twice the flips were identical {HH, TT} and twice they were different {HT, TH}. This is seen in Figure 2.1.

The probability that the two flips were the same is equal to the number of favorable outcomes (2) divided by the total outcomes (4), $2/4 = 1/2$

$$\text{Probability} = \frac{\text{Favorable outcomes}}{\text{Total outcomes}} = \frac{2}{4} = \frac{1}{2}$$

Probability is based on the long-run average. Thus, over many iterations or repeated trials of flipping two coins, you would expect about $1/2$ of the time for the two flips to be the same. Before we can determine whether we expect to win or lose with our bet in this flipping the coin game, we need to understand the concept of expected value.

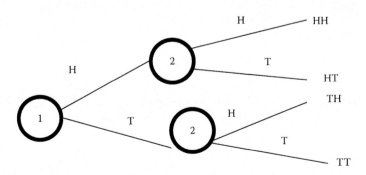

FIGURE 2.1
Flipping a fair coin sample space as a tree.

2.2.1 Expected Value

First, we define a random variable as a rule that assigns a number to every outcome of a sample.

We use $E[X]$, which is stated as the expected value of X. We define $E[X]$ as follows:

Expected value, $E[X]$, is the *mean* or *average value*.

Further, we provide the following two formulas: in the discrete case, $E[X] = \Sigma_{i=1}^{n} x_i p(x_i)$ and in the continuous case, $E[X] = \int_{-\infty}^{+\infty} x * f(x)dx$.

There are numerous ways to calculate the average value. We present a few common methods that you could use in decision theory.

If you had 2 quiz grades, an 82 and a 98, almost intuitively you would add the two numbers and divide by 2, giving an average of 90.

Average scores: Two scores that were earned were 82 and 98. Compute the average.

$$E[X] = \frac{(82+98)}{2} = 90$$

If after 5 quizzes, you had three 82s and two 98s, you would add them and divide by 5.

$$\text{Average} = \frac{3(82)+2(98)}{5} = 88.4$$

Rearranging the terms, we obtain

$$\text{Average} = \frac{3}{5}(82) + \frac{2}{5}(98)$$

In this form, we have two payoffs, 82 and 98, each multiplied by the weights, 3/5 and 2/5. This is analogous to the definition of expected value.

Suppose a game has outcomes $a_1, a_2, ..., a_n$, each with a payoff $w_1, w_2, ..., w_n$ and a corresponding probability $p_1, p_2, ..., p_n$ where $p_1 + p_2 + ... + p_n = 1$ and $0 \le p_i \le 1$, then the quantity

$$E = w_1 p_1 + w_1 p_2 + ... + w_1 p_n$$

is the expected value of the game. Note that expected value is analogous to weighted average, but the weights must be probabilities ($0 \le p_i \le 1$) and the weights must sum to 1.

Weighted average: Let us assume that in a class there were 8 scores of 100, 5 scores of 95, 3 scores of 90, 2 scores of 80, and 1 score of 75. We compute the weighted average as follows:

$$E[X] = \overline{X} = \frac{\sum w_i x_i}{\sum w_i} = \frac{8 \cdot 100 + 5 \cdot 95 + 3 \cdot 90 + 2 \cdot 80 + 1 \cdot 75}{8 + 5 + 3 + 2 + 1} = \frac{1780}{19} = 93.68$$

Probabilistic mean: Finding the expected value when probabilities are involved:

X	0	1	2	3	4
Probability	0.2	0.3	0.32	0.11	0.07

Number of persons in a queue for service at a bank at any given time in a peak period.

$$E[X] = \sum X \cdot P(X = x)$$

where X is a discrete random variable. A random variable is a rule that assigns a number to every outcome in a sample space. We provide a definition as follows:

A random variable is a rule that assigns a number to every outcome in a sample space. Thus, it is a numerical measure of the outcome from a probability experiment, so its value is determined by chance. Random variables are denoted using capital letters such as X. The values that they take on are annotated by small letters, x. We would say the $P(X = x)$, probability that the random variable X takes on a value x.

$$E[X] = (0)*(0.2) + (1)*(0.3) + (2)*(0.32) + (3)*(0.11) + (4)*(0.07) = 1.55$$

There are, on average, 1.55 people in the queue at any given time during a peak period. If your store policy is less than two people in a queue on average, then your current operation is meeting the policy. If your policy is that at no time there will be two or more people in a queue, then clearly you are not meeting the stated policy as 11% of time, three persons are in a queue and 7% of the time, four persons are in a queue.

What if we are dealing with a monetary decision with two potential outcomes. If the probability of success, $P(s)$, is 1/6 and if successful, we make $60,000 and if we are unsuccessful in our monetary venture, we lose $6,000. We find the expected value as $E[X] = 60,000(1/6) - 6,000(5/6) = 10,000 - 5,000 = $5,000.

We also present the continuous version of this concept. Consider a random variable, X, that represents the lifting ability of a male person who is more than 30 years old on team A of the National Football League (NFL). We measure the weight in pounds and define X as follows: $325 < x < 525$. The probability distribution function (pdf) might be given as a uniform distribution modeled by 1/200 for $325 < x < 525$. The cumulative distribution function (CDF) for the probability in an interval is defined as an integral:

$$F(x) = \int_{x_1}^{x_2} \frac{1}{b-a} dx = P(x_1 \le x \le x_2)$$

and the expected value, $E[X]$ is defined as

$$E[X] = \int_a^b x \frac{1}{b-a} dx$$

In our example, the CDF to find the probability from any (x_1, x_2) is $(x_2 - x_1)/(525 - 325)$. Thus the probability that some lifts between 411 and 500 lb is $(500 - 411)/200 = 89/200 = 0.445$.

To find the expected lift weight for this example, we compute the integral as follows:

$$E[X] = \int_{325}^{525} x \frac{1}{525-325} dx = 425 \, \text{lb}$$

Example 2.1: Life Insurance

A term life insurance policy will pay a beneficiary a certain sum of money upon the death of the policy holder. These policies have premiums that must be paid annually. The company can sell either a $250,000 or a $300,000 policy. Suppose a life insurance company sells a $250,000 1-year-term life insurance policy to a 50-year-old female for $750. According to the National Vital Statistics Report (1997), Vol. 47, No. 28 (p. 74), the probability that the 50-year-old female will die is 2.88 per thousand or 0.00288; thus, the probability that the female will survive the year is $1 - 0.00288 = 0.99712$. Compute the expected value of this policy to the insurance company. If successful, should the company sell the $300,000 policy?

Solution: We compute the expected value and make a decision. $E[X] = 750 * 1 - (250,000 + 750)*(1 - 0.99712) = 750 - 722.16 = 27.84$. The life insurance company expects to make a little more than $27.84 per policy sold. As a matter of fact, if the company sells 10,000 policies, then they would expect to make $278,400. If $E[X] > 0$, then they should sell the policy. If $E[X] < 0$, then they should not sell the policy.

What about a $300,000 policy?

$E[X] = 750 * 1 - (300,000 + 750)*(1 - 0.99712) = 750 - 866.16 = -\$116.16$. As this expected value is negative, we should not sell the policy for $750. Charging any amount more than $866.16 would give the company a positive expected value.

Example 2.2: Flipping the Coin Game

Let us consider our flipping the coin game where we win if both coins are heads or both coins are tails and we lose if each coin is different. We are interested in the outcomes: winning and losing. Winning has a payoff of $5, whereas losing has a payoff of negative $1, the cost of the game. Winning (either a HH or TT result) has a probability of 1/2. Losing (outcomes of HT or TH) has a probability of 1/2. That is, 1/2 = 1 − 1/2, or by computing from the definition, there are two ways of losing and four possibilities, so the probability of losing is 2/4 = 1/2. Our expected value then is

$$E = \frac{1}{2}(\$5) + \frac{1}{2}(\$ - 1) = \$2$$

The interpretation of the result is that we win $2 over the long haul. If you have broken even with an expected value of $0, then the game is *fair*. This coin flipping game is not fair with $5 for winning and $1 for losing. If we either won $1 or lost $1 in the game, then the expected value would be $0 and the game would be considered fair.

$$E = \frac{1}{2}(\$1) + \frac{1}{2}(\$ - 1) = \$0$$

Example 2.3: Order Skis for Rentals

A large skiing group in Lake Tahoe runs a very profitable business renting skis and ski wear. The management is planning on next year's skiing season and wants to know how many skis to order? They can order a small amount, a medium amount, and a large amount of skis for next season. The account staff members have estimated the demand and the profit:

	Outcomes		
Alternatives	Light Snow ($p = 0.3$)	Moderate Snow ($p = 0.6$)	Heavy Snow ($p = 0.1$)
Small order	$4,000	$5,700	$7,200
Medium order	$2,400	$8,000	$10,000
Large order	−$1,600	$4,200	$12,000

$$E(\text{Small order}) = 4000(0.3) + 5700(0.6) + 7200(0.1) = \$5340$$

$$E(\text{Medium order}) = 2,400(0.3) + 8,000(0.6) + 10,000(0.1) = \$6,520$$

$$E(\text{Large order}) = -16,000(0.3) + 4,200(0.6) + 12,000(0.1) = \$3,240$$

Our decision, based on the expected value, is to make a medium order because it results in the largest expected value.

SENSITIVITY ANALYSIS

We see that a typical decision model depends on many assumptions. For instance in Example 2.3, the probabilities of the amount of snow are likely estimates based on past experiences. They are likely to change. How sensitive is the estimate of the probability of the amount of snow to the order decision? To the net profit of a correct order size?

The probabilities for snow were 0.1 for light snow, 0.4 for average snow, and 0.5 for heavy snow.

$$E(\text{Small order}) = \$6280$$

$$E(\text{Medium order}) = \$8440$$

$$E(\text{Large order}) = \$7520$$

We still would choose the medium order.

Exercises 2.2

Determine the expected value in exercises 1–6.

1. Let us assume that you have the following numerical grades in a course: 100, 92, 83, 95, and 100. Compute your average grade.

2. *Weighted means*: Let us assume that in a class there were 8 scores of 100, 5 scores of 96, 3 scores of 92, 2 scores of 84, and 1 score of 75. Find the average grade for the class.

3. Finding the expected value when probabilities are involved. The number of attempted ATM uses per person and their probabilities are listed in the following. Compute the expected value and interpret that value.

	1	2	3	4	5
Probability	0.45	0.38	0.10	0.05	0.02

4. Assume that the probability of success for a venture is $P(s) = 1/4$, and further assume that if we are successful, we make $5000 and if we are unsuccessful, we lose $500. Find the expected value.

5. A term life insurance policy will pay a beneficiary a certain sum of money upon the death of the policy holder. These policies have premiums that must be paid annually. Suppose a life insurance company sells a $200,000 1-year-term life insurance policy to a 49-year-old female for $650. According to the National Vital Statistics Report (1997), Vol. 47, No. 28, the probability the female will survive the year is 0.99791. Compute the expected value of this policy to the insurance company.

6. Assume that team B of the NFL has more than 30 males who lift weights between 311 and 619 lb, following the same uniform distribution. Compute the expected value weight lifted for team B.

2.3 Decision Theory and Simple Decision Trees

Decision theory, or decision analysis, is a collection of mathematical models that are designed to assist decision-makers when choosing among alternative courses of action (strategies). In this section, we consider models where the decision criterion is to maximize the expected value. In Section 2.3, we present alternative decision criteria to maximize the expected value.

We have already introduced the maximum expected value concept in the last section where we examined making a decision under chance when probabilities are known and when our decision is based on the expected value.

Example 2.4: Concessions at Sporting Events

Consider a firm that handles concessions for a sporting event. The firm's manager needs to know whether to stock up with coffee or coke-cola products. A local agreement restricts you to only one beverage. You estimate a $1500 profit of selling coke-cola products if it is cold and a $5000 profit of selling cola if it is warm. You also estimate a $4000 profit of selling coffee if it is cold and a $1000 profit of selling coffee if it is warm. The forecast says that there is a 30% of a cold front, otherwise the weather will be warm. What do you do?

For decisions under risk (probabilities estimated or known), we use the expected value.

$$E[\text{coke-cola products}] = 1500 * 0.3 + 5000 * 0.7 = \$3950$$

$$E[\text{coffee}] = 4000 * 0.3 + 1000 * 0.7 = \$1900$$

$$E[\text{coke-cola products}] > E[\text{coffee}]$$

Out decision is to sell coke-cola products.

SENSITIVITY ANALYSIS OF THE DECISION

Under what conditions should we sell coffee?

Assume that p_1 is the probability that the weather is warm and p_2 is the probability that the weather is cold where $p_1 + p_2 = 1$.

Set the two expected values equal and solve.

$$E[\text{cola}] = E[\text{coffee}]$$

$$1500 * p_2 + 5000 * p_1 = 4000 * p_2 + 1000 * p_1$$

Let $p_1 = 1 - p_2$ and substitute.

$$1500p_2 + 5000(1 - p_2) = 4000p_2 + 1000(1 - p_2)$$

$$1500p_2 + 5000 - 5000p_2 = 4000p_2 + 1000 - 1000p_2$$

$$4000 = 6500p_2$$

$$p_2 = \frac{4000}{6500} = 0.615 \text{ (to 3 decimal places)}$$

Decision: When the probability of cold weather is greater than 0.615, we sell coffee. Otherwise, we sell coke-cola products.

Example 2.5: Ski Resort Decisions

The financial success of a ski resort in Squaw Valley is dependent on the amount of early snow fall in the fall and winter months. If the snow fall is greater than 40 in., the resort always has a successful ski season. If the snow is between 30 and 40 in., the resort has a moderate season, and if the snow fall is less than 30 in., the season is poor and the resort will lose money. The seasonal snow probabilities from the weather service are displayed in the following table with the expected revenue that was historically calculated for the previous 10 seasons. A hotel chain has offered to lease the resort during the winter for $100,000. You must decide whether to operate yourself or lease the resort. What decision should you make?

	States of Nature		
	Snow > 40″ $p_1 = 0.40$	30″ < Snow < 40″ $p_2 = 0.20$	Snow < 30″ $p_3 = 0.40$
Financial return if we operate	$280,000	$100,000	−$40,000
Lease	$100,000	$100,000	$100,000

Possible solution:

$$E[\text{Operate}] = 0.4(280,000) + 0.2(100,000) - 0.4(40,000) = \$116,000$$

$$E[\text{Lease}] = 0.4(100,000) + 0.2(100,000) + 0.4(100,000) = \$100,000$$

Under these conditions, we would recommend to operate ourselves because $E[\text{Operate}] > E[\text{Lease}]$. This solution was based on long-term forecasts of snow based on many assumptions. As this is a one-time decision, other key assumptions might be included that could change our decision.

DECISION TREES

Decision trees are especially informative when a sequence of decisions must be made. The notation we will use is as follows: Decision node or fork is a square:

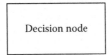

Courses of actions are lines disseminating from the decision node with a decision branch for each alternative course of action (strategy):

An uncertainty node or event fork that reflects chance (the state of nature) with an outcome branch for each possible outcome at that uncertainty node is identified by a circle:

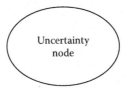

A terminal node with a consequence branch that shows the payoff for that outcome using a triangle:

Example 2.6: Contractor and Building a Parking Lot

Here are the data:

Large Parking Lot	Small Parking Lot
Win contract: $150,000	Win contract: $140,000 Profit
Lose contract: $−10,000	Lose contract: $−5,000
Probability of award of contract: 30%	Probability of award of contract: 35%

In this case, we have only one decision and our decision criterion is to compute the expected value of each alternative course of action as we did in Example 2.5 and choose the larger. That is, we compute the expected

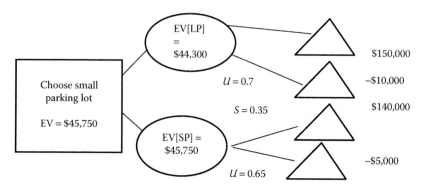

FIGURE 2.2
Decision tree for Example 2.6.

value of each uncertainty node, and then make the decision based on the higher expected value using Figure 2.2. Let S stand for successful, U stand for unsuccessful, LP for the large parking lot construction, and SP for the small parking lot.

Our decision for the company is to build the elementary schools.

As in this example, our procedure is to build the tree, then solve the decision tree by starting at the end, and replace each uncertainty node with an expected value. We make the decision before the uncertainty nodes by choosing the larger expected value. This is often referred to as the *fold back* method.

Exercises 2.3

1. Is a larger expected value always better? Consider Example 2.6 in Section 2.3, list some decisions that would cause the decision-maker to not choose the larger expected value.

2. Consider a construction firm that is deciding to specialize in building high schools or elementary schools, or a combination of both over the long haul. The construction company must submit a bid proposal that costs money to prepare, and there are no guarantees that they will be awarded the contract. If they bid on the high school, they have a 28% chance of getting the contract and they expect to make $62,500 net profit. However, if they do not get the contract, they lose $1350. If they bid on the elementary school, there is a 21% chance of getting the contract. For the elementary school, they would get $38,000 net profit. However if they do not get the contract, they lose $775. Build a decision tree to assist the company in their decision.

3. Company ABC has developed a new line of products. Top management is attempting to decide on both marketing and production strategies. Three strategies are considered and will be referred as *A* (aggressive), *B* (basic), and *C* (cautious). The conditions under

which the study will be conducted are S (strong) and W (weak) conditions. Management's best estimates for net profits (in millions of dollars) are given in the following table. Build a decision tree to assist the company to determine the best strategy.

Decision	Strong (with probability 48%)	Weak (with probability 52%)
A	33	−5
B	22	8
C	15	13

4. Assume the following probability distribution of daily demand for strawberries:

Daily demand	0	1	2	3	4
Probability	0.2	0.3	0.25	0.2	0.05

Also assume that unit cost = $2.50, selling price = $4.99 (i.e., profit on sold unit = $2.49), and salvage value on unsold units = $2 (i.e., loss on unsold unit = $1). We can stock either 0, 1, 2, 3, or 4 units. How many units should be stocked each day? Assume that units from one day cannot be sold the next day.

5. An oil company is considering making a bid on a shale oil development contract to be awarded by the government. The company has decided to bid $210 million. They estimate that they have a 70% chance of winning the contract bid. If the company wins the contract, the management has three alternatives for processing the shale. It can develop a new method for processing the oil, can use the existing method, or can ship the shale overseas for processing. The development cost of the new process is estimated at $30 million. The outcomes and probabilities associated in developing the new method are given as follows:

Event	Probability	Financial Outcome (in Millions)
Extremely successful	0.6	$350
Moderately successful	0.3	$250
Failure	0.1	$90

The existing methods cost $6.5 million to execute, and the outcomes and probabilities are given as follows:

Event	Probability	Financial Outcome (in Millions)
Extremely successful	0.5	$300
Moderately successful	0.25	$200
Failure	0.25	$40

The cost to ship overseas is $5 million. If it is shipped overseas, the contract guarantee is $230 million. Construct a decision tree and determine the best strategy.

6. The local TV station has $150,000 available for research and wants to decide whether to market a new advertising strategy for their station. The station is located in the city, but its viewers are statewide. They have three alternatives:

Alternative 1: Test locally with a small test group, then utilize the results of the local study to determine if a statewide study is needed.

Alternative 2: Immediately market with no studies.

Alternative 3: Immediately decide not to use the new strategy and keep everything *status quo*.

In the absence of a study, they believe that the new strategy has a 65% chance of success and a 35% chance of failure at the state level. If successful, the new strategy will bring $300,000 additional assets and if a failure, we will lose $100,000 in assets. If they do the study (which costs $30,000), there is a 60% chance of favorable outcome and a 40% chance of an unfavorable outcome. If the study shows that it is a local success, then there is an 85% chance that it will be a state success. If the study shows that it was a local failure, then there is only a 10% chance that it will be a state success. What should the local TV station do?

2.4 Sequential Decisions and Conditional Probability

In many cases, decisions must be made sequentially. Often decisions are based on other or previously made decisions. These are multistage decisions. We will examine the method and effects of sequential decisions. Often, a key part to decisions is to compute or use conditional probabilities.

Example 2.7: A Paving Company Decision with Multistage Choices

Consider a company that needs to decide whether or not to bid and build large or small parking lots. The three alternatives under consideration with the respective demand revenues and losses with the estimated demand probabilities are as follows:

	Outcomes		
	High Demand	**Moderate Demand**	**Low Demand**
Alternatives	($p_{hd} = 0.35$)	($p_{md} = 0.40$)	($p_{ld} = 0.25$)
Large parking lot	$200,000	$120,000	−$125,000
Small parking lot	$100,000	$55,000	−$25,000
No parking lots	$0	$0	$0

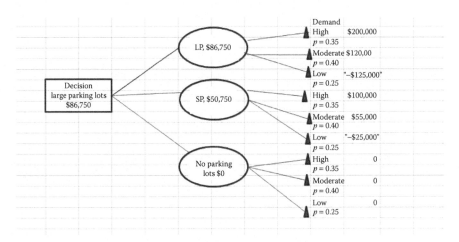

FIGURE 2.3
Decision tree for Example 2.7.

Initial analysis shows that we build the large warehouse, as shown in Figure 2.3. The large parking lot yields an expected value of $86,750, which is greater than the small parking lot expected value of $50,750, and building no parking lot yields an expected value of $0.

Example 2.8: Revisit Paving Company from Example 2.7 with New Options

Let us assume that before making a decision, the company has an option to hire a market research company for $4000. This company will survey the markets that are serviced by this company as to the attractiveness of the new outdoor play sets. The company knows that the market research does not provide perfect information but does provide updated information based on their sample survey. The company has additionally to decide whether or not to hire the market research team. If the research is conducted, the assumed probabilities of a success survey and an unsuccessful survey are 0.57 and 0.43, respectively. Further, as we have gained more information, our probabilities for the demand will chance. Given a successful survey outcome, the probability of high demand is 0.509, for moderate demand is 0.468, and/or for low demand is 0.023. Given an unsuccessful survey outcome, the probability of high demand is 0.023, for moderate demand is 0.543, and/or for low demand is 0.434. The tree diagram is shown in Figures 2.4 and 2.5.

Figure 2.4 shows the upper portion of tree diagram with the option to not hire a consulting firm. Figure 2.5 shows the lower portion of tree diagram with the option to hire the firm.

Decision: We conclude that we should hire and conduct the market survey to obtain the largest expected value of $92,066.40.

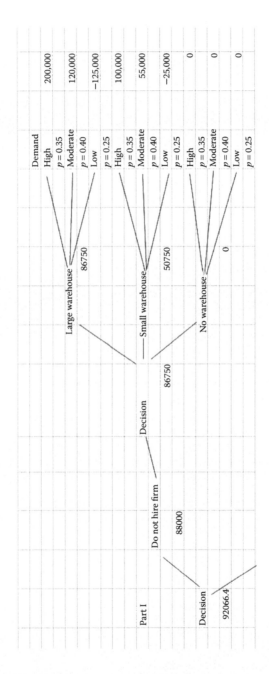

FIGURE 2.4
Excel screenshot part I.

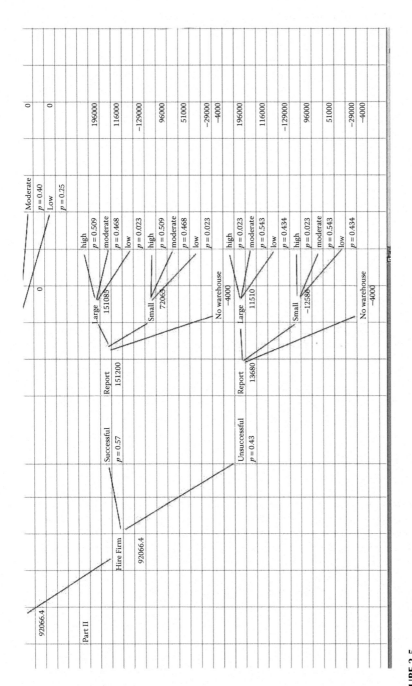

FIGURE 2.5
Excel screenshot part II.

Example 2.9: Computer Company

Acme Computer company manufactures memory chips in lots of
10 chips. From past experience, Acme knows that 80% of all lots contain 10% (1 out of 10) defective chips, and 20% of all lots contain 50%
(5 out of every 10) defective chips. If a good (that is 10% defective)
batch of chips is sent to the next stage of production, processing costs
of $1000 are incurred, and if a bad batch (50% defective) is sent to
the next stage of production, processing costs of $4000 are incurred.
Acme also has an alternative reworking batch before forwarding to
production at a cost of $1000. A reworked batch is sure to be a good
batch. Alternatively, for a cost of $100, Acme may test one chip from
each batch in an attempt to determine whether the batch is defective.
Determine how Acme should proceed to minimize the expected cost
per batch.

Solution: There are two states: G = batch is good and B = batch is bad.
Prior probabilities are given as $p(G) = 0.80$ and $p(B) = 0.20$.

Acme has an option to inspect each batch. The possible outcomes are
as follows: D is the defective chip and ND is the nondefective chip.

PROBABILITY

We introduce three new probability concepts: independence, conditional probability, and Bayes' theorem. Independence between events
occurs when the chance of one event happening does not affect the other
events chance of happening. Mathematically, we state that if we have
two events A and B, that events A and B are independent if and only if
$p(A \cap B) = p(A)*p(B)$. The second concept condition probability allows the
knowledge of one event to possibly influence the probability of a second
event. Mathematically, we state $p(A|B)$, the probability of event A given
event B has occurred, as follows in Equation 2.1:

$$p(A|B) = \frac{p(A \cap B)}{p(B)} \tag{2.1}$$

Bayes' theorem is illustrated with a simple Venn diagram, Figure 2.6.
We want to know the probability of event B, but all we know are the
intersections of B with event A. By summing all the known intersections, we are able to determine the probability of event B. With these
new concepts, we proceed with our example. The resulting formula is
given in Equation 2.2.

We are given the following conditional probabilities:

$$p(D|G) = 0.10\, p(ND|G) = 0.90\, p(D|B) = 0.50\, p(ND|B) = 0.50$$

To complete and find our values required, we need to determine some
additional probabilities. We will use two sets of formulas: conditional
probability (Equation 2.2) and Bayes' theorem (Equation 2.3).

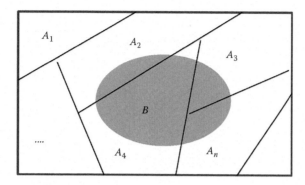

FIGURE 2.6
Venn diagram of Events B and A_i.

$$p(a|b) = \frac{p(a \cup b)}{p(b)} \qquad (2.2)$$

$$p(a_j|b) = \frac{p(a_j \cap b)}{p(b)} \qquad (2.3)$$

We start with a tree for Bayes' theorem initially as in Figure 2.7.

$$p(D) = p(D \cap gb) + p(D \cap bb) = 0.10 + 0.08 = 0.18$$

$$p(ND) = p(ND \cap gb) + p(ND \cap bb) = 0.72 + 0.10 = 0.82$$

We can now find our needed probabilities for our decision model.

$$p(bb|D) = \frac{p(bb \cap D)}{p(D)} = \frac{0.10}{0.18} = 0.55555$$

$$p(gb|D) = 1 - 0.5555555 = 0.45555555$$

$$p(bb|ND) = \frac{p(bb \cap ND)}{p(ND)} = \frac{0.10}{0.82} = 0.12195$$

$$p(gb|ND) = 1 - 0.12195 = 0.87805$$

Decision for computer company: Our decision is to test the chip as it has the minimum expected cost of $1580 as compared to $1600 to send the batch on as is or compared to $2000 to rework the batch.

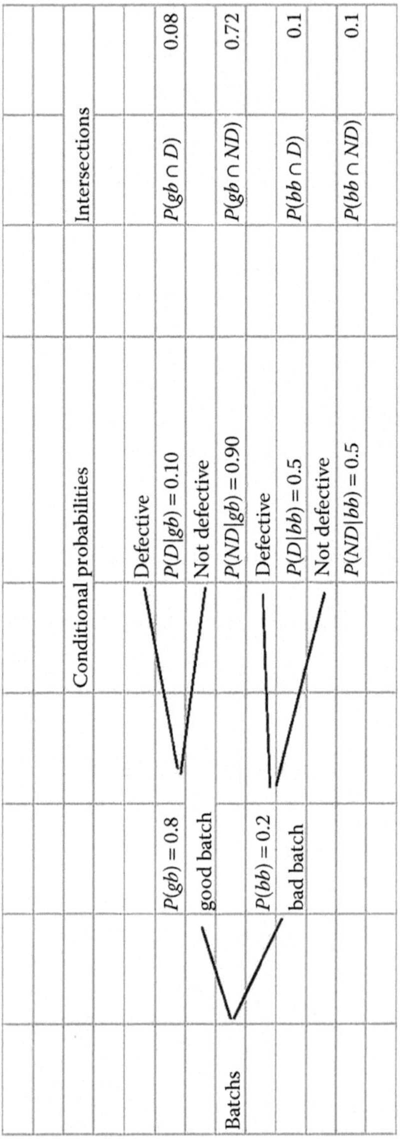

	Conditional probabilities	Intersections	
$P(gb) = 0.8$ good batch	Defective $P(D|gb) = 0.10$	$P(gb \cap D)$	0.08
	Not defective $P(ND|gb) = 0.90$	$P(gb \cap ND)$	0.72
$P(bb) = 0.2$ bad batch	Defective $P(D|bb) = 0.5$	$P(bb \cap D)$	0.1
Batchs	Not defective $P(ND|bb) = 0.5$	$P(bb \cap ND)$	0.1

FIGURE 2.7
Bayes' theorem tree diagram from Excel.

Exercises 2.4

1. Redo Example 2.7 with the following new tables of values:

	Outcomes		
	High Demand	Moderate Demand	Low Demand
Alternatives	($p_{hd} = 0.25$)	($p_{md} = 0.35$)	($p_{ld} = 0.40$)
Large warehouse	$250,000	$125,000	−$135,000
Small warehouse	$150,000	$65,000	−$35,000
No warehouse	$0	$0	$0

2. Acme Computer company manufactures memory chips in lots of 10 chips. From historical data, we built a frequency table of results:

	Defective Chip	No Defective Chip	Total
Good batch	40	660	700
Bad batch	120	180	300
Total	160	840	1000

If a good batch is sent on, costs of $1000 are incurred and if a bad batch is sent on, costs of $4000 are incurred. Acme can rework a batch of chips for $1000 and is sure to get a good batch. Alternatively, for a cost of $100, Acme can randomly test one chip per batch in an attempt to determine whether the batch is defective or good. Determine how Acme can minimize the expected costs per batch. Should they do any of the proposed inspections?

2.5 Decision Criteria under Risk and under Uncertainty

In the previous Sections 2.3 and 2.4, we used the criterion of maximizing the expected value to make decisions. Certainly, if you are repeating the decision many times over a long time, maximizing your expected value is very appealing. But, even then, there may be instances where you want to consider explicitly the *risk* involved. In this case, we will define risk as being able to estimate a probability of success and a probability of failure. For instance, in the example concerning the construction of small and large parking lots, consider the revised data where the net profit on a large parking lot increases, whereas its probability of award decreases and the data for the small parking lots remain the same as before:

Large Parking Lot (LP)	Small Parking Lot (SP)
Win contract: $260,000	Win contract: $140,000 profit
Lose contract: $−10,000	Lose contract: $−5,000
Probability of award of contract: 20%	Probability of award of contract: 35%

$$E(\text{LP}) = 0.20(\$280,000) + 0.80(-\$10,000) = \$48,000$$

and

$$E(\text{SP}) = 0.35(\$140,000) + 0.65(-\$5,000) = \$45,750$$

Following the criterion of maximizing the expected value, we would now choose to construct the large parking lots and over the long haul, we would make considerably more total profit. But what if this is a new company with limited capital in the short term, then there is *risk* involved. The numbers representing the expected value do not reveal—you cannot simply compare $48,000 with $45,750. First, there is a much lower probability of award for the larger parking lots than that of previously given. In the short term, are the higher probability of an award and the lower cost of failure of the smaller parking lots the better options for the new company? How do we measure risk in such cases?

And what about *one-time* decisions? Consider the former popular TV game *Deal or No Deal*. There are 26 suitcases containing an amount of money varying from $0.01 to $1,000,000. The contestant picks one of the suitcases. The contestant then begins to open a varying amount of suitcases. After opening the specified number of suitcases, the host makes the contestant an offer, which can either be accepted (deal) or rejected (no deal). Now let us suppose the contestant stays until the end, and we are down to the final two suitcases. One contains $0.01 and the other $1,000,000. The host offers the contestant $400,000. Should the contestant accept or reject the offer? If the contestant rejects the offer, the contestant would open the final two suitcases to see if he or she had picked the correct one. The expected value of continuing to play is

$$E = 0.5(1,000,000) + 0.5(\$0.01) = \$500,000.005$$

which is about 25% better than the offer of $400,000? Following the expected value criterion, the contestant would choose to play. But understand this: about half the time, the contestant would go home with only a penny. However, $400,000 is much better than $0.01 and although not as good as $500,00.005, perhaps it would be better to take the $400,000 offer. What would you do?

Now let us consider other decision criteria related to uncertainty. We define uncertainty as conditions for making decisions where we cannot estimate or know the related probabilities. Consider the following table that reflects (in $100,000) the final amount (after 5 years) of an initial investment of $100,000. Four different investment strategies A, B, C, and D and states of nature E, F, G, and H are available. The amount of the investment varies depending on the nature of the economy during the next 5 years. In this case, the investment is locked in for 5 years, so this is a one-time decision.

Example 2.10: Investments by Expected Value

Values for a 5-year investment of alternative investment strategies as a function of the nature of the economy:

		Fast, F	Moderate, M	Normal, N	Slow, S
		Nature of the Economy Investment			
	Fortune 500 stocks, A	2	3	1	1
You	Bonds, B	2	1	2	1
	Options, C	0.5	4	1	1
	Mutual funds, D	1	3	0.5	1

Case 1: One-time decisions and probabilities are known, and we want to maximize our expected outcome.

Maximize expected value criterion: Compute the expected value for each option. Choose the largest expected value. Let us assume that an accomplished econometrician has estimated a *subjective* probability. A subjective probability differs from our relative frequency definition used previously in that it is not the ratio of favorable outcomes to total outcomes because experimental data are not available, but rather it is the best estimate of a qualified expert. Let us assume that the probabilities for the states of the economy F, M, N, and S are 0.2, 0.4, 0.3, and 0.1, respectively. Then $E[A] = 2$, $E[B] = 1.5$, $E[C] = 2.1$, $E[D] = 1.65$. We would choose option C. Although the expected value does not reflect the risk involved, there may be instances where it is an appropriate criterion for a one-time decision. For example, suppose we are given a *one-time* chance to win $1000 on a *single roll* of a pair of dice, we would elect to bet on the number 7 because it has the highest probability of occurring and therefore the highest expected value. Others may choose differently.

Case 2: One-time decisions, probabilities unknown.

Laplace criterion: This decision criterion assumes that the unknown probabilities are equal. Therefore, we can simply average the payoffs (expected value) for each investment, or, equivalently, can choose the investment strategy with the highest sum because the weights are equal. In the previous example, the sums for each strategy are $A = 7$, $B = 6$, $C = 6.5$, and $D = 5.5$. So following the Laplace criterion, we would Choose A as the weighted average $(7/4) > (6.5/4) > (6/4) > (5.5/4)$. The Laplace method is simply the same as maximizing the expected value assuming the states of nature are equally likely.

Maximin criterion: Here, we want to compute the worst that can happen if we choose each strategy. We look at each row and find the minimum value. We now examine the columns of the row minimums and select the maximum of these minimum values. Hence the term *maximum (minima)* or *maximin*.

	Min	Maximin
A	1	Tie
B	1	Tie
C	0.5	
D	0.5	

The maximin strategy is *pessimistic* in that it chooses the strategy based only on the worst case for each strategy under consideration, completely neglecting the better cases that could result. When considering an investment that you wish to be conservative, such as investing funds for the college education of your children, and a minimax strategy gets the job done, it may be the decision criterion for which you are looking.

Maximax criterion: There may be instances where we want to obtain the best result possible. We examine each row and obtain the maximum in each row. Then, we examine the columns of max's and choose the largest value. Thus, our maximax, for our example, is to choose option C.

	Max	Maximax
A	3	
B	2	
C	4	4
D	3	

Obviously, this strategy is *optimistic* because it considers only the best case while neglecting the risk of each strategy. If all your investment needs were taken care of, if you want to invest some *extra money*, and if you are willing to suffer the risk involved for the possibility of a big gain, the optimistic maximax criterion may be of interest to you.

Coefficient of optimism criterion: This is a very subjective criterion that combines the pessimistic maximin criterion with the optimistic maximax criterion. You simply choose a coefficient of optimism $0 < x < 1$. For each strategy, you then compute

$$x(\text{row maximum}) + (1-x)(\text{row minimum})$$

Then you choose the row with the highest weighted value. For example, letting $x = 4/5$ in our example, we have

$$A = 4/5\ (3) + 1/5\ (1) = 13/5$$

$$B = 4/5\ (2) + 1/5\ (1) = 9/5$$

$$C = 4/5\ (4) + 1/5\ (0.5) = 16.5/5$$

$$D = 4/5\ (3) + 1/5\ (0.5) = 12.5/5$$

Thus we would choose Strategy C. Note, when $x = 0$, the coefficient of optimism criterion becomes the maximin criterion, and when $x = 1$, it becomes maximax criterion.

Summary: Note that for the investment strategies in Example 2.10, for the given economic conditions presented, we chose A, B, and C by one of the methods!

Example 2.11: Investment Strategy

Assume that we have 100,000 to invest (one time). We have three alternatives: stocks, bonds, or savings. We place the estimated returns in a payoff matrix.

Alternatives	Conditions		
Investments	Fast Growth (Risk)	Normal Growth	Slow Growth
Stocks	$10,000	$6,500	−$4,000
Bonds	8,000	$6,000	1,000
Savings	5,000	$5,000	5,000

Laplace criterion: Assumes equal probabilities. As there are three conditions, the chance of each is 1/3. We compute the expected value.

$$E[\text{Stocks}] = \frac{1}{3}(100,000) + \frac{1}{3}(6,500) + \frac{1}{3}(-4,000) = 4,167$$

$$E[\text{Bonds}] = \frac{1}{3}(8000) + \frac{1}{3}(6000) + \frac{1}{3}(1000) = 5000$$

$$E[\text{Savings}] = \frac{1}{3}(5000) + \frac{1}{3}(5000) + \frac{1}{3}(5000) = 5000$$

We have a tie, so we can do either bonds or savings under the Laplace criterion.

Maximin criterion: This assumes that the decision-maker is pessimistic about the future. According to this criterion, the minimum returns for each alternative are compared, and we choose the maximum of these minimums.

Stocks	−4000
Bonds	1000
Savings	5000

The maximum of these is savings at 5000.

Maximax criterion: This assumes an optimistic decision-maker. We take the maximum of each alternative and then the maximum of those.

Stocks	10,000
Bonds	8,000
Savings	5,000

The maximum of these is 10,000 with stocks.

Coefficient of optimism criterion: This is a compromise of the maximin and the maximax. It requires a coefficient of optimism called x that is between 0 and 1.

In this example, we assume that the coefficient of optimism is 0.6.

	Maximin	Maximax
Stocks	−4,500	10,000
Bonds	1,000	8,000
Savings	5,000	5,000

We compute the expected values using:

$$x * \text{Maximax value} + (1-x) * \text{Maximin value}$$

We illustrate:

$$EH[\text{Stocks}] = 0.6(10,000) + 0.4(-4,500) = 4,400$$

$$EH[\text{Bonds}] = 0.6(8000) + 0.4(1000) = 5200$$

$$EH[\text{Savings}] = 0.6(5000) + 0.4(5000) = 5000$$

5200 is the best answer, so with our choice of $\alpha = 0.6$ we pick bonds. Let us summarize our decision as follows:

Criterion	Choice
Laplace	Savings or bonds
Maximin	Savings
Maximax	Stocks
Coefficient of optimism	Bonds with $x = 0.6$

Thus, as there is no unanimous decision, we must consider the characteristics of the decision-maker. First, a modeler finds all the possible decisions and then considers the decision maker's characteristics or needs when making their recommendations.

Exercises 2.5

1. Given the following payoff matrix. Show all work to answer parts (a) and (b).

	States of Nature			
	$p = 0.30$	$p = 0.35$	$p = 0.20$	$p = 0.15$
Alternatives	#1	#2	#3	#4
A	1250	1000	500	400
B	950	1500	1000	500
C	800	1200	600	1000

   a. Which alternative do we choose under risk?

   b. Find the opportunity loss table and compute the expected opportunity loss for each alternative. What decision do you make?

2. We are considering one of three alternatives A, B, or C under *uncertainty* conditions. The payoff matrix is as follows:

	Conditions		
Alternative	#1	#2	#3
A	3500	4500	6300
B	1500	9500	2200
C	4500	4500	3500

   Determine the best plan by each of the following criteria and show work:

   a. Laplace

   b. Maximin

   c. Maximax

   d. Coefficient of optimism (assume that $x = 0.65$)

3. We have a choice of two investment strategies: stocks and bonds. The returns for each under two possible economic conditions are as follows:

States of Nature	
$p_1 = 0.68$	$p_2 = 0.32$

Alternative	Condition 1	Condition 2
Stocks	$11,500	−$4,300
Bonds	$7,700	$1,400

a. Compute the expected value and select the best alternative.

b. What probabilities for conditions 1 and 2 would have to exist to be indifferent toward stocks and bonds?

4. Given the following payoff matrix:

		Conditions	
Alternative	#1	#2	#3
A	$1400	2400	900
B	850	1250	950
C	650	650	650

Determine the best plan by each of the following criteria and show work:

a. Laplace

b. Maximin

c. Maximax

d. Coefficient of optimism (assume that $x = 0.55$)

5. A local investor is three alternative considering real estate investments, a hotel, a restaurant, and a convenience store. The hotel and the convenience store will be adversely or favorably affected depending on their closeness to gasoline stations, whereas the restaurant will be assumed to be relatively stable. The payoff is given as follows:

	Conditions		
Alternative	#1 Gas Close Distance	#2 Gas Medium Distance	#3 Gas Far Away Distance
Hotel	$25,000	10,000	−8,000
Convenience store	4,000	8,000	−12,000
Restaurant	5,000	6,000	6,000

Determine the best plan by each of the following criteria and show work:

a. Laplace

b. Maximin

c. Maximax

d. Coefficient of optimism (assume that $x = 0.48$)

2.6 EXCEL Add-Ins

Add-ins in Excel or any software to assist with decision analysis is very useful. We recently downloaded the add-in from SourceForge.net (https://sites. google.com/site/simpledecisiontree/).

We recommend to download software that you want to use and to read all the literature concerning the software before attempting to use it.

We will resolve the following concession problem with that software as an illustration.

Consider a firm that handles concessions for a sporting event. The firm's manager needs to know whether to stock up with coffee or coke-cola products. A local agreement restricts you to only one beverage. You estimate a $1500 profit of selling coke-cola products if it is cold and a $5000 profit of selling cola if it is warm. You also estimate a $4000 profit of selling coffee if it is cold and a $1000 profit of selling coffee if it is warm. The forecast says there is a 30% of a cold front, otherwise the weather will be warm. What do you do?

Our decision is to sell coke product and to get an estimated profit of $3950, shown in Figure 2.8. In this example, we set up the probabilities in the sheet that we, the user, may change them for the sensitivity analysis.

We find at a probability of cold of about 0.616, Figure 2.9, we would change our decision and sell coffee products.

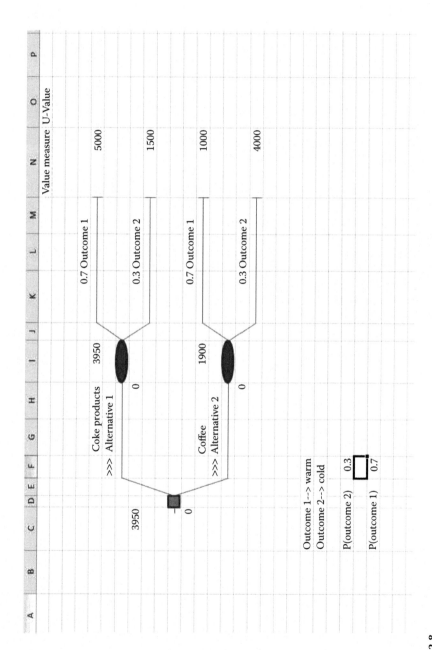

FIGURE 2.8
Screenshot for decision tree technology for concession problem.

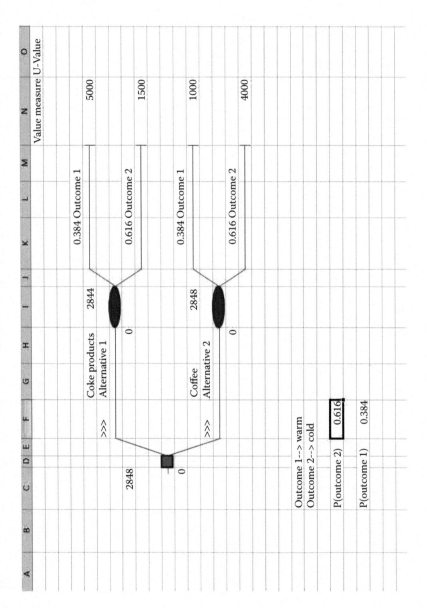

FIGURE 2.9
Screenshot for sensitivity analysis using the Add-In.

References and Suggested Further Readings

Bellman, R. 1957. *Dynamic Programming*. Mineola, NY: Dover Publications.

Bellman, R. and S. Dreyfus. 1962. *Applied Dynamic Programming*. Princeton, NJ: Princeton University Press.

Giordano, F., W. Fox, and S. Horton. 2014. *A First Course in Mathematical Modeling*. Boston, MA: Cengage Publishing, Chapter 9.

Hillier, F. and M. Hillier. 2010. *Introduction to Management Science: A Modeling and Case Studies Approach with Spreadsheets* (4th ed.). New York: McGraw-Hill.

Lawrence, J. and B. Pasterneck, 2002. *Applied Management Science: Modeling, Spreadsheet Analysis, and Communication for Decision Making* (2nd ed.). New York: Wiley Publishers.

Lee, S., L. Moore, and B. Taylor. 1985. *Management Science* (2nd ed.). Boston, MA: Allyn and Bacon Publishers, Chapter 4.

National Vital Statistics Report. 1997. Vol. 47, No. 28, p. 74. https://www.cdc.gov/nchs/data/nvsr/nvsr47/nvs47_28.pdf.

Winston, W. L. 1994. *Operations Research: Applications and Algorithms* (3rd ed.). Boston, MA: Duxbury Press, Chapter 13.

3

Mathematical Programming Models:
Linear, Integer, and Nonlinear Optimization

OBJECTIVES

1. Formulate mathematical programming problems.
2. Distinguish between types of mathematical programming problems.
3. Use appropriate technology to solve the problem.
4. Understand the importance of sensitivity analysis.

3.1 Introduction

Recall that the Emergency Service Coordinator (ESC) for a county is interested in locating the county's three ambulances to maximize the residents that can be reached within 8 minutes in emergency situations. The county is divided into six zones, and the average time required to travel from one region to the next under semiperfect conditions are summarized in Table 3.1.

The population in zones 1, 2, 3, 4, 5, and 6 is given in Table 3.2.

In Chapter 1, we presented the problem statement and basic assumptions:

Problem Statement: Determine the location for placement of the ambulances to maximize coverage within the allotted time.

Assumptions: Time travel between zones is negligible. Times in the data are averages under ideal circumstances.

Here, we further assume that employing an optimization technique would be worthwhile. We will begin with assuming a linear model, and then we might enhance the model with integer programming.

TABLE 3.1

Average Travel Times from Zone i to Zone j in Perfect Conditions

	1	2	3	4	5	6
1	1	8	12	14	10	16
2	8	1	6	18	16	16
3	12	18	1.5	12	6	4
4	16	14	4	1	16	12
5	18	16	10	4	2	2
6	16	18	4	12	2	2

TABLE 3.2

Populations in Each Zone

1	50,000
2	80,000
3	30,000
4	55,000
5	35,000
6	20,000
Total	270,000

Perhaps, consider planning the shipment of needed items from the warehouses where they are manufactured and stored to the distribution centers where they are needed.

There are three warehouses at different cities: Detroit, Pittsburgh, and Buffalo. They have 250, 130, and 235 tons of paper accordingly. There are four publishers in Boston, New York, Chicago, and Indianapolis. They ordered 75, 230, 240, and 70 tons of paper to publish new books. Transportation costs, in dollars, of one ton of paper are listed in the following table:

From\To	Boston (BS)	New York (NY)	Chicago (CH)	Indianapolis (IN)
Detroit (DT)	15	20	16	21
Pittsburgh (PT)	25	13	5	11
Buffalo (BF)	15	15	7	17

Management wants you to minimize the shipping costs while meeting demand. This problem involves the allocation of resources and can be modeled as a linear programming problem as we will discuss.

In engineering management, the ability to optimize results in a constrained environment is crucial to success. In addition, the ability to perform critical sensitivity analysis, or *what if analysis*, is extremely important for decision-making. Consider starting a new diet, which needs to healthy. You go to a nutritionist that gives you lots of information on foods. They recommend sticking to six different foods: bread, milk, cheese, fish, potato, and yogurt and provides you a table of information including the average cost of the items:

	Bread	Milk	Cheese	Potato	Fish	Yogurt
Cost ($)	2.0	3.5	8.0	1.5	11.0	1.0
Protein (g)	4.0	8.0	7.0	1.3	8.0	9.2
Fat (g)	1.0	5.0	9.0	0.1	7.0	1.0
Carbohydrates (g)	15.0	11.7	0.4	22.6	0.0	17.0
Calories (cal)	90	120	106	97	130	180

We go to a nutritionist and he or she recommends that our diet should contain not less than 150 cal, not more than 10 g of protein, not less than 10 g of carbohydrates, and not less than 8 g of fat. Also, we decide that our diet should have *minimal cost*. In addition, we conclude that our diet should include at least 0.5 g of fish and not more than 1 cup of milk. Again this is an allocation of resources problem where we want the optimal diet at minimum cost. We have six unknown variables that define weight of the food. There is a lower bound for fish as 0.5 g. There is an upper bound for milk as 1 cup. To model and solve this problem, we can use linear programming.

Modern linear programming was the result of a research project undertaken by the U.S. Department of Air Force under the title of Project Scientific Computation of Optimum Programs (SCOOP). As the number of fronts in the Second World War increased, it became more and more difficult to coordinate troop supplies effectively. Mathematicians looked for ways to use the new computers being developed to perform calculations quickly. One of the SCOOP team members, George Dantzig, developed the simplex algorithm for solving simultaneous linear programming problems. This simplex method has several advantageous properties: it is very efficient, allowing its use for solving problems with many variables; it uses methods from linear algebra, which are readily solvable.

In January 1952, the first successful solution to a linear programming problem was found using a high-speed electronic computer on the National Bureau of Standards Eastern Automatic Computer (SEAC) machine. Today, most linear programming problems are solved via high-speed computers. Computer specific software, such as LINDO, EXCEL SOLVER, GAMS, have been developed to help in the solving and analysis of linear programming problems. We will use the power of LINDO to solve our linear programming problems in this chapter.

To provide a framework for our discussions, we offer the following basic model in Equation 3.1:

Maximize (or minimize) $f(X)$

Subject to

$$g_i(X) \begin{Bmatrix} \geq \\ = \\ \leq \end{Bmatrix} b_i \text{ for all } i \qquad (3.1)$$

Now let us explain this notation. The various components of the vector **X** are called the decision variables of the model. These are the variables that can be controlled or manipulated. The function, $f(X)$, is called the objective function. By subject to, we connote that there are certain side conditions, resource requirement, or resource limitations that must be met. These conditions are called constraints. The constant b_i represents the level of the associated constraint $g(X_i)$ and is called the right-hand side (RHS) in the model.

Linear programming is a method for solving linear problems, which occur very frequently in almost every modern industry. In fact, areas using linear programming are as diverse as defense, health, transportation, manufacturing, advertising, and telecommunications. The reason for this is that in most situations, the classic economic problem exists—you want to maximize output, but you are competing for limited resources. The *linear* in linear programming means that in the case of production, the quantity produced is proportional to the resources used and also the revenue generated. The coefficients are constants, and no products of variables are allowed.

In order to use this technique, the company must identify a number of constraints that will limit the production or transportation of their goods; these may include factors such as labor hours, energy, and raw materials. Each constraint must be quantified in terms of one unit of output, as the problem-solving method relies on the constraints being used.

An optimization problem that satisfies the following five properties is said to be a linear programming problem:

1. There is a unique objective function, $f(X)$.
2. Whenever a decision variable, X, appears in either the objective function or a constraint function, it must appear with an exponent of 1, possibly multiplied by a constant.
3. No terms contain products of decision variables.
4. All coefficients of decision variables are constants.
5. Decision variables are permitted to assume fractional and integer values.

Linear problems, by the nature of the many unknowns, are very hard to solve by human inspection, but methods have been developed to use the power of computers to do the hard work.

3.2 Formulating Mathematical Programming Problems

A linear programming problem is a problem that requires an objective function to be maximized or minimized subject to resource constraints. The key to formulating a linear programming problem is recognizing the decision variables. The objective function and all constraints are written in terms of these decision variables.

The conditions for a mathematical model to be a linear program (LP) were as follows:

- All variables must be continuous (i.e., they can take on fractional values).
- There must be a single objective (minimize or maximize).
- The objective and constraints must be linear, that is, any term must either be a constant or a constant multiplied by an unknown.
- The decision variables must be nonnegative.

LPs are important—this is because of the following:

- Many practical problems can be formulated as LPs.
- There exists an algorithm (called the *simplex* algorithm) that enables us to solve LPs numerically relatively easily.

We will return to the simplex algorithm later for solving LPs, but for the moment we will concentrate on formulating LPs.

Some of the major application areas to which LP can be applied are as follows:

- Blending
- Production planning
- Oil refinery management
- Distribution
- Financial and economic planning
- Manpower planning
- Blast furnace burdening
- Farm planning

We consider in the following some specific examples of the types of problem that can be formulated as LPs. Note here that the key to formulating LPs is *practice*. However a useful hint is that common objectives for LPs are to *minimize cost* or *maximize profit*.

Example 3.1: Manufacturing

Consider the following problem statement: A company wants to can two new different drinks for the holiday season. It takes 2 hours to can one gross of Drink A, and it takes 1 hour to label the cans. It takes 3 hours to can one gross of Drink B, and it takes 4 hours to label the cans. The company makes $10 profit on one gross of Drink A and a $20 profit on one gross of Drink B. Given that we have 20 hours to devote for canning the drinks and 15 hours to devote for labeling cans per week, how many cans of each type of drink should the company package to maximize profits?

Problem Identification: Maximize the profit of selling these new drinks. *Define variables*:

X_1 is the number of gross cans produced for Drink A per week
X_2 is the number of gross cans produced for Drink B per week

Objective Function:

$$Z = 10X_1 + 20X_2$$

Constraints:

1. Canning with only 20 hours available per week

$$2X_1 + 3X_2 \leq 20$$

2. Labeling with only 15 hours available per week

$$X_1 + 4X_2 \leq 15$$

3. Nonnegativity restrictions
   $X_1 \geq 0$ (nonnegativity of the production items)
   $X_2 \geq 0$ (nonnegativity of the production items)

The complete formulation:

Maximize $Z = 10X_1 + 20X_2$
Subject to

$$2X_1 + 3X_2 \leq 20$$
$$X_1 + 4X_2 \leq 15$$
$$X_1 \geq 0$$
$$X_2 \geq 0$$

We will see in the next section how to solve these two-variable problems graphically.

Example 3.2: Financial Planning

A bank makes four kinds of loans to its personal customers, and these loans yield the following annual interest rates to the bank:

- First mortgage—14%
- Second mortgage—20%
- Home improvement—20%
- Personal overdraft—10%

The bank has a maximum foreseeable lending capability of $250 million and is further constrained by the policies:

1. First, mortgages must be at least 55% of all mortgages issued and at least 25% of all loans issued (in $ terms)

2. Second, mortgages cannot exceed 25% of all loans issued (in $ terms)
3. To avoid public displeasure and the introduction of a new windfall tax, the average interest rate on all loans must not exceed 15%

Formulate the bank's loan problem as an LP so as to maximize interest income while satisfying the policy limitations.

Note here that these policy conditions, while potentially limiting the profit that the bank can make, also limit its exposure to risk in a particular area. It is a fundamental principle of risk reduction that risk is reduced by spreading money (appropriately) across different areas.

FINANCIAL PLANNING FORMULATION

Note here that as in all formulation exercises, we are translating a verbal description of the problem into an equivalent mathematical description.

A useful tip when formulating LPs is to express the variables, constraints, and objective in words before attempting to express them in mathematics.

VARIABLES

Essentially, we are interested in the amount (in dollars) that the bank has loaned to customers in each of the four different areas (not in the actual number of such loans). Hence let x_i = amount loaned in area i in millions of dollars (where $i = 1$ corresponds to first mortgages, $i = 2$ corresponds to second mortgages, etc.) and note that each $x_i \geq 0$ ($i = 1, 2, 3, 4$).

Note here that it is conventional in LPs to have all variables ≥ 0. Any variable (X, say) that can be positive *or* negative can be written as $X_1 - X_2$ (the difference of two new variables) where $X_1 >= 0$ and $X_2 >= 0$.

CONSTRAINTS

1. Limit on amount lent

$$x_1 + x_2 + x_3 + x_4 \leq 250$$

2. Policy condition 1

$$x_1 \geq 0.55(x_1 + x_2)$$

that is, first mortgages ≥ 0.55 (total mortgage lending) and also

$$x_1 \geq 0.25(x_1 + x_2 + x_3 + x_4)$$

that is, first mortgages ≥ 0.25 (total loans)
3. Policy condition 2

$$x_2 \leq 0.25(x_1 + x_2 + x_3 + x_4)$$

4. Policy condition 3—we know that the total annual interest is 0.1 $4x_1 + 0.20x_2 + 0.20x_3 + 0.10x_4$ on total loans of $(x_1 + x_2 + x_3 + x_4)$. Hence the constraint relating to policy condition (3) is

$$0.14x_1 + 0.20x_2 + 0.20x_3 + 0.10x_4 \leq 0.15(x_1 + x_2 + x_3 + x_4)$$

OBJECTIVE FUNCTION

To maximize interest income (which is given earlier), that is,

$$\text{Maximize } Z = 0.14x_1 + 0.20x_2 + 0.20x_3 + 0.10x_4$$

Example 3.3: Blending and Formulation

Consider the example of a manufacturer of animal feed who is producing feed mix for dairy cattle. In our simple example, the feed mix contains two active ingredients. One kg of feed mix must contain a minimum quantity of each of four nutrients as follows:

Nutrient	A	B	C	D
Gram	90	50	20	2

The ingredients have the following nutrient values and cost

	A	B	C	D	Cost/kg
Ingredient 1 (gram/kg)	100	80	40	10	40
Ingredient 2 (gram/kg)	200	150	20	0	60

What should be the amounts of active ingredients in 1 kg of feed mix that minimizes cost?

BLENDING PROBLEM SOLUTION

Variables

In order to solve this problem, it is best to think in terms of 1 kg of feed mix. That 1 kg is made up of two parts: ingredient 1 and ingredient 2:

x_1 is the amount (kg) of ingredient 1 in 1 kg of feed mix.
x_2 is the amount (kg) of ingredient 2 in 1 kg of feed mix.

where $x_1 \geq 0, x_2 \geq 0$
  Essentially, these variables (x_1 and x_2) can be thought of as the recipe telling us how to make up 1 kg of feed mix.

Constraints

- Nutrient constraints

  $100x_1 + 200x_2 >= 90$ (nutrient A)
  $80x_1 + 150x_2 >= 50$ (nutrient B)
  $40x_1 + 20x_2 >= 20$ (nutrient C)
  $10x_1 >= 2$ (nutrient D)

- Balancing constraint (an *implicit* constraint due to the definition of the variables)

$$x_1 + x_2 = 1$$

Objective Function

Presumably to minimize cost, that is,

$$\text{Minimize } Z = 40x_1 + 60x_2$$

This gives us our complete LP model for the blending problem.

Example 3.4: Production Planning Problem

A company manufactures four variants of the same table and in the final part of the manufacturing process, there are assembly, polishing, and packing operations. For each variant, the time required for these operations is shown in the following table (in minutes) as is the profit per unit sold.

	Assembly	Polish	Pack	Profit ($)
Variant 1	2	3	2	1.50
2	4	2	3	2.50
3	3	3	2	3.00
4	7	4	5	4.50

Given the current state of the labor force, the company estimates that, each year, they have 100,000 minutes of assembly time, 50,000 minutes of polishing time, and 60,000 minutes of packing time available. How many of each variant should the company make per year and what is the associated profit?

VARIABLES

Let
x_i be the number of units of variant $i (i = 1, 2, 3, 4)$ made per year
where $x_i > 0$ $i = 1, 2, 3, 4$

CONSTRAINTS

Resources for the operations of assembly, polishing, and packing

$$2x_1 + 4x_2 + 3x_3 + 7x_4 <= 100{,}000 \text{ (assembly)}$$
$$3x_1 + 2x_2 + 3x_3 + 4x_4 < = 50{,}000 \text{ (polishing)}$$
$$2x_1 + 3x_2 + 2x_3 + 5x_4 < = 60{,}000 \text{ (packing)}$$

OBJECTIVE FUNCTION

Maximize $Z = 1.5x_1 + 2.5x_2 + 3.0x_3 + 4.5x_4$

Example 3.5: Shipping

Consider planning the shipment of needed items from the warehouses where they are manufactured and stored to the distribution centers where they are needed as shown in the introduction. There are three warehouses at different cities: Detroit, Pittsburgh, and Buffalo. They have

250, 130, and 235 tons of paper accordingly. There are four publishers in Boston, New York, Chicago, and Indianapolis. They ordered 75, 230, 240, and 70 tons of paper to publish new books.

Transportation costs, in dollars, of 1 ton of paper:

From\To	Boston (BS)	New York (NY)	Chicago (CH)	Indianapolis (IN)
Detroit (DT)	15	20	16	21
Pittsburgh (PT)	25	13	5	11
Buffalo (BF)	15	15	7	17

Management wants you to minimize the shipping costs while meeting demand.

We define x_{ij} to be the travel from city i (1 is Detroit, 2 is Pittsburg, 3 is Buffalo) to city j (1 is Boston, 2 is New York, 3 is Chicago, and 4 is Indianapolis).

Minimize $Z = 15x_{11} + 20x_{12} + 16x_{13} + 21x_{14} + 25x_{21} + 13x_{22} + 5x_{23} + 11x_{24} + 15x_{31} + 15x_{32} + 7x_{33} + 17x_{34}$

Subject to

$$x_{11} + x_{12} + x_{13} + x_{14} \leq 250 \text{ (availability in Detroit)}$$
$$x_{21} + x_{22} + x_{23} + x_{24} \leq 130 \text{ (availability in Pittsburg)}$$
$$x_{31} + x_{32} + x_{33} + x_{34} \leq 235 \text{ (availability in Buffalo)}$$
$$x_{11} + x_{21} + x_{31} \geq 75 \text{ (demand Boston)}$$
$$x_{12} + x_{22} + x_{32} \geq 230 \text{ (demand New York)}$$
$$x_{13} + x_{23} + x_{334} \geq 240 \text{ (demand Chicago)}$$
$$x_{14} + x_{24} + x_{34} \geq 70 \text{ (demand Indianapolis)}$$
$$x_{ij} \geq 0$$

INTEGER PROGRAMMING AND MIXED-INTEGER PROGRAMMING

For integer and mixed-integer programming, we will take advantage of technology. We will not present the branch and bound technique, but we suggest that a thorough review of the topic can be found in Winston (2002) or other similar mathematical programming texts.

Perhaps in Example 3.5, shipping, we decide that all shipments must be integer shipment and no partial shipments are allowed. That would cause us to solve Example 3.5 as an integer programming problem. Assignment problems, transportation problems, and assignments with binary constraints are among the most used integer and binary integer problems.

NONLINEAR PROGRAMMING

It is not our plan to present material on how to formulate or solve nonlinear programs. Often, we have nonlinear objective functions or nonlinear constraints. Suffice it to say that we will recognize these and use technology to assist in the solution. Excellent nonlinear programming information, methodology, and algorithms can be gained from our recommended suggested reading. Many problems are, in fact, nonlinear. We will provide a few examples later in the chapter. We point out that often numerical algorithms such as one-dimensional golden section or two-dimensional gradient search methods are sued to solve nonlinear problems.

Exercises 3.2

Formulate the following problems:

1. A company wants to can two different drinks for the holiday season. It takes 3 hours to can one gross of Drink A, and it takes 2 hours to label the cans. It takes 2.5 hours to can one gross of Drink B, and it takes 2.5 hours to label the cans. The company makes $15 profit on one gross of Drink A and an $18 profit on one gross of Drink B. Given that we have 40 hours to devote for canning the drinks and 35 hours to devote for labeling cans per week, how many cans of each type of drink should the company package to maximize profits?

2. The Mariners Toy Company wishes to make three models of ships to maximize their profits. They found that a model steamship takes the cutter 1 hour, the painter 2 hours, and the assembler 4 hours of work; it produces a profit of $6.00. The sailboat takes the cutter 3 hours, the painter 3 hours, and the assembler 2 hours. It produces a $3.00 profit. The submarine takes the cutter 1 hour, the painter 3 hours, and the assembler 1 hour. It produces a profit of $2.00. The cutter is only available for 45 hours per week, the painter for 50 hours, and the assembler for 60 hours. Assume that they sell all the ships that they make, formulate this LP to determine how many ships of each type that Mariners should produce.

3. In order to produce 1000 tons of nonoxidizing steel for BMW engine valves, at least the following units of manganese, chromium, and molybdenum will be needed weekly: 10 units of manganese, 12 units of chromium, and 14 units of molybdenum (1 unit is 10 lb). These materials are obtained from a dealer who markets these metals in three sizes: small (S), medium (M), and large (L). One S case costs $9 and contains two units of manganese, two units of chromium, and one unit of molybdenum. One M case costs $12 and contains two units of manganese, three units of chromium, and one unit of molybdenum. One L case costs $15 and contains one unit of manganese, one unit of chromium, and five units of molybdenum. How many cases of each kind (S, M, L) should be purchased weekly so that we have enough manganese, chromium, and molybdenum at the smallest cost?

4. The Super Bowl advertising agency wishes to plan an advertising campaign in three different media: television, radio, and magazines.

The purpose or goal is to reach as many potential customers as possible. Results of a marketing study are given in the following:

	Day Time TV	Prime Time TV	Radio	Magazines
Cost of advertising unit	$40,000	$75,000	$30,000	$15,000
Number of potential customers reached per unit	400,000	900,000	500,000	200,000
Number of woman customers reached per unit	300,000	400,000	200,000	100,000

The company does not want to spend more than $800,000 on advertising. It further requires (1) at least 2 million exposures to take place among woman, (2) TV advertising to be limited to $500,000, (3) at least three advertising units to be bought on day time TV and two units on prime time TV, and (4) the number of radio and magazine advertisement units should each be between 5 and 10 units.

5. A tomato cannery has 5,000 lb of grade A tomatoes and 10,000 lb of grade B tomatoes from which they will make whole canned tomatoes and tomato paste. Whole tomatoes must be composed of at least 80% grade A tomatoes, whereas tomato paste must be made with at least 10% grade A tomatoes. Whole tomatoes sell for $0.08 per pound and grade B tomatoes sell for $0.05 per pound. Maximize revenue of the tomatoes.

   Hint: Let X_{wa} be pounds of grade A tomatoes used as whole tomatoes and X_{wb} be pounds of grade B tomatoes used as whole tomatoes. The total number of whole tomato cans produced is the sum of $X_{wa} + X_{wb}$ after each is found. Also remember that a percent is a fraction of the whole times 100%.

6. The McCow Butchers is a large-scale distributor of dressed meats for Myrtle Beach restaurants and hotels. Ryan's steak house orders meat for meatloaf (mixed ground beef, pork, and veal) for 1000 lb according to the following specifications:

   a. Ground beef must be not less than 400 lb and not more than 600 lb.

   b. The ground pork must be between 200 and 300 lb.

   c. The ground veal must weigh between 100 and 400 lb.

   d. The weight of the ground pork must be not more than one and one half (3/2) times the weight of the veal.

   The contract calls for Ryan's to pay $1200 for the meat. The cost per pound for the meat is $0.70 for hamburger, $0.60 for pork, and $0.80 for the veal. How can this be modeled?

7. *Portfolio investments*: A portfolio manager who is in charge of a bank wants to invest $10 million. The securities available for purchase, as well as their respective quality ratings, maturate, and yields, are shown in the following table:

Bond Name	Bond Type	Moody's Quality Scale	Bank's Quality Scale	Years to Maturity	Yield at Maturity (%)	After-Tax Yield (%)
A	Municipal	Aa	2	9	4.3	4.3
B	Agency	Aa	2	15	5.4	2.7
C	Govt 1	Aaa	1	4	5	2.5
D	Govt 2	Aaa	1	3	4.4	2.2
E	Local	Ba	5	2	4.5	4.5

The bank places certain policy limitations on the portfolio manager's actions:

a. Government and agency bonds must total at least $4 million.

b. The average quality of the portfolios cannot exceed 1.4 on the bank's quality scale. Note that a low number means high quality.

c. The average years to maturity must not exceed 5 years.

Assume that the objective is to maximize after-tax earnings on the investment.

8. Suppose a newspaper publisher must purchase three kinds of paper stock. The publisher must meet their demand but desire to minimize their costs in the process. They decide to use an economic lot size model to assist them in their decisions. Given an economic order quantity model (EOQ) with constraints where the total cost is the sum of the individual quantity costs:

$$C(Q_1, Q_2, Q_3) = C(Q_1) + C(Q_2) + C(Q_3)$$

$$C(Q_i) = \frac{a_i d_i}{Q_i} + \frac{h_i Q_i}{2}$$

where:
  d is the order rate
  h is the cost per unit time (storage)
  $Q/2$ is the average amount on hand
  a is the order cost

The constraint is the amount of storage area available to the publisher so that he or she can have three kinds of paper on hand for use. The items cannot be stacked but can be laid side by side. They are constrained by the available storage area, *S*.

The following data are collected:

	Type I	Type II	Type III
d	32 rolls/week	24	20
a	$25	$18	$20
h	$1/roll/week	$1.5	$2.0
s	4 sq ft/roll	3	2

You have 200 sq ft of storage space available. Formulate the problem.

9. Suppose, you want to use the Cobb–Douglass function $P(L,K) = A\,L^a\,K^b$ to predict output in thousands based on amount of capital and labor used. Suppose you know that the prices of capital and labor per year are $10,000 and $7,000, respectively. Your company estimates the values of *A* as 1.2, *a* as 0.3, and *b* as 0.6. Your total cost is assumed to be $T = PL*L + Pk*k$, where *PL* and *Pk* are the prices of capital and labor. There are three possible funding levels: $63,940, $55,060, or $71,510. Formulate the problem to determine which budget yields the best solution for your company.

10. The manufacturer of a new plant is planning the introduction of two new products: a 19-inch stereo color set with a manufacturer's suggested retail price (MSRP) of $339 and a 21-inch stereo color set with a MSRP of $399. The cost to the company is $195 per 19-inch set and $225 per 21-inch set, plus an additional $400,000 in fixed costs of initial parts, initial labor, and machinery. In a competitive market in which they desire to sell the sets, the number of sales per year will affect the average selling price. It is estimated that for each type of set, the average selling price drops by one cent for each additional unit sold. Furthermore, sales of 19-inch sets will affect the sales of 21-inch sets and vice versa. It is estimated that the average selling price for the 19-inch set will be reduced by an additional 0.3 cents for each 21-inch set sold, and the price for the 21-inch set will decrease by 0.4 cents for each 19-inch set sold. We desire to provide them the optimal number of units of each type set to produce and to determine the expected profits. Recall that profit is revenue minus cost, $P = R - C$. Formulate the model to maximize profits. Ensure that you have accounted for all revenues and costs. Define all your variables.

11. In problem 10, we assumed that the company has the potential to produce any number of TV sets per year. Now we realize that there is a limit on production capacity. Consideration of these two products came about, because the company plans to discontinue manufacturing of its black and white sets, thus providing excess capacity at

its assembly plants. This excess capacity could be used to increase production of other existing product lines, but the company feels that these new products will be more profitable. It is estimated that the available production capacity will be sufficient to produce 10,000 sets per year (about 200 per week). The company has ample supply of 19-inch and 21-inch color tubes, chassis, and other standard components; however, circuit assemblies are in short supply. In addition, the 19-inch TV requires different circuit assemblies than the 21-inch TV. The supplier can deliver 8000 boards per year for the 21-inch model and 5000 boards per year for the 19-inch model. Taking this new information into account, what should the company now do? Formulate this problem.

3.3 Graphical Linear Programming

Many applications in business and economics involve a process called optimization. In optimization problems, you are asked to find the minimum or the maximum result. This section illustrates the strategy in graphical simplex method of linear programming. We will restrict ourselves in this graphical context to two dimensions. Variables in the simplex method are restricted to positive variables (e.g., $x \geq 0$).

A two-dimensional linear programming problem consists of a linear objective function and a system of linear inequalities called resource constraints. The objective function gives the linear quantity that is to be maximized (or minimized). The constraints determine the *set of feasible solutions*. Understanding the two-variable case helps in the understanding of more complicated programming problems. Let us illustrate a two-variable example.

Example 3.6: Memory Chips for CPUs

Let us start with a manufacturing example. Suppose a small business wants to know how many of two types of high-speed computer chips to manufacture weekly to maximize their profits. First, we need to define our decision variables. Let

x_1 be the number of high-speed chip type A to produce weekly
x_2 be the number of high-speed chip type B to produce weekly

The company reports a profit of \$140 for each type A chip and \$120 for each type B chip sold. The production line reports the information in Table 3.3.

TABLE 3.3

Memory Chip Data for CPUs

	Chip A	Chip B	Quantity Available
Assembly time (hours)	2	4	1400
Installation time (hours)	4	3	1500
Profit	140	120	

The constraint information from the table becomes inequalities that are written mathematically as follows:

$2x_1 + 4x_2 \leq 1400$ (assembly time)
$4x_1 + 3x_2 \leq 1500$ (installation time)
$x_1 > 0, x_2 \geq$

The profit equation is

$$\text{Profit } Z = 140x_1 + 120x_2$$

THE FEASIBLE REGION

We use the constraints of the linear program:

$2x_1 + 4x_2 \leq 1400$ (assembly time)
$4x_1 + 3x_2 \leq 1500$ (installation time)
$x_1 > 0, x_2 \geq 0$

The constraints of a linear program, which include any bounds on the decision variables, essentially shape the region in the x–y plane that will be the domain for the objective function before any optimization being performed. Every inequality constraint that is part of the formulation divides the entire space defined by the decision variables into two parts: the portion of the space containing points that violate the constraint and the portion of the space containing points that satisfy the constraint.

It is very easy to determine which portion will contribute to shaping the domain. We can simply substitute the value of some point in either *half-space* into the constraint. Any point will do, but the origin is particularly appealing. As there is only one origin, if it satisfies the constraint, then the *half-space* containing the origin will contribute to the domain of the objective function.

When we do this for each of the constraints in the problem, the result is an area representing the intersection of all the *half-spaces* that satisfied the constraints individually. This intersection is the domain for the objective function for the optimization. As it contains points that satisfy all the constraints simultaneously, these points are considered feasible to the problem. The common name for this domain is the *feasible region*.

Consider our constraints:

$2x_1 + 4x_2 \leq 1400$ (assembly time)
$4x_1 + 3x_2 \leq 1500$ (installation time)
$x_1 > 0, x_2 \geq 0$

For our graphical work, we use the constraints: $x_1 \geq 0$, $x_2 \geq 0$ to set the region. Here, we are strictly in the x_1–x_2 plane (the first quadrant).

Let us first take constraint #1 (assembly time) in the first quadrant: $2x_1 + 4x_2 \leq 1400$ shown in Figure 3.1.

First, we graph each constraint as equality, one at a time. We choose a point, usually the origin to test the validity of the inequality constraint. We shade all the areas where the validity holds. We repeat this process for all constraints to obtain Figure 3.2.

Figure 3.2 shows a plot of (1) the assembly hour's constraint and (2) the installation hour's constraint in the first quadrant. Along with the nonnegativity restrictions on the decision variables, the intersection of the half-spaces defined by these constraints is the feasible region. This area represents the domain for the objective function optimization.

We region shaded in our feasible region.

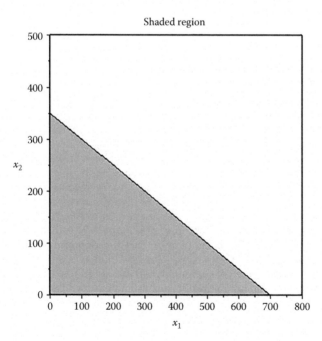

FIGURE 3.1
Shaded inequality.

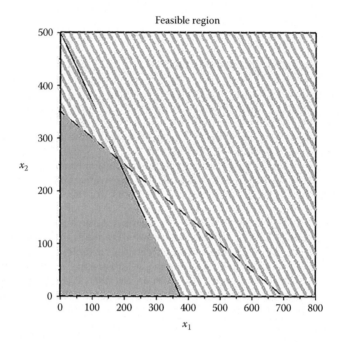

FIGURE 3.2
Plot of (1) the assembly hour's constraint and (2) the installation hour's constraint in the first quadrant.

SOLVING A LINEAR PROGRAMMING PROBLEM GRAPHICALLY

We have decision variables defined and an objection function that is to be maximized or minimized. Although all points inside the feasible region provide feasible solutions, the solution, if one exists, occurs according to the fundamental theorem of linear programming:

If the optimal solution exists, then it occurs at a corner point of the feasible region.

Notice the various corners formed by the intersections of the constraints in example. These points are of great importance to us. There is a cool theorem (did not know there were any of these, huh?) in linear optimization that states, "if an optimal solution exists, then an optimal corner point exists." The result of this is that any algorithm searching for the optimal solution to a linear program should have some mechanism of heading toward the corner point where the solution will occur. If the search procedure stays on the outside border of the feasible region while pursuing the optimal solution, it is called an *exterior point* method. If the search procedure cuts through the inside of the feasible region, it is called an *interior point* method.

Thus, in a linear programming problem, if there exists a solution, it must occur at a corner point of the set of feasible solutions (these are the vertices of the region). Note that in Figure 3.2, the corner points of the feasible region are the four coordinates, and we might use algebra to find these: (0,0), (0,350) (375,0), and (180,260).

How did we get the point (180,260)? This point is the intersection of the lines: $2x + 4y = 1400$ and $4x + 3y = 1500$. We use matrix algebra and solve for (x,y) from

$$\begin{bmatrix} 2 & 4 \\ 4 & 3 \end{bmatrix} \begin{bmatrix} x \\ y \end{bmatrix} = \begin{bmatrix} 1400 \\ 1500 \end{bmatrix}$$

Now, that we have all the possible solution coordinates for (x,y), we need to know which is the optimal solution. We evaluate the objective function at each point and choose the best solution.

Assume that our objective function is to maximize $Z = 140x + 120y$. We can set up a table of coordinates and corresponding Z values as follows:

Coordinate of corner point	$Z = 140x + 120y$
(0,0)	$Z = 0$
(0,350)	$Z = 42,000$
(180,260)	$Z = 56,400$
(375,0)	$Z = 52,500$
Best solution is (180,260)	**$Z = 56,400$**

Graphically, we see the result by plotting the objective function line, $Z = 140x + 120y$, with the feasible region. Determine the parallel direction for the line to maximize (in this case) Z. Move the line parallel until it crosses the last point in the feasible set. That point is the solution. The line that goes through the origin at a slope of $-7/6$ is called the ISO-profit line. We have provided this in Figure 3.3.

Here is a short cut to sensitivity analysis using the KTC conditions. We set up the function, L, using the form:

$$L = f(x) + l_1[b_1 - g_1(x)] + l_2(b_2 - g_2(x) + \ldots$$

For our example, this becomes

$$140x_1 + 120x_2 + l_1(1400 - 2x_1 + 4x_2) + l_2(1500 - 4x_1 + 3x_2)$$

We take the partial derivatives of L with respect to x_1, x_2, l_1, and l_2. For sensitivity analysis, we only care about the partial derivatives with respect to the l's. Thus, we will solve the following two equations and two unknowns:

$$140 - 2l_1 - 4l_2$$
$$120 - 4l_1 - 3l_2$$

We find $l_1 = 6$ and $l_2 = 32$.

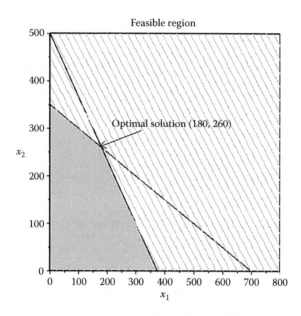

FIGURE 3.3
ISO-profit lines added.

We will see later with technology that these are shadow prices. We find here that a one unit increase in the second resource provides a larger increase to Z than a unit increase in the resource for the first constraint ($32\Delta > 6\Delta$).

We summarize the steps for solving a linear programming problem involving only two variables:

1. Sketch the region corresponding to the system of constraints. The points satisfying all constraints make up the feasible solution.
2. Find all the corner points (or intersection points in the feasible region).
3. Test the objective function at each corner point and select the values of the variables that optimize the objective function. For bounded regions, both a maximum and a minimum will exist. For an unbounded region, if a solution exists, it will exist at a corner.

Example 3.7: Minimization Linear Program

Minimize $Z = 5x + 7y$
Subject to $2x + 3y \geq 6$

$$3x - y \leq 15$$
$$-x + y \leq 4$$
$$2x + 5y \leq 27$$
$$x \geq 0$$
$$y \geq 0$$

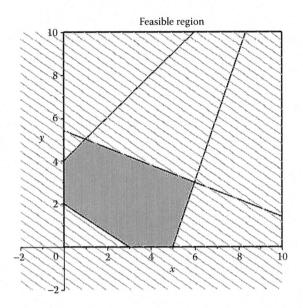

FIGURE 3.4
Feasible region for minimization example.

The corner points in Figure 3.4 are (0,2), (0,4,) (1,5), (6,3), (5,0), and (3,0). See if you can find all these corner points.

If we evaluate $Z = 5x + 7y$ at each of these points, we find the following:

Corner point	$Z = 5x + 7y$ (Minimize)
(0,2)	$Z = 14$
(1,5)	$Z = 40$
(6,3)	$Z = 51$
(5,0)	$Z = 25$
(3,0)	$Z = 15$
(0,4)	$Z = 28$

The minimum value occurs at (0,2) with a Z value of 14. Notice in our graph that the ISO-profit lines will last across the point (0,2) as it moves out of the feasible region in the direction that minimizes Z.

Example 3.8: An Unbounded Linear Program

Let us examine the concept of an unbounded feasible region. Look at the constraints:

$x + 2y \geq 4$
$3x + y \geq 7$
$x \geq 0$ and $y \geq 0$

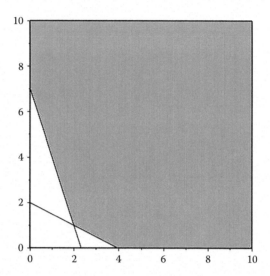

FIGURE 3.5
Unbounded feasible region.

Note that the corner points are (0,7), (2,1), and (4,0), and the region is unbounded as shown in Figure 3.5. If our solution is to minimize $Z = x + y$, then our solution is (2,1) with $Z = 3$. Determine why there is no solution to the LP to maximize $Z = x + y$?

Exercises 3.3

Find the maximum and minimum solution. Assume that we have $x \geq 0$ and $y \geq 0$ for each problem.

1. $Z = 2x + 3y$

   Subject to

$$2x + 3y \geq 6$$
$$3x - y \leq 15$$
$$-x + y \leq 4$$
$$2x + 5y \leq 27$$

2. $Z = 6x + 4y$

   Subject to

$$-x + y \leq 12$$
$$x + y \leq 24$$
$$2x + 5y \leq 80$$

3. $Z = 6x + 5y$

   Subject to

$$x + y \geq 6$$
$$2x + y \geq 9$$

4. $Z = x - y$

   Subject to

$$x + y \geq 6$$
$$2x + y \geq 9$$

5. $Z = 5x + 3y$

   Subject to

$$2x + 0.6y \leq 24$$
$$2x + 1.5y \leq 80$$

Projects 3.3

For each scenario:

1. List the decision variables and define them.
2. List the objective function.
3. List the resources that constrain this problem.
4. Graph the *feasible region*.
5. Label all intersection points of the feasible region.
6. Plot the objective function in a different color (highlight the objective function line, if necessary) and label it as the ISO-profit line.
7. Clearly indicate on the graph the point that is the optimal solution.
8. List the coordinates of the optimal solution and the value of the objective function.
9. Answer all scenario specific questions.
10. With the rising cost of gasoline and increasing prices to consumers, the use of additives to enhance performance of gasoline is being considered. Consider two additives: Additive 1 and Additive 2. The following conditions must hold for the use of additives:

    - Harmful carburetor deposits must not exceed 1/2 lb per car's gasoline tank.
    - The quantity of Additive 2 plus twice the quantity of Additive 1 must be at least 1/2 lb per car's gasoline tank.
    - 1 lb of Additive 1 will add 10 octane units per tank, and 1 lb of Additive 2 will add 20 octane units per tank. The total number of octane units added must not be less than six.
    - Additives are expensive and cost $1.53/lb for Additive 1 and $4.00/lb for Additive 2.

We want to determine the quantity of each additive that will meet the aforementioned restrictions and will minimize their cost.

1. Assume now that the manufacturer of additives has the opportunity to sell you a nice TV with special deal to deliver at least 0.5 lb of Additive 1 and at least 0.3 lb of Additive 2. Use graphical LP methods to help recommend whether you should buy this TV offer. Support your recommendation.

2. Write a one-page cover letter to your boss of the company that summarizes the results that you found.

2. A farmer has 30 acres on which to grow tomatoes and corn. Each 100 bushels of tomatoes requires 1000 gal of water and 5 acres of land. Each 100 bushels of corn requires 6000 gal of water and 2½ acres of land. Labor costs are $1 per bushel for both corn and tomatoes. The farmer has 30,000 gal of water and $750 in capital available. He knows that he cannot sell more than 500 bushels of tomatoes or 475 bushels of corn. He estimates a profit of $2 on each bushel of tomatoes and $3 of each bushel of corn. How many bushels of each should he raise to maximize profits?

   a. Assume now that farmer has the opportunity to sign a nice contract with a grocery store to grow and deliver at least 300 bushels of tomatoes and at least 500 bushels of corn. Use graphical LP methods to help recommend a decision to the farmer. Support your recommendation.

   b. If the farmer can obtain an additional 10,000 gal of water for a total cost of $50, is it worth to obtain the additional water? Determine the new optimal solution caused by adding this level of resource.

   c. Write a one-page cover letter to your boss that summarizes the result that you found.

3. *Fire Stone Tires* headquartered in Akron, Ohio has a plant in Florence, SC that manufactures two types of tires: SUV 225 radials and SUV 205 radials. Demand is high because of the recent recall of tires. Each 100 SUV 225 radials requires 100 gal of synthetic plastic and 5 lb of rubber. Each 100 SUV 205 radials requires 60 gal of synthetic plastic and 2½ lb of rubber. Labor costs are $1 per tire for each type of tire. The manufacturer has weekly quantities available of 660 gal of synthetic plastic, $750 in capital, and 300 lb of rubber. The company estimates a profit of $3 on each SUV 225 radial and $2 of each SUV 205 radial. How many of each type of tire should the company manufacture in order to maximize their profits?

   a. Assume now that manufacturer has the opportunity to sign a nice contract with a tire outlet store to deliver at least 500 SUV

225 radial tires and at least 300 SUV 205 radial tires. Use graphical LP methods to help recommend a decision to the manufacturer. Support your recommendation.

b. If the manufacturer can obtain an additional 1000 gal of synthetic plastic for a total cost of $50, is it worth it to obtain this amount? Determine the new optimal solution caused by adding this level of resource.

c. If the manufacturer can obtain an additional 20 lb of rubber for $50, should they obtain the rubber? Determine the new solution caused by adding this amount.

d. Write a one-page cover letter to your boss of the company that summarizes the results that you found.

4. Consider a toy maker that carves wooden soldiers. The company specializes in two types: Confederate soldiers and Union soldiers. The estimated profit for each is $28 and $30, respectively. A Confederate soldier requires 2 units of lumber, 4 hours of carpentry, and 2 hours of finishing in order to complete the soldier. A Union soldier requires 3 units of lumber, 3.5 hours of carpentry, and 3 hours of finishing to complete. Each week the company has 100 units of lumber delivered. The workers can provide at most 120 hours of carpentry and 90 hours of finishing. Determine the number of each type of wooden soldiers to produce maximize weekly profits.

3.4 Mathematical Programming with Technology

3.4.1 Linear Programming

Technology is critical to solving, analyzing, and performing sensitivity analysis on linear programming problems. Technology provides a suite of powerful, robust routines for solving optimization problems, including linear programs (LPs). Technology that we illustrate includes Excel, LINDO, and LINGO as these appear to be used often in engineering. We also examined GAMS, which we found powerful but too cumbersome to discuss here. We tested all these other software packages and found them all useful.

We show the computer chip problem first with technology.

Profit $Z = 140x_1 + 120x_2$

Subject to

$$2x_1 + 4x_2 \leq 1400 \text{ (assembly time)}$$
$$4x_1 + 3x_2 \leq 1500 \text{ (installation time)}$$
$$x_1 \geq 0, x_2 \geq 0$$

Using Excel

Put the problem formulation into Excel. Note that you must have formulas in terms of cells for the objective function and the constraints as shown in Figure 3.6.

Highlight the objective function, open the Solver, select the solving method, SimplexLP. This is shown in Figure 3.7.

Insert the decision variables into the By Changing Variable Cells as shown in Figure 3.8.

Enter the constraints by evoking the Add command as shown in Figure 3.9.

Enter the constraints by using the Add Constraint for each constraint and select Add again until all constraints are added as shown in Figure 3.10.

Linear Programming

Decision Variables

		Decision Variables	
x1= number of high speed chip type A to produce weekly	180	
x2= number of high speed chip type B to produce week	260	

Objective function Z=140x1+120x2		=140*C8+120*C9	

Constraints		Used	RHS
(1) 2 x1 + 4x2 <= 1400		=2*C8+4*C9	1400
(2) 4 x1 + 3 x2 <=1500		=4*C8+3*C9	1500
(3) x1, x2 >=0			

Linear Programming

Decision Variables

		Decision Variables
x1= number of high speed chip type A to produce weekly	180
x2= number of high speed chip type B to produce weekly	260

Objective function Z=140x1+120x2	56400

Constraints	Used	RHS
(1) 2 x1 + 4x2 <= 1400	1400	1400
(2) 4 x1 + 3 x2 <-1500	1500	1500
(3) x1, x2 >=0		

FIGURE 3.6
Screenshot of input with formulas.

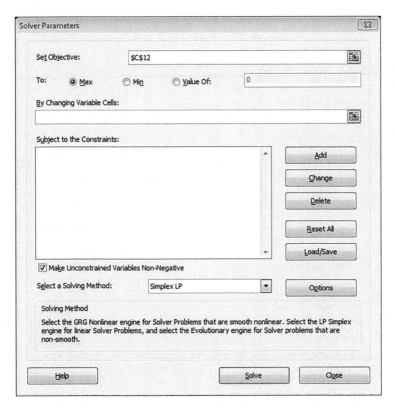

FIGURE 3.7
Screenshot of opening solver and putting in the objective function from Figure 3.6.

Next we select the Solve command and save both the answer and sensitivity analysis worksheets as shown in Figure 3.11.

We view the spreadsheet solution shown in Figure 3.12.

Next, we view solution and analysis reports. The answer report is shown as Figure 3.13 and the sensitivity report as Figure 3.14.

As expected, we have the same answers as we found earlier.

We present the following example via each technology:

Example 3.9:

Maximize $Z = 28x_1 + 30x_2$
   Subject to

$$20x_1 + 30x_2 \leq 690$$
$$5x_1 + 4x_2 \leq 120$$
$$x_1, x_2, \geq 0$$

Using Excel, we set up this LP as shown in Figure 3.15.

FIGURE 3.8
Screenshot of inserting the decision variables into the solver.

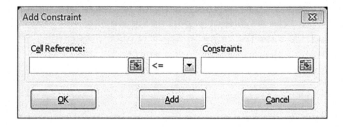

FIGURE 3.9
Screenshot of ADD constraint.

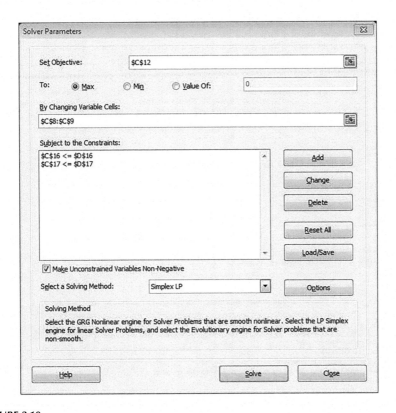

FIGURE 3.10
Screenshot of final model with constraints entered.

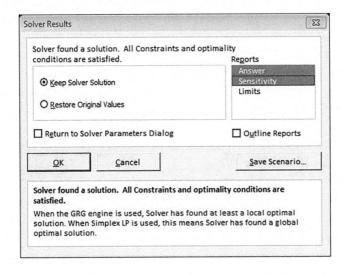

FIGURE 3.11
Screenshot of solver results and saving answer and sensitivity worksheets.

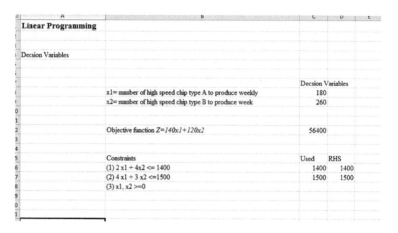

FIGURE 3.12
Screenshot of solver's solution to Example 3.6.

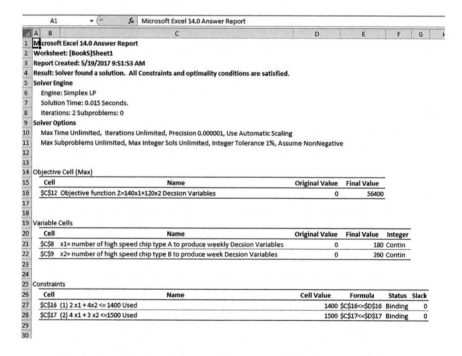

FIGURE 3.13
The answer report to Example 3.6.

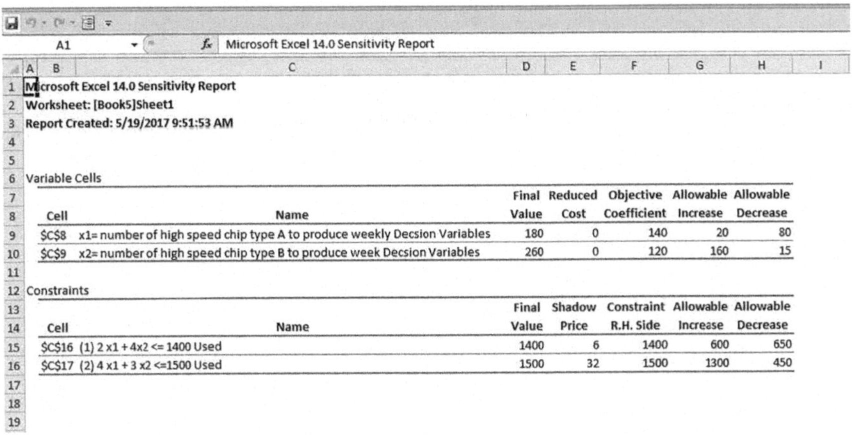

Microsoft Excel 14.0 Sensitivity Report
Worksheet: [Book5]Sheet1
Report Created: 5/19/2017 9:51:53 AM

Variable Cells

Cell	Name	Final Value	Reduced Cost	Objective Coefficient	Allowable Increase	Allowable Decrease
$C$8	x1= number of high speed chip type A to produce weekly Decsion Variables	180	0	140	20	80
$C$9	x2= number of high speed chip type B to produce week Decsion Variables	260	0	120	160	15

Constraints

Cell	Name	Final Value	Shadow Price	Constraint R.H. Side	Allowable Increase	Allowable Decrease
$C$16	(1) 2 x1 + 4x2 <= 1400 Used	1400	6	1400	600	650
$C$17	(2) 4 x1 + 3 x2 <=1500 Used	1500	32	1500	1300	450

FIGURE 3.14
The sensitivity report to Example 3.6.

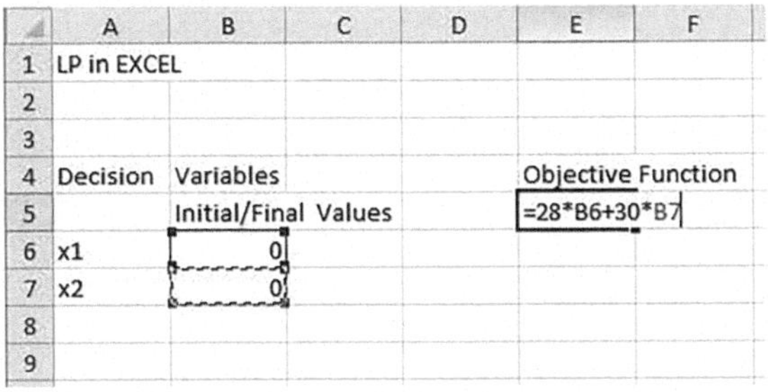

	A	B	C	D	E	F
1	LP in EXCEL					
2						
3						
4	Decision	Variables			Objective Function	
5		Initial/Final Values			=28*B6+30*B7	
6	x1	0				
7	x2	0				
8						
9						

FIGURE 3.15
Setting up our Example 3.9 in Excel.

We add the constraints as shown in Figure 3.16.

We highlight the objective function and open the Solver as shown in Figure 3.17.

We add the constraints into Solver as shown in Figure 3.18.

Now that we have the full Set up, we click Solve and obtain the output as shown in Figure 3.19.

We obtain the answers as $x_1 = 12$, $x_2 = 15$, $Z = 786$.

In addition, we obtain reports from Excel. Two key reports are the answer report and the sensitivity report.

We view these two reports as shown in Figures 3.20 and 3.21.

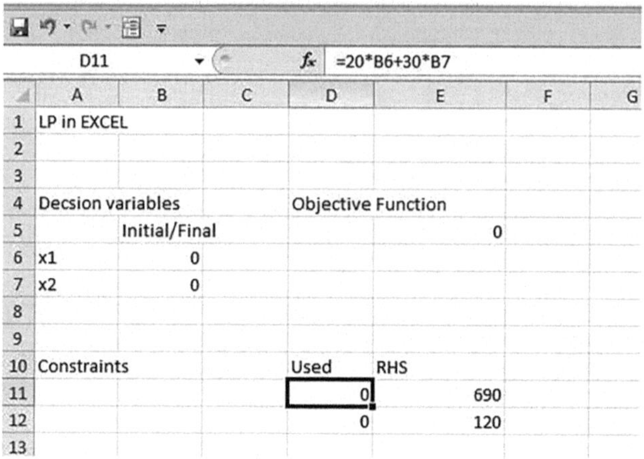

FIGURE 3.16
Excel's setup for Example 3.9.

FIGURE 3.17
Screenshot opening solver with inputs and decision variables for Example 3.9.

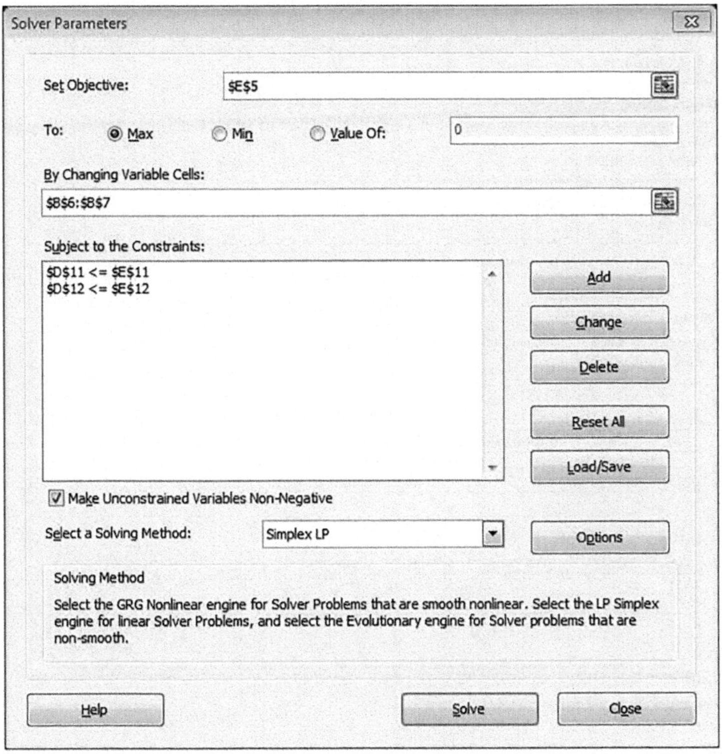

FIGURE 3.18
Screenshot with constraints added to Example 3.9.

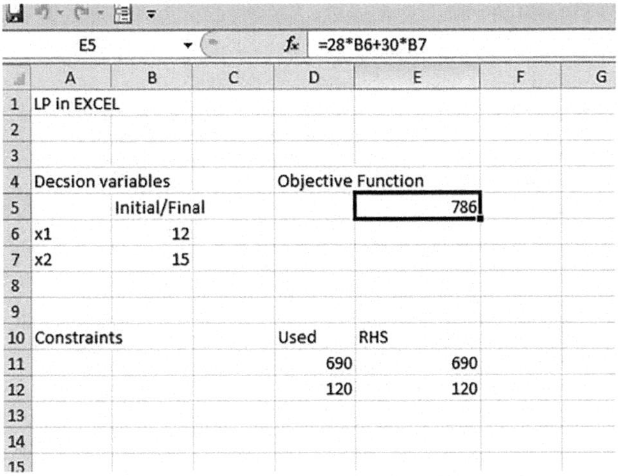

FIGURE 3.19
Screenshot of output for Example 3.9.

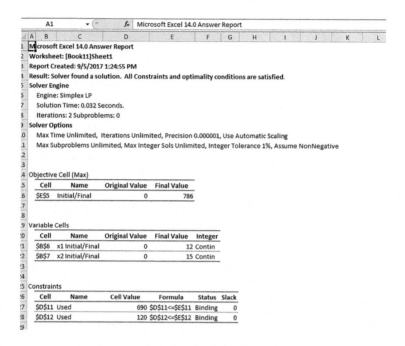

FIGURE 3.20
The answer report for Example 3.9.

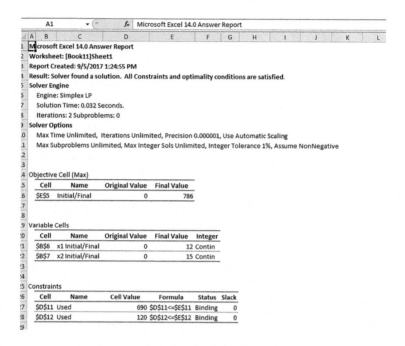

FIGURE 3.21
The sensitivity report for Example 3.9.

We find that our solution is $x_1 = 9$, $x_2 = 24$, $P = \$972$. From the standpoint of sensitivity analysis, Excel is satisfactory in that it provides shadow prices.

Limitation: No tableaus are provided making it difficult to find alternate solutions.

For further discussions on alternate optimal solution detection and shadow prices, please refer to linear programming or mathematical programming texts in the references section.

3.4.1.1 Using LINDO

This is the format to type in the formulation directly into LINDO.

```
MAX 25 X1 + 30 X2
   SUBJECT TO
           2) 20 X1 + 30 X2 <= 690
           3)  5 X1 +  4 X2 <= 120
   END
```

THE TABLEAU

	ROW	(BASIS)	X1	X2	SLK	2 SLK	3
	1 ART		-25.000	-30.000	0.000	0.000	0.000
	2 SLK	2	20.000	30.000	1.000	0.000	690.000
	3 SLK	3	5.000	4.000	0.000	1.000	120.000
ART	ART		-25.000	-30.000	0.000	0.000	0.000

LP OPTIMUM FOUND AT STEP 2

OBJECTIVE FUNCTION VALUE

1) 750.0000

VARIABLE	VALUE	REDUCED COST
X1	12.000000	0.000000
X2	15.000000	0.000000

ROW	SLACK OR SURPLUS	DUAL PRICES
2)	0.000000	0.714286
3)	0.000000	2.142857

NO. ITERATIONS=   2

```
RANGES IN WHICH THE BASIS IS UNCHANGED:

                         OBJ COEFFICIENT RANGES
    VARIABLE      CURRENT       ALLOWABLE       ALLOWABLE
                   COEF         INCREASE        DECREASE
         X1     25.000000      12.500000        5.000000
         X2     30.000000       7.500000       10.000000

                        RIGHTHAND SIDE RANGES
      ROW        CURRENT       ALLOWABLE       ALLOWABLE
                   RHS          INCREASE        DECREASE
       2        690.000000    210.000000      209.999985
       3        120.000000     52.499996       28.000000


THE TABLEAU
      ROW   (BASIS)      X1        X2  SLK   2  SLK   3
        1 ART          0.000     0.000   0.714    2.143  750.000
        2       X2     0.000     1.000   0.071   -0.286   15.000
        3       X1     1.000     0.000  -0.057    0.429   12.000
```

3.4.1.2 Using LINGO

We type the formulation into LINGO and Solve.

```
MODEL:

MAX = 25 * x1 + 30 * x2;

20 * x1 + 30 * x2 <= 690;

5 * x1 + 4 * x2 <= 120;

x1>=0;
x2>=0;

END
```

```
Variable             Value         Reduced Cost
                      X1          12.00000            0.0000000
                      X2          15.00000            0.0000000

                     Row     Slack or Surplus       Dual Price
                      1          750.0000             1.000000
                      2          0.0000000            0.7142857
                      3          0.0000000            2.142857
                      4          12.00000             0.0000000
                                 15.00000             0.0000000
```

3.4.1.3 MAPLE

MAPLE is a computer algebra package. It has an optimization package included that solves linear programming problems. The following is an example of a setup for the problem. Note that errors occur if you capitalize the first letter on the name. We enter the commands:

> $objectiveLP := 25 \cdot x_1 + 30 \cdot x_2;$

$objectiveLP := 25\, x_1 + 30\, x_2$

> $constraintsLP := \{20 \cdot x_1 + 30 \cdot x_2 \le 690, 5 \cdot x_1 + 4 \cdot x_2 \le 120, x_1 \ge 0, x_2 \ge 0\}$

$constraintsLP := \{0 \le x_1, 0 \le x_2, 5\, x_1 + 4\, x_1 \le 120, 20\, x_2 + 30\, x_2 \le 690\}$

We then call the optimization packages and in this case maximize the linear programming problem. There are two MAPLE approaches one with simplex and either maximize or minimize command and the other with LPSolve with either maximize or minimize as shown in the following to obtain our same answers.

> $with(Optimization) : with(simplex):$
> $maximize(objectiveLP, constraintsLP, NONNEGATIVE);$
$\{x_1 = 12, x_2 = 15\}$
> $LPSolve(objectiveLP, constraintsLP, maximize);$
$[750., [x_1 = 12., x_2 = 15.]]$

The basic LP package in MAPLE is not equipped to provide tableaus or sensitivity analysis directly. Fishback (2010) wrote a nice book on linear programming in Maple. This is a step-by-step process in which the user has to understand the simplex procedure.

Here are the commands for our problem, and tableaus are provided.

> *restart;*

> *with(LiniearAlgebra) :*

> $c := Vector[row]([25, 30]);$

$c := [25 \ 30]$

> $A := Matrix(2, 2, [20, 30, 5, 4]);$

$$A := \begin{bmatrix} 20 & 30 \\ 5 & 4 \end{bmatrix}$$

> $b := \langle 690, 120 \rangle;$

$$b := \begin{bmatrix} 690 \\ 120 \end{bmatrix}$$

> $b_1 := \langle 0, b \rangle;$

$$b_1 := \begin{bmatrix} 0 \\ 690 \\ 120 \end{bmatrix}$$

> $n := 2 : m := 2 :$

> $x := array(1 \ ..n) : s := array(1 \ ..m);$

$s := array(1 \ ..2, [\ ])$

> *Labels* $:= Matrix(1, 2 + n + m, [z, seq(x[i], i = 1 \ ..n), seq(s[j], j = 1 \ ..m), RHS]);$

$$Labels := \begin{bmatrix} z \ x_1 \ x_2 \ s_1 \ s_2 \ RHS \end{bmatrix}$$

> *#LP*

> $LPMatrix := \langle UnitVector(1, m + 1) | \langle -c, A \rangle | \langle ZeroVector[row](m), IdentityMatrix(m) \rangle | \langle 0, b \rangle \rangle;$

$$LPMatrix := \begin{bmatrix} 1 & -25 & -30 & 0 & 0 & 0 \\ 0 & 20 & 30 & 1 & 0 & 690 \\ 0 & 5 & 4 & 0 & 1 & 120 \end{bmatrix}$$

> $LP_1 := LPMatrix;$

$$LP_1 := \begin{bmatrix} 1 & -25 & -30 & 0 & 0 & 0 \\ 0 & 20 & 30 & 1 & 0 & 690 \\ 0 & 5 & 4 & 0 & 1 & 120 \end{bmatrix}$$

> *Tableau* := proc(*M*);return(⟨*Lables*, *M*⟩)end:

>

$RowRatios := proc(M, c)$ local k : for k from 2
to *nops* (convert(*Column*(*M*, *c* + 1), *list*)) do if $M[k, c + 1] = 0$
then *print* (*cat*("Row", convert(*k* − 1, *string*), "*Undefined*"))
else *print* (*cat* ("Row", *convert*(*k* − 1, *string*), "Ratio",

$$convert \left(evalf \left(\frac{M[k, nops(convert(Row(M, k), list))]}{M[k, c + 1]} \right), \right.$$

string))) end if; end do; end;

>

> $Iterate := proc(M, r, c)$ $RowOperation(M, r + 1, (M[r + 1, c + 1])^{(-1)},$
$inplace = true) : Pivot(M, r + 1, c + 1, inplace = true) :$
return(*Tableau*(*M*)) :end;

$Iterate := proc(M, r, c)$
$LinearAlgebra:-RowOperation(M, r + 1, 1/M[r + 1, c + 1],$
$inplace = true);$
$LinearAlgebra:-Pivot(M, r + 1, c + 1, inplace = true);$
return *Tableau*(*M*)
end proc

> *Tableau*(*LPMatrix*);

$$\begin{bmatrix} z & x_1 & x_2 & s_1 & s_2 & RHS \\ 1 & -25 & -30 & 0 & 0 & 0 \\ 0 & 20 & 30 & 1 & 0 & 690 \\ 0 & 5 & 4 & 0 & 1 & 120 \end{bmatrix}$$

>

$RowRatios := proc(M, c)$ local k :
for k from 2 to *nops* (convert(*Column*(*M*, *c* + 1), *list*)) do
if $M[k, c + 1] = 0$ then *print* (*cat*("Row", convert(k − 1, *string*),
"*Undefined*"))
else *print*(*cat*("Row", *convert*(k − 1, *string*), "Ratio",
convert(*evalf* (*M*[*k*, *nops*(convert(*Row*(*M*, *k*), *list*))]/*M*[*k*, *c* + 1]),
string)))
end if; end do;end:

> *RowRatios*(*LPMatrix*, 2);

*Row*1*Ratio*23.

*Row2Ratio*30.

> *Iterate(LPMatrix, 1, 2);*

$$\begin{bmatrix} z & x_1 & x_2 & s_1 & s_2 & RHS \\ 1 & -5 & 0 & 1 & 0 & 690 \\ 0 & \frac{2}{3} & 1 & \frac{1}{30} & 0 & 23 \\ 0 & \frac{7}{3} & 0 & -\frac{2}{15} & 1 & 28 \end{bmatrix}$$

> *RowRatios(LPMatrix, 1);*

*Row1Ratio*34.50000000

*Row2Ratio*12.

> *Iterate(LPMatrix, 2, 1);*

$$\begin{bmatrix} z & x_1 & x_2 & s_1 & s_2 & RHS \\ 1 & 0 & 0 & \frac{5}{7} & \frac{15}{7} & 750 \\ 0 & 0 & 1 & \frac{1}{14} & -\frac{2}{7} & 15 \\ 0 & 1 & 0 & -\frac{2}{35} & \frac{3}{7} & 12 \end{bmatrix}$$

3.4.1.4 *Integer and Nonlinear Programming*

Integer: Integer programming in Excel requires only that you identify the variables as integers in the constraint set. Your choices are binary integers {0,1} or integers. We state that the Solver does not identify the methodology used.

Nonlinear Programming: There are many forms of nonlinear problems in optimization. Maple and Excel are both useful in obtaining solutions.

We will illustrate the use of technology in the case study examples later.

Exercises 3.4

Solve the exercises and projects in Section 3.3 using appropriate and available technology.

3.5 Case Studies in Mathematical Programming

Example 3.10: Supply Chain Operations

In our case study, we present linear programming for supply chain design. We consider producing a new mixture of gasoline. We desire to minimize the total cost of manufacturing and distributing the new mixture. There is a supply chain involved with a product that must be modeled. The product is made up of components that are produced separately.

Crude Oil Type	Compound A (%)	Compound B (%)	Compound C (%)	Cost/ Barrel	Barrel Avail (000 of barrels)
X10	35	25	35	$26	15,000
X20	50	30	15	$32	32,000
X30	60	20	15	$55	24,000

Demand information is as follows:

Gasoline	Compound A (%)	Compound B (%)	Compound C (%)	Expected Demand (000 of barrels)
Premium	≥ 55	≤ 23		14,000
Super		≥ 25	≤ 35	22,000
Regular	≥ 40		≤ 25	25,000

Let i = crude type 1, 2, 3 (X10, X20, X30, respectively)
Let j = gasoline type 1, 2, 3 (premium, super, regular, respectively)

We define the following decision variables:

G_{ij} = amount of crude i used to produce gasoline j
For example, G_{11} = amount of crude X10 used to produce Premium gasoline
G_{12} = amount of crude type X20 used to produce Premium gasoline
G_{13} = amount of crude type X30 used to produce Premium gasoline
G_{12} = amount of crude type X10 used to produce Super gasoline
G_{22} = amount of crude type X20 used to produce Super gasoline
G_{32} = amount of crude type X30 used to produce Super gasoline
G_{13} = amount of crude type X10 used to produce Regular gasoline
G_{23} = amount of crude type X20 used to produce Regular gasoline
G_{33} = amount of crude type X30 used to produce Regular gasoline

LP formulation

Minimize Cost $= \$86(G_{11} + G_{21} + G_{31}) + \$92(G_{12} + G_{22} + G_{32}) + \$95(G_{13} + G_{23} + G_{33})$
Subject to: Demand

$$G_{11} + G_{21} + G_{31} > 14{,}000 \text{ (Premium)}$$
$$G_{12} + G_{22} + G_{32} > 22{,}000 \text{ (Super)}$$
$$G_{13} + G_{23} + G_{33} > 25{,}000 \text{ (Regular)}$$

Availability of products

$$G_{11} + G_{12} + G_{13} < 15{,}000 \text{ (crude 1)}$$
$$G_{21} + G_{22} + G_{23} < 32{,}000 \text{ (crude 2)}$$
$$G_{31} + G_{32} + G_{33} < 24{,}000 \text{ (crude 3)}$$

Product mix in mixture format

$(0.35G_{11} + 0.50G_{21} + 0.60G_{31})/(G_{11} + G_{21} + G_{31}) > 0.55$ (X10 in Premium)
$(0.25G_{11} + 0.30G_{21} + 0.20G_{31})/(G_{11} + G_{21} + G_{31}) < 0.23$ (X20 in Premium)
$(0.35G_{13} + 0.15G_{23} + 0.15G_{33})/(G_{13} + G_{23} + G_{33}) > 0.25$ (X20 in Regular)
$(0.35G_{13} + 0.15G_{23} + 0.15G_{33})/(G_{13} + G_{23} + G_{33}) < 0.35$ (X30 in Regular)
$(0.35G_{12} + 0.50G_{22} + 0.60G_{23})/(G_{12} + G_{22} + G_{32}) < 0.40$ (Compound X10 in Super)
$(0.35G_{12} + 0.15G_{22} + 0.15G_{32})/(G_{12} + G_{22} + G_{32}) < 0.25$ (Compound X30 in Super)

The solution was found using LINDO and we noticed an alternate optimal solution:
Two solutions are found yielding a minimum cost of $1,904,000.

Decision variable	Z = $1,940,000	Z = $1,940,000
G_{11}	0	1,400
G_{12}	0	3,500
G_{13}	14,000	9,100
G_{21}	15,000	1,100
G_{22}	7,000	20,900
G_{23}	0	0
G_{31}	0	12,500
G_{32}	25,000	7,500
G_{33}	0	4,900

Depending on whether we want to additionally minimize delivery (across different locations) or maximize sharing by having more distribution point involved then we have choices.

We present one of the solutions with LINDO:

```
LP OPTIMUM FOUND AT STEP 7

        OBJECTIVE FUNCTION VALUE

    1)      1904000.

VARIABLE            VALUE        REDUCED COST
      P1          0.000000          0.000000
      R1      15000.000000          0.000000
      E1          0.000000          0.000000
      P2          0.000000          0.000000
      R2       7000.000000          0.000000
      E2      25000.000000          0.000000
      P3      14000.000000          0.000000
      R3          0.000000          0.000000
      E3          0.000000          0.000000

     ROW    SLACK OR SURPLUS      DUAL PRICES
      2)          0.000000          9.000000
      3)          0.000000          3.000000
      4)      10000.000000          0.000000
      5)          0.000000        -35.000000
      6)          0.000000        -35.000000
      7)          0.000000        -35.000000
      8)        700.000000          0.000000
      9)       3500.000000          0.000000
     10)       1400.000000          0.000000
     11)       2500.000000          0.000000
     12)       2500.000000          0.000000
     13)        420.000000          0.000000

NO. ITERATIONS= 7

RANGES IN WHICH THE BASIS IS UNCHANGED:

                        OBJ COEFFICIENT RANGES
VARIABLE       CURRENT       ALLOWABLE       ALLOWABLE
                  COEF        INCREASE        DECREASE
      P1     26.000000        INFINITY        0.000000
      R1     26.000000        0.000000        INFINITY
      E1     26.000000        INFINITY        0.000000
      P2     32.000000        0.000000        0.000000
      R2     32.000000        0.000000        0.000000
      E2     32.000000        0.000000       35.000000
      P3     35.000000        0.000000        3.000000
      R3     35.000000        INFINITY        0.000000
      E3     35.000000        INFINITY        0.000000
```

RIGHTHAND SIDE RANGES

ROW	CURRENT RHS	ALLOWABLE INCREASE	ALLOWABLE DECREASE
2	15000.000000	4200.000000	0.000000
3	32000.000000	4200.000000	0.000000
4	24000.000000	INFINITY	10000.000000
5	14000.000000	10000.000000	14000.000000
6	22000.000000	0.000000	4200.000000
7	25000.000000	0.000000	4200.000000
8	0.000000	700.000000	INFINITY
9	0.000000	3500.000000	INFINITY
10	0.000000	INFINITY	1400.000000
11	0.000000	2500.000000	INFINITY
12	0.000000	INFINITY	2500.000000
13	0.000000	INFINITY	420.000000

Source: Fox, W. P. and F. Garcia., Chapter 9, Modeling and linear programming in engineering management, in F. P. G. Márquez and B. Lev (Eds.), *Engineering Management*, Rijeka, Croatia, InTech, pp. 216–225, 2013.

Example 3.11: Recruiting Raleigh Office

Although this is a simple model, it was adopted by the U.S. Army Recruiting Command for operations. The model determines the optimal mix of prospecting strategies that a recruiter should use in a given week. The two prospecting strategies initially modeled and analyzed are phone and email prospecting. The data came from the Raleigh Recruiting Company, U.S. Army Recruiting Command in 2006. On average, each phone lead yields 0.041 enlistments, and each email lead yields 0.142 enlistments. Forty recruiters who are assigned to the Raleigh recruiting office prospected a combined 19,200 minutes of work per week via phone and e-mail. The company's weekly budget is $60,000.

	Phone (x_1)	Email (x_2)
Prospecting time (Minutes)	60 minutes per lead	1 minute per lead
Budget (dollars)	$10 per lead	$37 per lead

The decision variables are as follows:

x_1 is the number of phone leads.
x_2 is the number of email leads.
Maximize $Z = 0.041x_1 + 0.142x_2$
Subject to

$$60x_1 + 1x_2 \leq 19{,}200 \text{ (prospecting minutes available)}$$
$$10x_1 + 37x_2 \leq 60{,}000 \text{ (budget dollars available)}$$
$$x_1, x_2 \geq 0 \text{ (nonnegativity)}$$

If we examine all the intersections point, we find a suboptimal point, $x_1 = 294.29$, $x_2 = 154.082$, achieving 231.04 recruitments.

We examine the sensitivity analysis report,

Microsoft Excel 14.0 Sensitivity Report

Worksheet: [Book4]Sheet1

Report Created: 5/5/2015 2:21:25 PM

Variable Cells

Cell	Name	Final Value	Reduced Cost	Objective Coefficient	Allowable Increase	Allowable Decrease
$B$3	x_1	294.2986425	0	0.041	8.479	0.002621622
$B$4	x_2	1542.081448	0	0.142	0.0097	0.141316667

Constraints

Cell	Name	Final Value	Shadow Price	Constraint R.H. Side	Allowable Increase	Allowable Decrease
$C$10		19,200	4.38914E-05	19,200	340,579	17,518.64865
$C$11		60,000	0.003836652	60,000	648,190	56,763.16667
$C$12		294.2986425	0	1	293.2986425	1E+30
$C$13		1542.081448	0	1	1541.081448	1E+30

First, we see that we maintain a mixed solution over a fairly large range of values for the coefficient of x_1 and x_2. Further the shadow prices provide additional information. A one unit increase in prospecting minutes available yields an increase of approximately 0.00004389 in recruits, whereas an increase in budget of $1 yields an additional 0.003836652 recruits. At initial look it appears as though, we might be better off with an additional $1 in resource.

Let us assume that it costs only $0.01 for each additional prospecting minute. Thus, we could get 100*0.00004389 or a 0.004389 increase in recruits for the same unit cost increase. In this case, we would be better off obtaining the additional prospecting minutes.

Source: McGrath, G., Email marketing for the U.S. Army and Special Operations Forces Recruiting, Master's Thesis, Naval Postgraduate School, 2007.

Exercises 3.5

1.

In the Supply Chain Case Study, Resolve with the following Data Table Crude Oil Type	Compound A (%)	Compound B (%)	Compound C (%)	Cost/ Barrel	Barrel Available (000 of Barrels)
X10	45	35	45	$26.50	18,000
X20	60	40	25	$32.85	35,000
X30	70	30	25	$55.97	26,000

Demand information is as follows:

Gasoline	Compound A (%)	Compound B (%)	Compound C (%)	Expected Demand (000 of Barrels)
Premium	≥ 55	≤ 23		14,000
Super		≥ 25	≤ 35	22,000
Regular	≥ 40		≤ 25	25,000

2. In the Raleigh recruiting case study, assume that the data have been updated as follows as resolve.

	Phone (x_1)	Email (x_2)
Prospecting time (Minutes)	45 minutes per lead	1.5 minute per lead
Budget (dollars)	$15 per lead	$42 per lead

3.6 Examples for Integer, Mixed-Integer, and Nonlinear Optimization

Example 3.12: Emergency Services

Recall this as example from Chapter 1. Here we formulate and present a solution.

Solution: We assume that due to the nature of the problem requiring a facility location that we should decide to employ integer programming to solve our problem.

Decision Variables:

$$y_i = \begin{cases} 1 \text{ if node is covered} \\ 0 \text{ if node not covered} \end{cases}$$

$$x_j = \begin{cases} 1 \text{ if ambulance is located in } j \\ 0 \text{ if not located in } j \end{cases}$$

m is the number of ambulances available
h_i is the population to be served at demand node i
t_{ij} is the shortest time from node j to node I in perfect conditions
i is the set of all demand nodes
j is the set of nodes where ambulances can be located

Model formulation:

Maximize $Z = 50{,}000y_1 + 80{,}000y_2 + 30{,}000y_3 + 55{,}000y_4 + 35{,}000y_5 + 20{,}000y_6$

Subject to

$$x_1 + x_2 \geq y_1$$
$$x_1 + x_2 + x_3 \geq y_2$$
$$x_3 + x_5 + x_6 \geq y_3$$
$$x_3 + x_4 + x_6 \geq y_4$$
$$x_4 + x_5 + x_6 \geq y_5$$
$$x_3 + x_5 + x_6 \geq y_6$$
$$x_1 + x_2 + x_3 + x_4 + x_5 + x_6 = 3$$

all variables are binary integers

Solution and analysis: We find that we can cover all 270,000 potential patients with three ambulances posted in locations 1, 3, and 6. We can cover all 270,000 potential patients with only two ambulances posted in locations 1 and 6. If we had only one ambulance, we can cover at most 185,000 patients with the ambulance located in location 3. We will have 85,000 not covered. For management, they have several options that meet demand. They might use the option that is least costly.

Example 3.13: Optimal Path to Transport Hazardous Material

Federal Emergency Management Agency (FEMA) is requesting a two-part analysis. They are concerned about the transportation of nuclear waste from the Savannah River nuclear plant to the appropriate disposal site. After the route is found, FEMA requests analysis as to the location and composition of cleanup sites. In this example, we only discuss the optimal path portion of the model using generic data.

Consider a model whose requirement is to find the route from node A to node B that minimizes the probability of a vehicle accident. A primary concern is the I-95 and I-20 corridor where both interstate meet and converge in Florence, SC.

To simplify the ability of the use of technology, we transform the model to maximize the probability of not having an accident.

Maximize $f(x_{12}, x_{13}, \dots x_{9,10}) = (1-p_{12}*x_{12})*(1-p_{13}x_{13})*\dots(1-p_{9,10}\, x_{9,10})$
Subject to

$$-x_{12} - x_{13} - x_{14} = -1$$
$$x_{12} - x_{24} - x_{26} = 0$$
$$x_{13} - x_{34} - x_{35} = 0$$
$$x_{14} + x_{24} + x_{34} - x_{45} - x_{46} - x_{48} = 0$$
$$x_{35} + x_{45} - x_{57} = 0$$
$$x_{26} + x_{46} - x_{67} - x_{68} = 0$$
$$x_{57} + x_{67} - x_{78} - x_{7,10} = 0$$
$$x_{48} + x_{68} + x_{78} - x_{8,10} = 0$$
$$x_{79} - x_{9,10} = 0$$
$$x_{7,10} + x_{8,10} + x_{9,10} = 1$$

Nonnegativity

From/To	Route		Prob of accident	No accident		Node	Input-Output	Demand
12	1		0.003	0.997	0.997	1		-1
13	0		0.004	0.996	1	2		0
14	0		0.002	0.998	1	3		0
24	0		0.01	0.99	1	4		0
26	1		0.006	0.994	0.994	5		0
34	0		0.002	0.998	1	6		0
35	0		0.01	0.99	1	7		0
45	0		0.002	0.998	1	8		0
46	0		0.004	0.996	1	9		0
48	0		0.009	0.991	1	10		1
57	0		0.001	0.999	1			
67	0		0.01	0.99	1			
68	1		0.001	0.999	0.999			
78	0		0.004	0.996	1			
79	0		0.001	0.999	1			
710	0		0.005	0.995	1			
810	1		0.001	0.999	0.999			
910	0		0.006	0.994	1			
		Objective functtion			0.989037		constraints	rhs
							-1	-1
							0	0
							0	0
	Solution	Safe corridors to travel					0	0
	1 to 4						0	0
	2 to 6						0	0
	6 to 8						0	0
	8 to 10						0	0
							0	0
							1	1

Example 3.14: Minimum Variance of Expected Investment Returns

A new company has $5000 to invest, but the company needs to earn about 12% interest. A stock expert has suggested three mutual funds {A, B, and C} in which the company could invest. Based on previous year's returns, these funds appear relatively stable. The expected return, variance on the return, and covariance between funds are shown in the following table:

Expected value	A	B	C
	0.14	0.11	0.10
Variance	A	B	C
	0.2	0.08	0.18
Covariance	AB	AC	BC
	0.05	0.02	0.03

Formulation:
We use laws of expected value, variance, and covariance in our model. Let x_j be the number of dollars invested in funds j ($j = 1, 2, 3$).

Minimize $V_1 = \text{var}(Ax_1 + Bx_2 + Cx_3) = x_1^2 \,\text{Var}\,(A) + x_2^2 \,\text{Var}\,(B) + x_3^2 \,\text{Var}(C) + 2x_1x_2 \,\text{Cov}(AB) + 2x_1x_3 \,\text{Cov}(AC) + 2x_2x_3 \,\text{Cov}(BC) = 0.2x_1^2 + 0.08x_2^2 + 0.18x_3^2 + 0.10x_1x_2 + 0.04x_1x_3 + 0.06x_2x_3.$

Our constraints include the following:

1. The expectation to achieve at least the expected return of 12% from the sum of all the expected returns:

$$0.14x_1 + 0.11x_2 + 0.10x_3 \geq (0.12 \times 5000) \text{ or}$$
$$0.14x_1 + 0.11x_2 + 0.10x_3 \geq 600$$

2. The sum of all investments must not exceed the $5000 capital.

$$x_1 + x_2 + x_3 \leq \$5000$$

The optimal solution via LINGO is
$x_1 = 1904.80$, $x_2 = 2381.00$, $x_3 = 714.20$, $z = \$1880942.29$ or a standard deviation of $1371.50.

The expected return is $0.14(1904.8) + 0.11(2381) + 0.1(714.2)/5000 = 12\%$

This example was used as a typical standard for investment strategy.

Source: Fox, W. P. and Garcia, F., Chapter 9, Modeling and linear programming in engineering management, in Márquez, F. P. G. and Lev, B. (Eds.), *Engineering Management*, Rijeka, Croatia, InTech, pp. 216–225, 2013.

Example 3.15: Cable Instillation

Consider a small company that is planning to install a central computer with cable links to five new departments with a schematic shown in Figure 3.22. According to their floor plan, the peripheral computers for the five departments will be situated as shown by the dark circles in the

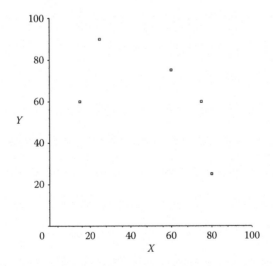

FIGURE 3.22
Computer links to departments.

figure. The company wishes to locate the central computer so that the minimal amount of cable will be used to link to the five peripheral computers. Assuming that cable may be strung over the ceiling panels in a straight line from a point above any peripheral to a point above the central computer, the distance formula may be used to determine the length of cable needed to connect any peripheral to the central computer. Ignore all lengths of cable from the computer itself to a point above the ceiling panel immediately over that computer. That is, work only with lengths of cable strung over the ceiling panels.

The coordinates of the locations of the five peripheral computers are listed as follows:

X	Y
15	60
25	90
60	75
75	60
80	25

GRID COORDINATES OF FIVE DEPARTMENTS

Assume that the central computer will be positioned at coordinates (m,n) where m and n are *integers* in the grid representing the office space. Determine the coordinates (m,n) for placement of the central computer that minimizes the total amount of cable needed. Report the total number of feet of cable needed for this placement along with the coordinates (m,n).

THE MODEL

This is an unconstrained optimization model. We want to minimize the sum of the distances from each department to the placement of the central computer system. The distances represent cable lengths assuming that a straight line is the shortest distance between two points. Using the distance formula,

$$d = \sqrt{(x - X_1)^2 + (y - Y_1)^2}$$

where d represents the distance (cable length in feet) between the location of the central computer (x,y) and the location of the first peripheral computer (X_1,Y_1). As we have five departments, we define

$$\text{dist} = \sum_{i=1}^{5} \sqrt{(x - X_i)^2 + (y - Y_i)^2}$$

Using the gradient search method on the Excel solver, we find that our solution is, *distance* = 157.66 ft when the central computer is placed at coordinates (56.82, 68.07).

Exercises 3.6

1. Your company is considering for investments. Investment 1 yields a net present value (NPV) of $17,000; investment 2 yields a NPV of $23,000; investment 3 yields a NPV of $13,000; and investment 4 yields a NPV of $9,000. Each investment requires a current cash flow of investment 1, $6000; investment 2, $8000; investment 3, $5000; and investment 4, $4000. At present, $21,000 is available for investment. Formulate and solve as an integer programming problem assuming that you can only invest at most one time in each investment.

2. Your company is considering for investments. Investment 1 yields a NPV of $17,000; investment 2 yields a NPV of $23,000; investment 3 yields a NPV of $13,000; and investment 4 yields a NPV of $9,000. Each investment requires a current cash flow of Investment 1, $6000; investment 2, $8000; investment 3, $5000; and investment 4, $4000. At present $21,000 is available for investment. Formulate and solve as an integer programming problem assuming that you can only invest more than once in any investment.

3. For the cable installation example, assume that we are moving the computers around to the following coordinates and resolve.

X	Y
10	50
35	85
60	77
75	60
80	35

Projects 3.6

Find multiple available nonlinear software packages. Using Example 3.14, solve with each package. Compare speed and accuracy.

3.7 Simplex Method in Excel

With problems more than two variables, an algebraic method may be used. This method is called the simplex method. The *simplex method*, developed by George Dantzig in 1947, incorporates both *optimality* and *feasibility* tests to find the optimal solution(s) to a LP (if an optimal solution exists).

An *optimality test* shows whether or not an intersection point corresponds to a value of the objective function better than the best value found so far.

A *feasibility test* determines whether the proposed intersection point is feasible. It does not violate any of the constraints.

The simplex method starts with the selection of a corner point (usually the origin if it is a feasible point) and then, in a systematic method, moves to adjacent corner points of the feasible region until the optimal solution is found, or it can be shown that no solution exists.

We will use our computer chip example to illustrate:

Maximize Profit $Z = 140x_1 + 120x_2$

$$2x_1 + 4x_2 \le 1400 \text{ (assembly time)}$$
$$4x_1 + 3x_2 \le 1500 \text{ (installation time)}$$
$$x_1 \ge 0, x_2 \ge 0$$

3.7.1 Steps of the Simplex Method

1. *Tableau format*: Place the LP in tableau Format, as explained in the following:

   Maximize Profit $Z = 140x_1 + 120x_2$
   $2x_1 + 4x_2 \le 1400$ (assembly time)
   $4x_1 + 3x_2 \le 1500$ (installation time)
   $x_1 \ge 0, x_2 \ge 0$

To begin the simplex method, we start by converting the inequality constraints (of the form $\le$) to equality constraints. This is accomplished by adding a unique, nonnegative variable, called a slack variable, to each constraint. For example, the inequality constraint $2x_1 + 4x_2 \le 1400$ is converted to an equality constraint by adding the slack variable S_1 to obtain:

$$2x_1 + 4x_2 + S_1 = 1400, \text{ where } S_1 \ge 0.$$

The inequality $2x_1 + 4x_2 \le 1400$ states that the sum $2x_1 + 4x_2$ is less than or equal to 1400. The slack variable *takes up the slack* between the values used for x_1 and x_2 and the value 1400. For example, if $x_1 = x_2 = 0$, then $S_1 = 14,000$. If $x_1 = 240, x_2 = 0$, then $2(240) + 4(0) + S_1 = 1400$, so $S_1 = 920$. A unique slack variable must be added to each inequality constraint.

Maximize $Z = 140x_1 + 240x_2$

Subject to

$$2x_1 + 4x_2 + S_1 = 1400$$
$$4x_1 + 3x_2 + S_2 = 1500$$
$$x_1 \ge 0, x_2 \ge 0, S_1 \ge 0, S_2 \ge 0$$

Adding slack variables makes the constraint set a system of linear equations. We write these with all variables on the left-hand side of the equation and all constants on the right-hand side.

We will even rewrite the objective function by moving all variables to the left-hand side.

Maximize $Z = 120x_1 + 140x_2$ is written as

$$Z - 140x_1 - 120x_2 = 0$$

Now, these can be written in the following form:

$$Z - 140x_1 - 120x_2 = 0$$
$$2x_1 + 4x_2 + S_1 = 1400$$
$$4x_1 + 3x_2 + S_2 = 1500$$
$$x_1 \geq 0, x_2 \geq 0, S_1 \geq 0, S_2 \geq 0$$

or more simply in a matrix. This matrix is called the simplex tableau.

Z	x_1	x_2	S_1	S_2		RHS
1	−140	−120	0	0	=	0
0	2	4	1	0	=	1400
0	4	3	0	1	=	1500

Because we are working in Excel, we will take advantage of a few commands, MINVERSE and MMULT to update the tableau.

2. *Initial extreme point*: The simplex method begins with a known extreme point, usually the origin (0,0) for many of our examples. The requirement for a basic feasible solution gives rise to special simplex methods such as big M and two-phase simplex, which can be studied in a linear programming course.

The tableau previously shown that contains the corner point (0,0) is our initial solution.

Z	x_1	x_2	S_1	S_2		RHS
1	−140	−120	0	0	=	0
0	2	4	1	0	=	1400
0	4	3	0	1	=	1500

We read this solution as follows:

$$x_1 = 0$$
$$x_2 = 0$$

$$S_1 = 1400$$
$$S_2 = 1500$$
$$Z = 0$$

As a matter of fact, we see that the column for variables Z, s_1, and s_2 form a 3×3 identity matrix. These three are referred to as basic variables. Let us continue to define a few of these variables further. We have five variables $\{Z, x_1, x_2, S_1, S_2\}$ and three equations. We can have at most three solutions. Z will always be a solution by convention of our tableau. We have two nonzero variables among $\{x_1, x_2, S_1, S_2\}$. These nonzero variables are called the *basic variables*. The remaining variables are called the *nonbasic variables*. The corresponding solutions are called the *basic feasible solutions* (FBS) and correspond to corner points. The complete step of the simplex method produces a solution that corresponds to a corner point of the feasible region. These solutions are read directly from the tableau matrix.

We also note the basic variables are variables that have a column consisting of one 1 and the rest zeros in their column. We will add a column to label these as shown in the following table:

	Basic Variable			Basic variable	Basic variable		
	Z	x_1	x_2	S_1	S_2		RHS
Z	1	−140	−120	0	0	=	0
S_1	0	2	4	1	0	=	1400
S_2	0	4	3	0	1	=	1500

3. *Optimality test*: We need to determine if an adjacent intersection point improves the value of the objective function. If not, the current extreme point is optimal. If an improvement is possible, the optimality test determines which variable currently in the independent set (having value zero) should *enter* the dependent set as a basic variable and become nonzero. For our maximization problem, we look at the Z-row (The row marked by the basic variable Z). If any coefficient in that row is negative, then we

select the variable whose coefficient is the most negative as the entering variable.

In the Z-row, the coefficients are as follows:

	Z	x_1	x_2	S_1	S_2
Z	1	−140	−120	0	0

The variable with the most negative coefficient is x_1 with value −140. Thus, x_2 wants to become a basic variable. We can have only three basic variables in this example (because we have three equations) so one of the current basic variables $\{S_1, S_2\}$ must be replaced by x_1. Let us proceed to see how we determine which variable exists being a basic variable.

4. *Feasibility test*: To find a new intersection point, one of the variables in the basic variable set must *exit* to allow the entering variable from Step 3 to become basic. The feasibility test determines which current dependent variable to choose for exiting, ensuring that we stay inside the feasible region. We will use the minimum positive ratio test as our feasibility test. The minimum positive ratio test is the $\text{Min}(\text{RHS}_j/a_j > 0)$. Make a quotient of the RHS_j/a_j.

			Most Negative Coefficient (−30)					Ratio Test
	Z	x_1	x_2	S_1	S_2		RHS	Quotient
Z	1	−140	−120	0	0	=	0	
S_1	0	2	4	1	0	=	1400	1400/2 = 700
S_2	0	4	3	0	1	=	1500	1500/4 = 3 75

Note that we will always disregard all quotients with either 0 or negative values in the denominator. In our example, we compare {700,375} and select the smallest nonnegative value. This gives the location of the matrix pivot that we will perform. However, matrix pivots in Excel are not easy, so we will use the updated matrix B by swapping the second column with the column of the variable x_2. Then, we invert B to obtain B^{-1}. Then, we multiply the original tableau by B^{-1}.

Tableau 0

Basic Variable	x_1	x_2	Basic variable S_1	Basic variable S_2	RHS	Ratio test
Z	-140	-120	0	0	0	—
S_1	2	4	1	0	1400	700
S_2	4	3	0	1	1500	375 *

B

Z	S_1	S_2
1	0	0
0	1	0
0	0	1

Tableaus 1

Basic Variable	x_1	x_2	S_1	S_2	RHS	Ratio test
Z	0	-15	0	35	52500	
s_1	0	2.5	1	-0.5	650	260 *
x_1	1	0.75	0	0.25	375	500

Updated B

Z	S_1	x_1			B^{-1}		
1	0	-140			1	0	35
0	1	2			0	1	-0.5
0	0	4			0	0	0.25

Tableau 2

Basic Variable	x_1	x_2	S_1	S_2	RHS	
Z	0	0	6	32	56400	No negatives->STOP
x_2	0	1	0.4	-0.2	260	
x_1	1	2.22E-16	-0.3	0.4	180	

Updated B

Z	x_2	x_1			B^{-1}		
1	-120	-140			1	6	32
0	4	2			0	0.4	-0.2
0	3	4			0	-0.3	0.4

In three iterations of the simplex, we have found our solution.
The final solution is read as follows:

Basic variables

$$x_2 = 260$$
$$x_1 = 180$$
$$Z = 56{,}400$$

Nonbasic variables

$$S_1 = S_2 = 0$$

The final tableau is also important.

Tableau	2							
		Z	x1	x2	S1	S2	RHS	
	Z	1	0	0	6	32	56400	No negatives->STOP
	x2	0	0	1	0.4	-0.2	260	
	x1	0	1	2.22E-16	-0.3	0.4	180	

We look for possible alternate optimal solutions by looking in the
Z-row for costs of 0 for nonbasic variables. Here, there are none.
We also examine the cost coefficient for the nonbasic variables and
recognize them as reduced costs or shadow prices. In this case, the
shadow prices are 6 and 32, respectively. Again if the cost of an additional unit of each constraint is the same, then adding an additional
unit of constraint 2 produces the largest increase in Z (32 > 6).

Exercises 3.7

Resolve exercises from Section 3.3 using the tableau method in Excel or
Maple.

References and Suggested Further Reading

Albright, B. 2010. *Mathematical Modeling with EXCEL*. Burlington, MA: Jones and
Bartlett Publishers, Chapter 7.

Apaiah, R. and E. Hendrix. 2006. Linear programming for supply chain design: A
case on Novel protein foods. PhD Thesis, Wageningen University, Netherlands.

Balakrishnan, N., B. Render, and R. Stair. 2007. *Managerial Decision Making* (2nd ed.).
Upper Saddle River, NJ: Prentice Hall.

Bazarra, M. S., J. J. Jarvis, and H. D. Sheralli. 1990. *Linear Programming and Network
Flows*. Hoboken, NJ: John Wiley & Sons.

Ecker, J. and M. Kupperschmid. 1988. *Introduction to Operations Research.* New York: John Wiley & Sons.

Fishback, P. E. 2010. *Linear and Nonlinear Programming with Maple: An Interactive, Applications-Based Approach.* Boca Raton, FL: CRC Press, http://www.mathplace.org/C064X/main.pdf.

Fox, W. 2012. *Mathematical Modeling with Maple.* Boston, MA: Cengage Publishers, Chapters 7–10.

Fox, W. P. and F. Garcia. 2013. Chapter 9, Modeling and linear programming in engineering management. In F. P. G. Márquez and B. Lev (Eds.), *Engineering Management.* Rijeka, Croatia: InTech, pp. 216–225.

Giordano, F., W. Fox, and S. Horton. 2014. *A First Course in Mathematical Modeling* (5th ed.). Boston, MA: Cengage Publishers, Chapter 7.

Hiller, F. and G. J. Liberman. 1990. *Introduction to Mathematical Programming.* Hightstovm, NJ: McGraw Hill Publishing Company.

McGrath, G. 2007. Email marketing for the U.S. Army and Special Operations Forces Recruiting. Master's Thesis, Naval Postgraduate School.

Winston, W. L. 2002. *Introduction to Mathematical Programming Applications and Algorithms* (4th ed.). Pacific Grove, CA: Duxbury Press.

Winston, W. L. 1994. *Operations Research: Applications and Algorithms* (3rd ed.). Belmont, CA: Duxbury Press.

4

Introduction to Multi-Attribute Decision-Making in Business Analytics

OBJECTIVES

1. Know the types of multi-attribute decision techniques.
2. Know the basic solution methodologies.
3. Know the weighting schemes.
4. Know which technique or techniques to use.
5. Know the importance of sensitivity analysis.
6. Know the importance of technology in the solution process.

Risk analysis for homeland security: Consider providing support to the Department of Homeland Security (DHS). The department has only a limited number of assets and a finite amount of time to conduct investigations; thus, priorities might be established. The risk assessment office has collected the data for the morning meeting shown in Table 4.1. Your operations research team must analyze the information and must provide a priority list to the risk assessment team for that meeting with DHS.

Problem statement: Build a model that ranks the threats in a priority order.

Assumptions: We have past decision that will give us insights into the decision-maker's process. We have data only on reliability, approximate number of deaths, approximate costs to fix or rebuild, location, destructive influence, and on the number of intelligence gathering tips. These will be the criteria for our analysis. The data are accurate and precise. We solve this problem later in the chapter.

TABLE 4.1

DHS Risk Assessment Data

Threat Alternatives/Criterion	Reliability of Threat Assessment	Approximate Associated Deaths (000)	Cost to Fix Damages (in millions)	Location Density (in millions)	Destructive Psychological Influence	Number of Intelligence Related Tips
Dirty bomb threat	0.40	10	150	4.5	9	3
Anthrax-bio terror threat	0.45	0.8	10	3.2	7.5	12
District of Columbia-road and bridge network threat	0.35	0.005	300	0.85	6	8
New York subway threat	0.73	12	200	6.3	7	5
District of Columbia metro threat	0.69	11	200	2.5	7	5
Major bank robbery	0.81	0.0002	10	0.57	2	16
Federal Aviation Administration threat	0.70	0.001	5	0.15	4.5	15

4.1 Introduction

Multiple-attribute decision-making (MADM) concerns in making decisions when there are multiple but a finite list of alternatives and criteria. This differs from analysis where we have alternatives and only one criterion such as cost. We address problems as in the DHS scenario where we have seven alternatives and six criteria that impact the decision.

Consider a problem where management needs to prioritize or rank order alternative choices such as identifying key nodes in a business network, picking a contractor or subcontractor, choosing an airport, ranking recruiting efforts, ranking banking facilities, and ranking schools or colleges. How does one proceed to accomplish this analytically?

In this chapter, we will present four methodologies to rank order or prioritize alternatives based on multiple criteria. These four methodologies include the following:

1. Data envelopment analysis (DEA)
2. Simple average weighting (SAW)
3. Analytical hierarchy process (AHP)
4. Technique of order preference by similarity to ideal solution (TOPSIS)

For each method, we describe the method and provide a methodology, discuss some strengths and limitations to the method, discuss tips for conducting sensitivity analysis, and present several illustrative examples.

4.2 Data Envelopment Analysis

4.2.1 Description and Uses

DEA is a *data input–output-driven* approach for evaluating the performance of entities called decision-making units (DMUs) that convert multiple inputs into multiple outputs (Cooper et al., 2000). The definition of a DMU is generic and very flexible so that any entity to be ranked might be a DMU. DEA has been used to evaluate the *performance* or *efficiencies* of hospitals, schools, departments, U.S. Air Force wings, U.S. Armed Forces recruiting agencies, universities, cities, courts, businesses, banking facilities, countries, regions, Special Operations Forces (SOF) airbases, key nodes in networks, and the list goes on. According to Cooper et al. (2000), DEA has been used to gain insights into activities that were not obtained by other quantitative or qualitative methods.

Charnes et al. (1978) described DEA as a mathematical programming model applied to observational data. It provides a new way of obtaining

empirical estimates of relationship among the DMUs. It has been formally defined as a methodology directed to frontiers rather than central
tendencies.

4.2.2 Methodology

The model, in simplest terms, may be formulated and solved as a linear
programming problem (Callen, 1991; Winston, 1995). Although several formulations for DEA exist, we seek the most straightforward formulation
in order to maximize an efficiency of a DMU as constrained by inputs
and outputs as shown in Equation 4.1. As an option, we might normalize
the metric inputs and outputs for the alternatives if the values are poorly
scaled within the data. We will call this data matrix, $\mathbf{X}$, with entries x_{ij}. We
define an efficiency unit as E_i for $i = 1,2,\ldots$, nodes. Let w_i be the weights or
coefficients for the linear combinations. Further, we restrict any efficiency
from being larger than one. Thus, the largest efficient DMU will be 1. This
gives the following linear programming formulation for single output but
multiple inputs:

$$\text{Max } E_i$$

subject to

$$\sum_{i=1}^{n} w_i x_{ij} - E_i = 0, j = 1,2,\ldots \tag{4.1}$$

$$E_i \leq 1 \text{ for all } i$$

For multiple inputs and outputs, we recommend the formulations provided
by Winston (1995) and Trick (2014) using Equation 4.2.

For any DMU_0, let X_i be the inputs and Y_i be the outputs. Let X_0 and Y_0 be
the DMUs being modeled.

$$\text{Min } \theta$$

subject to

$$\sum \lambda_i X_i \leq \theta X_0 \tag{4.2}$$

$$\sum \lambda_i Y_i \leq Y_0$$

$$\lambda_l > 0$$

Nonnegativity

4.2.3 Strengths and Limitations to Data Envelopment Analysis

DEA can be a very useful tool when used wisely according to Trick (1996). A few of the strengths that make DEA extremely useful are (Trick, 1996): (1) DEA can handle multiple input and multiple output models, (2) DEA does not require an assumption of a functional form relating inputs to outputs, (3) DMUs are directly compared against a peer or combination of peers, and (4) Inputs and outputs can have very different units. For example, X_1 could be in units of lives saved and X_2 could be in units of dollars without requiring any a priori trade-off between the two.

The same characteristics that make DEA a powerful tool can also create limitations to the process and analysis. An analyst should keep these limitations in mind when choosing whether or not to use DEA. A few additional limitations include the following:

1. As DEA is an extreme point technique, noise in the data such as measurement error can cause significant problems.

2. DEA is good at estimating *relative* efficiency of a DMU, but it converges very slowly to *absolute* efficiency. In other words, it can tell you how well you are doing compared to your peers but not compared to a *theoretical maximum*.

3. As DEA is a nonparametric technique, statistical hypothesis tests are difficult and are the focus of ongoing research.

4. As a standard formulation of DEA with multiple inputs and outputs creates a separate linear program for each DMU, large problems can be computationally intensive.

5. Linear programming does not ensure that all weights are considered. We find that the values for weights are only for those that optimally determine an efficiency rating. If having all criteria weighted (inputs, outputs) is essential to the decision-maker, then do not use DEA.

4.2.4 Sensitivity Analysis

Sensitivity analysis is always an important element in analysis. According to Neralic (1998), an increase in any output cannot make a solution worse rating nor a decrease in inputs alone can worsen an already achieved efficiency rating. As a result, in our examples we only decrease outputs and increase inputs as just described (Neralic, 1998). We will illustrate some sensitivity analysis, as applicable, in our illustrative examples in Section 4.2.5.

4.2.5 Illustrative Examples

Example 4.1: Manufacturing with DEA

Consider the following manufacturing process where we have three DMUs each of which has two inputs and three outputs as shown in the data table:

DMU	Input #1	Input #2	Output #1	Output #2	Output #3
1	5	14	9	4	16
2	8	15	5	7	10
3	7	12	4	9	13

Source: Winston, W., *Introduction to Mathematical Programming*, Duxbury Press, Pacific Grove, CA, pp. 322–325, 1995.

As no units are given and as the scales are similar, we decide not to normalize the data. We define the following decision variables:

t_i is the value for a single unit of output of DMU$_i$, for $i = 1,2,3$
w_i is the cost or weight for one unit of input of DMU$_i$, for $i = 1,2$
Efficiency$_i$ = DMU$_i$ = (total value of i's outputs)/(total cost of i's inputs), for $i = 1,2,3$

The following modeling assumptions are made as follows:

1. No DMU will have an efficiency of more than 100%.
2. If any efficiency is less than 1, then it is inefficient.
3. We will scale the costs so that the cost of the inputs equals 1 for each linear program. For example, we will use $5w_1 + 14w_2 = 1$ in our program for DMU$_1$.
4. All values and weights must be strictly positive, so we use a constant such as 0.0001 in lieu of 0.

To calculate the efficiency of DMU$_1$, we define the linear program using Equation 4.2 as

Maximize DMU$_1 = 9t_1 + 4t_2 + 16t_3$
Subject to

$$-9t_1 - 4t_2 - 16t_3 + 5w_1 + 14w_2 > 0$$
$$-5t_1 - 7t_2 - 10t_3 + 8w_1 + 15w_2 > 0$$
$$-4t_1 - 9t_2 - 13t_3 + 7w_1 + 12w_2 > 0$$
$$5w_1 + 14w_2 = 1$$
$$t_i > 0.0001, i = 1,2,3$$
$$w_i > 0.0001, i = 1,2$$
Nonnegativity

To calculate the efficiency of DMU$_2$, we define the linear program using Equation 4.2 as

Maximize $DMU_2 = 5t_1 + 7t_2 + 10t_3$
Subject to

$$-9t_1 - 4t_2 - 16t_3 + 5w_1 + 14w_2 > 0$$
$$-5t_1 - 7t_2 - 10t_3 + 8w_1 + 15w_2 > 0$$
$$-4t_1 - 9t_2 - 13t_3 + 7w_1 + 12w_2 > 0$$
$$8w_1 + 15w_2 = 1$$
$$t_i > 0.0001, i = 1,2,3$$
$$w_i > 0.0001, i = 1,2$$
Nonnegativity

To calculate the efficiency of DMU_3, we define the linear program using Equation 4.2 as

Maximize $DMU_3 = 4t_1 + 9t_2 + 13t_3$
Subject to

$$-9t_1 - 4t_2 - 16t_3 + 5w_1 + 14w_2 > 0$$
$$-5t_1 - 7t_2 - 10t_3 + 8w_1 + 15w_2 > 0$$
$$-4t_1 - 9t_2 - 13t_3 + 7w_1 + 12w_2 > 0$$
$$7w_1 + 12w_2 = 1$$
$$t_i > 0.0001, i = 1,2,3$$
$$w_i > 0.0001, i = 1,2$$
Nonnegativity

The linear programming solutions show the efficiencies as $DMU_1 = DMU_3 = 1$, $DMU_2 = 0.77303$.

Interpretation: DMU_2 is operating at 77.303% of the efficiency of DMU_1 and DMU_3. Management could concentrate on some improvements or best practices from DMU_1 or DMU_3 for DMU_2. An examination of the dual prices for the linear program of DMU_2 yields $\lambda_1 = 0.261538$, $\lambda_2 = 0$, and $\lambda_3 = 0.661538$. The average output vector for DMU_2 can be written as

$$0.261538 \begin{bmatrix} 9 \\ 4 \\ 16 \end{bmatrix} + 0.661538 \begin{bmatrix} 4 \\ 9 \\ 13 \end{bmatrix} = \begin{bmatrix} 5 \\ 7 \\ 12.785 \end{bmatrix}$$

and the average input vector can be written as

$$0.261538 \begin{bmatrix} 5 \\ 14 \end{bmatrix} + 0.661538 \begin{bmatrix} 7 \\ 12 \end{bmatrix} = \begin{bmatrix} 5.938 \\ 11.6 \end{bmatrix}$$

In our data, output #3 is 10 units. Thus, we may clearly see that the inefficiency is in output #3 where 12.785 units are required. We find that they are short of 2.785 units ($12.785 - 10 = 2.785$). This helps focus on treating the inefficiency found for output #3.

Sensitivity Analysis: Sensitivity analysis in a linear program is sometimes referred to as *what if* analysis. Let us assume that without management requiring some additional training for DMU_2, that

DMU_2's output #3 would drop from 10 to 9 units of output while the input #2 hours increase from 15 to 16 hours. We find that these changes in the *technology coefficients* are easily handled in resolving the LPs. As DMU_2 is affected, we might only modify and solve the LP concerning DMU_2. We find with these changes that DMU's efficiency is now only 74% as effective as DMU_1 and DMU_3.

Example 4.2: Social Networks and Ranking Nodes

Consider the Kite Social Network (Krackhardt, 1990) shown in Figure 4.1.

ORA© (Carley, 2011), a social network software, was used to obtain the metrics for this network. A subset of the output is shown in Table 4.1. We restricted the metrics presented: total centrality (TC), eigenvector centrality (EC), in-closeness (IC), out-closeness (OC), information centrality (INC), and betweenness (Betw), whose definitions can be found in recent social network literature (Fox and Everton, 2013, 2014) (Table 4.2).

We formulate the linear program from Equation 4.1 to measure the efficiency of the nodes. We define the decision variables:

u_i is the efficiency of node i, $i = 1,2,3,...10$
w_j is the weight of input j, $j = 1,2,3,4,5$

Maximize u_1
Subject to
$A = 0$
$u_i < 1$ for $i = 1,2,3,...,10$

where

$$
A = \begin{bmatrix}
0.180555556\, w_1 + 0.175080826\, w_2 + 0.091993186\, w_3 + 0.10806175\, w_4 + 0.108849307\, w_5 + 0.202247191\, w_6 - u_1 \\
0.138888889\, w_1 + 0.137527978\, w_2 + 0.099659284\, w_3 + 0.100343053\, w_4 + 0.113090189\, w_5 + 0.15526047\, w_6 - u_2 \\
0.125\, w_1 + 0.137527978\, w_2 + 0.110732538\, w_3 + 0.089193825\, w_4 + 0.113090189\, w_5 + 0.104187947\, w_6 - u_3 \\
0.111111111\, w_1 + 0.114399403\, w_2 + 0.099659284\, w_3 + 0.100343053\, w_4 + 0.100932994\, w_5 + 0.019407559\, w_6 - u_4 \\
0.111111111\, w_1 + 0.114399403\, w_2 + 0.099659284\, w_3 + 0.100343053\, w_4 + 0.100932994\, w_5 + 0.019407559\, w_6 - u_5 \\
0.083333333\, w_1 + 0.093757772\, w_2 + 0.099659284\, w_3 + 0.100343053\, w_4 + 0.097540288\, w_5 - u_6 \\
0.083333333\, w_1 + 0.093757772\, w_2 + 0.099659284\, w_3 + 0.100343053\, w_4 + 0.097540288\, w_5 - u_7 \\
0.083333333\, w_1 + 0.104202935\, w_2 + 0.099659284\, w_3 + 0.100343053\, w_4 + 0.108849307\, w_5 + 0.317671093\, w_6 - u_8 \\
0.055555556\, w_1 + 0.024123352\, w_2 + 0.099659284\, w_3 + 0.100343053\, w_4 + 0.088493073\, w_5 + 0.181818182\, w_6 - u_9 \\
0.027777778\, w_1 + 0.005222581\, w_2 + 0.099659284\, w_3 + 0.100343053\, w_4 + 0.070681368\, w_5 - u_{10}
\end{bmatrix}
$$

The linear programming solution is as follows:

	DV	
Susan	DMU_1	1
Steven	DMU_2	0.785511
Sarah	DMU_3	0.785511
Tom	DMU_4	0.653409
Claire	DMU_5	0.653409
Fred	DMU_6	0.535511
David	DMU_7	0.535511
Claudia	DMU_8	0.59517
Ben	DMU_9	0.137784
Jennifer	DMU_{10}	0.02983

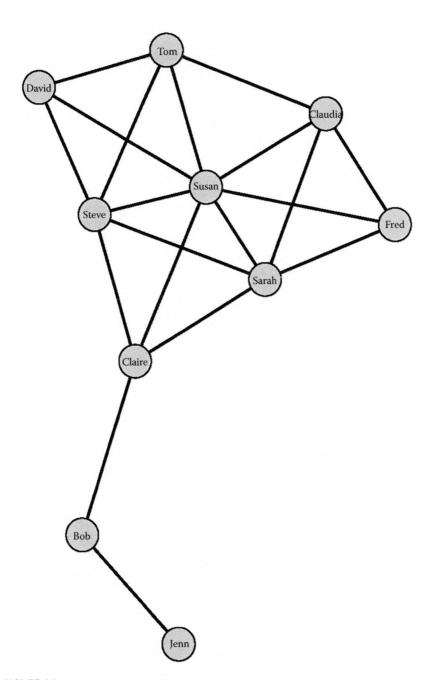

FIGURE 4.1
Kite network diagram from ORA. (From Carley, K. M., *Organizational Risk Analyzer* (*ORA*), Center for Computational Analysis of Social and Organizational Systems (CASOS), Carnegie Mellon University, Pittsburgh, PA, 2011.)

TABLE 4.2

ORA Metric Measures as Outputs for the Kite Network

TC	EC	IC	OC	INC	Betw
0.1806	0.1751	0.0920	0.1081	0.1088	0.2022
0.1389	0.1375	0.0997	0.1003	0.1131	0.1553
0.1250	0.1375	0.1107	0.0892	0.1131	0.1042
0.1111	0.1144	0.0997	0.1003	0.1009	0.0194
0.1111	0.1144	0.0997	0.1003	0.1009	0.0194
0.0833	0.0938	0.0997	0.1003	0.0975	0.0000
0.0833	0.0938	0.0997	0.1003	0.0975	0.0000
0.0833	0.1042	0.0997	0.1003	0.1088	0.3177
0.0556	0.0241	0.0997	0.1003	0.0885	0.1818
0.0278	0.0052	0.0997	0.1003	0.0707	0.0000

w_1	0
w_2	5.711648
w_3	0
w_4	0
w_5	0
w_6	0

Interpretation: We interpret the linear programming solution as follows: Player 1, u_1 = Susan, is rated most influential followed closely by Sarah and Steven. In addition, we see that the most important criterion in solving the optimal problem was the eigenvector centrality, w_2, of the network. The solution translated back into the original variables is found as

Susan = 1, Sarah = 0.78551, Steven = 0.78551, Claire = 0.6534, Tom = 0.6534, Fred = 0.5355, David = 0.5355, Claudia = 0.5951, Ben = 0.1377, and Jennifer = 0.02983 while $w_1 = w_3 = w_4 = w_5 = w_6 = 0$ and $w_2 = 5.7116$.

As the output metrics are network metrics calculated from ORA, we do not recommend any sensitivity analysis for this type of problem unless your goal is to improve the influence (efficiency) of another member of the network. If so, then finding the *dual prices* (*shadow prices*) would be required as shown in the first example.

Example 4.3: Recruiting

DEA to obtain efficiency in recruiting units is illustrated. Linear programming may be used to compare the efficiency of units, known as DMUs. The data envelopment method uses the following linear programming formulation to calculate its efficiencies. We want to measure the efficiency of 42 recruiting companies that are part of a recruiting brigade in the United States. The model uses six input measures and two output measures created from data that are obtained directly from the sixth brigade in 2014. The outputs are the percent fill-to-demand ratio for the unit and the percent language capability of the unit. The inputs are the number of recruiters and the percent of populations from which to recruit in a region. The main question was to determine if a larger percentage of recruiters' ability to speak languages other than English improved their units' ability to attract recruits. The goal is to identify those units that are not operating at the highest level so that improvement can be made to improve their efficiency. The data envelopment will calculate which of the companies, in this case DMU_1, DMU_2, ... , DMU_{42}, are more efficient when compared to the others (Figure 4.2).

The data needed for evaluating efficiency of the companies are provided in Table 4.3.

The weighted sum of the company's populations, or the first five columns in Table 4.3, must be equal to 1.00. It does not account for other ethnicities.

The output matrix array is the set of coefficient vectors for the fill-to-demand and language-to-recruiter output variables. It also includes a portion of its output coefficients:

The linear programming formulation to implement the solutions of the DMUs is as follows:

Objective function:
Maximize DMU_1, DMU_2, ..., DMU_c

Subject to:

Constraint 1: $\begin{bmatrix} w_1 \\ \cdots \\ w_c \end{bmatrix} - \begin{bmatrix} T_1 \\ \cdots \\ T_c \end{bmatrix} \geq 0$; limits the resource of outputs to that of inputs

Constraint 2: $\begin{bmatrix} DMU_1 \\ \cdots \\ DMU_c \end{bmatrix} - \begin{bmatrix} T_1 \\ \cdots \\ T_c \end{bmatrix} = 0$; the efficiencies cannot be greater than outputs

Constraint 3: $\begin{bmatrix} DMU_1 \\ \cdots \\ DMU_c \end{bmatrix} \leq 1$; limits of the efficiency to values less than or equal to 1

Constraint 4: $\begin{bmatrix} w_1 \\ \cdots \\ w_i \end{bmatrix} \geq 0.001$; limits the input weights to be greater than zero

Constraint 5: $\begin{bmatrix} t_1 \\ t_2 \end{bmatrix} \geq 0.001$; limits the output weights to be greater than zero

Constraint 6: $\begin{bmatrix} X_{1,\,input\,1} & \cdots & X_{1,\,input\,i} \\ \vdots & \ddots & \vdots \\ X_{c,\,input\,1} & \cdots & X_{c,\,input\,i} \end{bmatrix} \times \begin{bmatrix} w_1 \\ \cdots \\ w_i \end{bmatrix} = 1$;

The product of the input coefficient and decision variables must equal 1

FIGURE 4.2
Screenshot of recruiting generic formulation.

TABLE 4.3

Input and Output Coefficients for the DEA Approach (Array Named Matrix A)

6th REC BDE	Input Coefficients						Output Coefficients	
Company	%PopAPI	%PopAA	%PopH	%PopW	%PopNative	#Recruiters	Fill-to-Demand Ratio	Language-to-Recruiter Ratio
6F2—San Gabri El Vl	0.263722	0.031573	0.517156	0.185787	0.001763	38	0.872152	0.2105
6F3—Long Beach	0.087205	0.105755	0.672485	0.133068	0.001486	41	0.959648	0.2683
6F5—Sn Fernando Vl	0.080011	0.035218	0.533119	0.350004	0.001648	41	0.763623	0.0976
6F7—Coastal	0.157075	0.199102	0.295339	0.347086	0.001397	27	0.840470	0.2963
6f8—Los Angeles	0.136177	0.055508	0.546700	0.260140	0.001474	51	0.896214	0.3000
6h1—Eugene	0.036067	0.007294	0.093672	0.846818	0.016149	28	0.798541	0.0909
6h2—Vancouver	0.074656	0.038769	0.133544	0.745682	0.007349	49	0.888734	0.2083
6H3—Wi Lsonvi Lle	0.035723	0.010366	0.173345	0.769427	0.011140	25	0.763134	0.1000
6H5—Honolulu	0.600513	0.014254	0.122297	0.260485	0.002450	24	1.014687	0.0800
6H7—Guam	0.926119	0.010042	0.000000	0.063839	0.000000	24	1.016260	0.0000
6I0—Si Erra Nevada	0.047907	0.019145	0.250423	0.664284	0.018241	29	0.910788	0.0000
6I 1—Reddi Ng	0.040783	0.010726	0.151065	0.772217	0.025210	42	0.808599	0.0870
6I 3—Sacramento VI	0.081499	0.036714	0.216618	0.657542	0.007626	39	0.920515	0.0000
6I 4—San Joaqui N	0.113513	0.053919	0.445580	0.381530	0.005457	36	0.931590	0.0000
6I 5—Capi Tol	0.202628	0.107138	0.257041	0.427997	0.005197	52	0.888965	0.0000
6I 6—North Bay	0.086219	0.066036	0.319470	0.519875	0.008401	31	0.656557	0.0000
6J1—Ogden	0.018162	0.008748	0.129949	0.833434	0.009707	27	0.797701	0.1000
6J2—Salt Lake	0.043967	0.010922	0.149390	0.784530	0.011190	29	0.752606	0.0000
6J3—Butte	0.010439	0.002670	0.039882	0.885437	0.061573	25	0.714416	0.0000
6J4—Boi Se	0.017945	0.006657	0.165861	0.800868	0.008668	28	0.860633	0.0000
6J6—Las Vegas	0.102264	0.107663	0.309920	0.474534	0.005620	109	0.890533	0.2174

(Continued)

TABLE 4.3 (Continued)

Input and Output Coefficients for the DEA Approach (Array Named Matrix A)

6th REC BDE Company	Input Coefficients						Output Coefficients	
	%PopAPI	%PopAA	%PopH	%PopW	%PopNative	#Recruiters	Fill-to-Demand Ratio	Language-to-Recruiter Ratio
6J9—Bi G Horn	0.006928	0.003742	0.067161	0.844465	0.077704	17	0.754591	0.0435
6K1—Redlands	0.046855	0.080681	0.566912	0.300264	0.005287	43	0.969034	0.1111
6K2—Fullerton	0.201510	0.019564	0.492608	0.284349	0.001969	40	0.857143	0.1364
6K4—La Mesa	0.117789	0.053590	0.506028	0.317863	0.004730	39	0.715259	0.0690
6K5—Newport Beach	0.139674	0.013223	0.270650	0.574473	0.001979	32	0.743743	0.0800
6K6—San Marcos	0.044696	0.032839	0.508044	0.408251	0.006170	50	0.876591	0.0000
6K7—Ri Versi De	0.086515	0.090924	0.563245	0.256525	0.002791	55	0.855292	0.1163
6K8—San Di Ego	0.173149	0.050862	0.249331	0.523374	0.003284	35	0.757979	0.3500
6L1—Everett	0.086824	0.018779	0.109900	0.76868	0.015817	25	0.837500	0.3846
6L2—Seattle	0.201467	0.073000	0.102603	0.616732	0.006198	32	0.766444	0.1364
6L3—Spokane	0.023348	0.012660	0.055685	0.891007	0.017300	23	0.796624	0.1200
6L4—Tacoma	0.092312	0.068854	0.121083	0.704523	0.013227	27	0.990457	0.1818
6L5—Yaki Ma	0.016830	0.008463	0.338310	0.614801	0.021596	22	0.775581	0.3143
6L6—Alaska	0.074101	0.028043	0.061478	0.655457	0.180921	24	0.855114	0.2000
6L7—Olympi A	0.048443	0.018887	0.094673	0.814933	0.023064	29	0.901454	0.1034
6N1—Fresno	0.081141	0.043657	0.577334	0.292158	0.00571	54	0.792119	0.0238
6N2—Bakersfi Eld	0.036745	0.072447	0.546160	0.338127	0.006521	40	0.836003	0.0769
6N6—Gold Coast	0.055439	0.015520	0.471056	0.454545	0.003440	34	0.721532	0.1111
6N7—South Bay	0.321473	0.038378	0.254535	0.383649	0.001965	37	0.638575	0.1220
6N8—East Bay	0.225293	0.125870	0.299244	0.346959	0.002633	55	0.692921	0.1290
6N9—Monterey Bay	0.21374	0.023038	0.451766	0.309182	0.002274	29	0.726957	0.0909

In order to maximize the efficiency of the companies, or DMUs, the model formulation uses three set of decision variables. Excel Solver identifies the optimal values for the decision variables by solving a linear program, as shown in Figure 4.3, of which the objective is to maximize the efficiency of the companies.

Figure 4.3 shows how to implement the preceding DEA linear formulation using Excel Solver. The naming conventions in the Excel Solver screen, which is shown in Figure 4.3, represent the array of cells in which the data are found.

Decision variables contain an array of cells in an Excel column that has all 42 decision variables assigned as DMU_1, DMU_2, ..., DMU_c; six values for the w_1, w_2, ..., w_i; and two values for the outputs t_1 and t_2. Each formulation, in the *subject to the constraints* block, has similar naming conventions in order to simplify the location of the data in the Excel spreadsheet.

The efficiency results in Table 4.4 provide an opportunity to determine whether the data envelopment method for ethnic populations correlates with the actual recruiting numbers by ethnicity. Note that the DEA uses only the ethnic population distributions and the total number of recruiters; similarly, the outputs use the fill-to-demand and language-to-recruiter

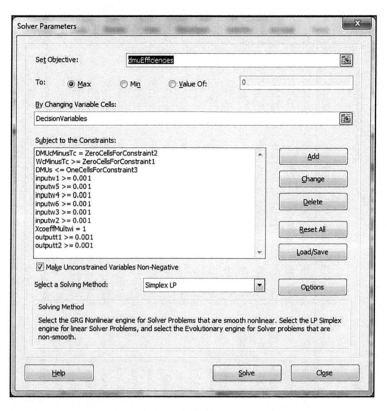

FIGURE 4.3
Linear program for the DEA problem using Excel Solver.

TABLE 4.4

Optimal DEA Efficiencies for the Sixth REC BDE's Companies

DMU Ranking[a]	DEA Efficiencies	Company
DMU_2	1.0000	6F3—Long Beach
DMU_{33}	1.0000	6L4—Tacoma
DMU_9	0.9886	6H5—Honolulu
DMU_{10}	0.9631	6H7—Guam
DMU_{23}	0.9558	6K1—Redlands
DMU_5	0.9506	6F8—Los Angeles
DMU_{30}	0.9235	6L1—Everett
DMU_{21}	0.9173	6J6—Las Vegas
DMU_7	0.9126	6H2—Vancouver
DMU_i	0.8976	6F2—San Gabriel Valley
DMU_4	0.8965	6F7—Coastal
DMU_{36}	0.8892	6L7—Olympia
DMU_{14}	0.8828	6I4—San Joaquin
DMU_{35}	0.8779	6L6—Alaska
DMU_{13}	0.8723	6I3—Sacramento Valley
DMU_{11}	0.8631	6I0—Sierra Nevada
DMU_{24}	0.8583	6K2—Fullerton
DMU_{28}	0.8498	6K7—Riverside
DMU_{15}	0.8424	6I5—Capitol
DMU_{34}	0.8411	6L5—Yakima
DMU_{29}	0.8365	6K8—San Diego
DMU_{27}	0.8307	6K6—San Marcos
DMU_{38}	0.8182	6N2—Bakersfield
DMU_{20}	0.8156	6J4—Boise
DMU_{12}	0.7956	6I1—Redding
DMU_{32}	0.7954	6L3—Spokane
DMU_{17}	0.7897	6J1—Ogden
DMU_6	0.7874	6H1—Eugene
DMU_{31}	0.7724	6L2—Seattle
DMU_{37}	0.7587	6N1—Fresno
DMU_8	0.757	6H3—Wilsonville
DMU_3	0.7566	6F5—Sn Fernando Vl
DMU_{26}	0.7318	6K5—Newport Beach
DMU_{22}	0.7298	6J9—Big Horn
DMU_{39}	0.7213	6N6—Gold Coast
DMU_{42}	0.7196	6N9—Monterey Bay
DMU_{18}	0.7132	6J2—Salt Lake
DMU_{25}	0.7011	6K4—La Mesa
DMU_{41}	0.7002	6N8—East Bay

(Continued)

TABLE 4.4 (*Continued*)

Optimal DEA Efficiencies for the Sixth REC BDE's Companies

DMU Ranking[a]	DEA Efficiencies	Company
DMU_{19}	0.677	6J3—Butte
DMU_{40}	0.6463	6N7—South Bay
DMU_{16}	0.6222	6I6—North Bay

[a] DMUs rank from highest to lowest.

ratios. However, the actual recruiting data—the number of recruits by ethnicity—are neither part of the inputs nor the outputs of the DEA method. The DEA method accounts for the company's performance in the form of the fill-to-demand ratio and indirectly, the P2P metrics. We will also show that the correlation between the recruiting efficiencies of the DEA and the P2P metrics suggests that the DEA model can be used to allocate recruiters with secondary languages. The decision-making criteria for allocating recruiters would be a bottom-up approach. In other words, the units at the bottom of the DEA ranking in Table 4.4 would be the ones to first receive new assignments of recruiters with secondary languages.

Analysis: The most efficient companies, those achieving a DEA score of 100%, are Long Beach, within the Los Angeles Battalion (BN) and Tacoma from the Seattle BN. The least efficient companies include North Bay from the Sacramento BN, achieving 62.2%, and South Bay from the Fresno BN, achieving 64.6%.

There are many other factors for improving recruitment numbers; nevertheless, DEA can be used as a tool to assess changes in conditions such as evolving demographic data, to reallocate recruiting center areas of operation, to update rankings based on new recruiting production and fill-to-demand ratios, or to assess changes in the recruiter's manning, or language-to-recruiter ratios.

Source: Figueroa, S., Improving recruiting in the 6th recruiting brigade through statistical analysis and efficiency measures, Master's Thesis, Naval Postgraduate School, Monterey, CA, 2014.

Exercises 4.2

1. Given the following input–output table for three hospitals where inputs are number of beds and labor hours in thousands per month and outputs, all measured in hundreds, are patient days for patients under 14, patient days for patients between 14 and 65, and patient days for patients more than 65. Determine the efficiency of the three hospitals.

	Inputs		Outputs		
Hospital	1	2	1	2	3
1	5	14	9	4	16
2	8	15	5	7	10
3	7	12	4	9	13

2. Resolve problem 1 with the following inputs and outputs.

	Inputs		Outputs		
Hospital	1	2	1	2	3
1	4	16	6	5	15
2	9	13	10	6	9
3	5	11	5	10	12

3. Consider ranking four bank branches in a particular city. The inputs are as follows:

Input 1 = labor hours in hundreds per month
Input 2 = space used for tellers in hundreds of square feet
Input 3 = supplies used in dollars per month
Output 1 = loan applications per month
Output 2 = deposits made in thousands of dollars per month
Output 3 = checks that processed thousands of dollars per month

The following data table is for the bank branches:

Branches	Input 1	Input 2	Input 3	Output 1	Output 2	Output 3
1	15	20	50	200	15	35
2	14	23	51	220	18	45
3	16	19	51	210	17	20
4	13	18	49	199	21	35

4. What *best practices* might you suggest to the branches that are less efficient in problem 3?

4.3 Weighting Methods

4.3.1 Modified Delphi Method

The Delphi method is a reliable way of obtaining the opinions of a group of experts on an issue by conducting several rounds of interrogative communications. This method was first developed in the U.S. Air Force in the 1950s, mainly for market research and sales forecasting (Chan et al., 2001). This modified method is basically a way to obtain inputs from experts and then to average their scores.

The panel consists of a number of experts who are chosen based on their experience and knowledge. As mentioned earlier, panel members remain anonymous to each other throughout the procedure to avoid the negative impacts of criticism on the innovation and creativity of panel members. The Delphi method should be conducted by a director. One can use the Delphi method for giving weights to the short-listed critical factors. The panel members should give weights to each factor and their reasoning. In this way, other panel members can evaluate the weights based on the reasons given and can accept, modify, or reject those reasons and weights. For example, consider a search region that has rows A–G and columns 1–6 as shown in Figure 4.4. A group of experts then places an x in the squares. In this example, each of 10 experts places 5 x's in the squares. We then total the number of x's in the squares and divide by the total of x placed, in this case 50.

We would find the weights as shown in Table 4.5.

	1	2	3	4	5	6
A						
B						
C						
D						
E						
F						
G						

FIGURE 4.4
Delphi example.

TABLE 4.5

Modified Delphi to Find Weights

Selection	Frequency	Relative Frequency or Weight
A1	3	3/50
B3	1	1/50
B4	1	1/50
B6	1	1/50
C4	1	1/50
C5	4	4/50
D6	8	8/50
E5	7	7/50
F5	8	8/50
G3	9	9/50
G4	7	7/50
All others	0	0/50
Total	50	50/50 = 1.0

4.3.2 Rank Order Centroid Method

This method is a simple way of giving weight to a number of items that are ranked according to their importance. The decision-makers usually can rank items much more easily than giving weight to them. This method takes those ranks as inputs and converts them to weights for each of the items. The conversion is based on the following formula:

$$w_i = \left(\frac{1}{M}\right)\sum_{n=i}^{M}\frac{1}{n}$$

1. List objectives in order from most important to least important
2. Use the aforementioned formulas for assigning weights

where M is the number of items and w_i is the weight of the i item. For example, if there are four items, the item ranked first will be weighted $(1+1/2+1/3+1/4)/4=0.52$, the second will be weighted $(1/2+1/3+1/4)/4=0.27$, the third $(1/3+1/4)/4=0.15$, and the last $(1/4)/4=0.06$. As shown in this example, the rank order centroid (ROC) is simple and easy to follow, but it gives weights that are highly dispersed (Chang, 2004). As an example, consider the same factors to be weighted (shortening schedule, agency control over the project, project cost, and competition). If they are ranked based on their importance and influence on decision as 1—shortening schedule, 2—project cost, 3—agency control over the project, and 4—competition, their weights would be 0.52, 0.27, 0.15, and 0.06, respectively. These weights almost eliminate the effect of the fourth factor, that is, among competitors. This could be an issue.

4.3.3 Ratio Method

The ratio method is another simple way of calculating weights for a number of critical factors. A decision-maker should first rank all the items according to their importance. The next step is giving weight to each item based on its rank. The lowest ranked item will be given a weight of 10. The weight of the rest of the items should be assigned as multiples of 10. The last step is normalizing these raw weights (Weber and Borcherding, 1993). This process is shown in the example in Table 4.6. Note that the weights should not necessarily jump 10 points from one item to the next. Any increase in the weight is based on the subjective judgment of the decision-maker and reflects the difference between the importance of the items. Ranking the items in the first step helps in assigning more accurate weights. An example of the ratio method is given in Table 4.6.

Normalized weights are simply calculated by dividing the raw weight of each item over the sum of the weights for all items. For example, normalized weight for the first item (shortening schedule) is calculated as 50/ (50 + 40 + 20 + 10) = 41.7%. The sum of normalized weights is equal to 100% (41.7 + 33.3 + 16.7 + 8.3 = 100), see Table 4.6.

4.3.4 Pairwise Comparison (Analytical Hierarchy Process)

In this method, the decision-maker should compare each item with the rest of the group and should give a preferential level to the item in each pairwise comparison (Chang, 2004). For example, if the item at hand is as important as the second one, the preferential level would be one. If it is much more important, its level would be 10. After conducting all the comparisons and after determining the preferential levels, the numbers will be added up and normalized. The results are the weights for each item. Table 4.7 can be used as a guide for giving a preferential level score to an item while comparing it with another one. The following example shows the application of the pairwise comparison procedure. Referring to the four critical factors identified earlier, let us assume that shortening the schedule, project cost, and agency control of the project are the most important parameters in the project delivery selection decision. Following the pairwise comparison, the decision-maker should pick one of these factors (e.g., shortening the schedule), compare it with the remaining factors, and should give a preferential level to it. For

TABLE 4.6

Ratio Method

Task/Item	Shorten Schedule	Project Cost	Agency Control	Competition
Ranking	1	2	3	4
Weighting	50	40	20	10
Normalizing	41.7%	33.3%	16.7%	8.3%

example, shortening the schedule is more important than project cost; in this case, it will be given a level of importance of 5.

The decision-maker should continue the pairwise comparison and should give weights to each factor. The weights, which are based on the preferential levels given in each pairwise comparison, should be consistent to the extent possible. The consistency is measured based on the matrix of preferential levels. The interested reader can find the methods and applications of consistency measurement in Temesi (2006). Table 4.7 provides the nine-point scale that we will use.

Table 4.8 shows the rest of the hypothetical weights and the normalizing process, the last step in the pairwise comparison approach.

Note that Column (5) is simply the sum of the values in Columns (1) through (4). In addition, note that if the preferential level of factor i to factor j is n, then the preferential level of factor j to factor i is simply $1/n$. The weights calculated for this exercise are 0.6, 0.1, 0.2, and 0.1, which add up to 1.0. Note that it is possible for two factors to have the same importance and weight.

TABLE 4.7

Saaty's Nine-Point Scale

Intensity of Importance in Pair-Wise Comparisons	Definition
1	Equal importance
3	Moderate importance
5	Strong importance
7	Very strong importance
9	Extreme importance
2,4,6,8	For comparing between the above
Reciprocals of above	In comparison of elements i and j if i is 3 compared to j, then j is $1/3$ compared to i

TABLE 4.8

Pairwise Comparison Example

	Shorten the Schedule (1)	Project Cost (2)	Agency Control (3)	Competition (4)	Total (5)	Weights (6)
Shorten the schedule	1	5	5/2	8	16.5	$16.5/27.225 = 0.60$
Project cost	1/5	1	½	1	2.7	$2.7/27/225 = 0.10$
Agency control	2/5	2	1	2	5.4	$5.4/27/225 = 0.20$
Competition	1/8	1	½	1	2.625	$2.625/27/225 = 0.10$
				Total	27.225	1

4.3.5 Entropy Method

Shannon and Weaver (1947) proposed the entropy concept, and this concept has been highlighted by Zeleny (1982) for deciding the weights of attributes. Entropy is the measure of uncertainty in the information using probability methods. It indicates that a broad distribution represents more uncertainty than a sharply peaked distribution does.

To determine the weights by the entropy method, the normalized decision matrix we call R_{ij} is considered. The equation used is

$$e_j = -k \sum_{i=1}^{n} R_{ij} \ln(R_{ij}) \tag{4.3}$$

where $k = 1/\ln(n)$ is a constant that guarantees that $0 < e_j < 1$. The value of n refers to the number of alternatives. The degree of divergence (d_j) of the average information contained by each attribute can be calculated as

$$d_j = 1 - e_j$$

The more divergent the performance rating R_{ij}, for all i and j, then the higher the corresponding d_j the more important the attribute B_j is considered to be.

The weights are found by Equation 4.4:

$$w_j = \frac{(1 - e_j)}{\sum (1 - e_j)} \tag{4.4}$$

Let us illustrate an example to obtain entropy weights.

Example 4.4: Entropy Weights for Criteria for Cars

1. The data are provided as follows:

	Cost	Safety	Reliability	Performace	MPG City	MPG HW	Interior and Style
a1	27.8	9.4	3	7.5	44	40	8.7
a2	28.5	9.6	4	8.4	47	47	8.1
a3	38.668	9.6	3	8.2	35	40	6.3
a4	25.5	9.4	5	7.8	43	39	7.5
a5	27.5	9.6	5	7.6	36	40	8.3
a6	36.2	9.4	3	8.1	40	40	8

2. Sum the columns

sums	184.168	57	23	47.6	245	246	46.9

3. Normalize the data. Divide each data element in a column by the sum of the column.

0.150949	0.164912	0.13043478	0.157563	0.17959184	0.162602	0.185501066
0.15475	0.168421	0.17391304	0.176471	0.19183673	0.191057	0.172707889
0.20996	0.168421	0.13043478	0.172269	0.14285714	0.162602	0.134328358
0.138461	0.164912	0.2173913	0.163866	0.1755102	0.158537	0.159914712
0.14932	0.168421	0.2173913	0.159664	0.14693878	0.162602	0.176972281
0.19656	0.164912	0.13043478	0.170168	0.16326531	0.162602	0.170575693

4. Use the entropy formula, where in the case $k = 6$.

$$e_j = -k \sum_{i=1}^{n} R_{ij} \ln(R_{ij})$$

	e1	e2	e3	e4	e5	e6	e7		k =	0.558111
	−0.28542	−0.29723	−0.2656803	−0.29117	−0.3083715	−0.29536	−0.31251265			
	−0.28875	−0.30001	−0.3042087	−0.30611	−0.31674367	−0.31623	−0.30330158			
	−0.32771	−0.30001	−0.2656803	−0.30297	−0.27798716	−0.29536	−0.26965989			
	−0.27376	−0.29723	−0.3317514	−0.29639	−0.30539795	−0.29199	−0.293142			
	−0.28396	−0.30001	−0.3317514	−0.29293	−0.28179026	−0.29536	−0.3064739			
	−0.31976	−0.29723	−0.2656803	−0.30136	−0.29589857	−0.29536	−0.3016761			

5. Find e_j,

0.993081	0.999969	0.98492694	0.999532	0.99689113	0.998825	0.997213162

6. Compute weights by formula:

	0.006919	3.09E−05	0.01507306	0.000468	0.00310887	0.001175	0.002786838	0.029561	
w	0.234044	0.001046	0.50989363	0.015834	0.1051674	0.039742	0.094273533		1

7. Check that weights sum to 1, as they did earlier.
8. Interpret weights and rankings.
9. Use these weights in further analysis.

Let us see the possible weights under another method.

Example 4.5: Pairwise Comparison Weighting for Cars

AHP: We will use template for Cars using pairwise comparison.

	Cost	Safety	Reliability	Performance	MPG City	MPG Highway	Interior and Style
	1	2	3	4	5	6	7
Cost	1	2	3	3	4	5	7
Safety	1/2	1	2	2	3	4	6
Reliability	1/3	1/2	1	2	3	4	5
Performance	1/3	1/2	1/2	1	2	4	6
MPG City	1/4	1/3	1/3	1/2	1	3	6
MPG Highway	1/5	1/4	1/4	1/4	1/3	1	3
Interior and Style	1/7	1/6	1/5	1/6	1/6	1/3	1

Results for the weights with a consistency ratio (CR) of 0.090 are as follows:

Cost	0.3612331
Safety	0.2093244
Reliability	0.14459
Performance	0.1166729
MPG City	0.0801478
MPG HW	0.0529871
Interior/Style	0.0350447

Example 4.6: Using the Ratio Method for Criteria Weights for Cars

Cost	Safety	Reliability	Performance	MPG City	MPG Highway	Interior and Style	
70	60	50	40	30	20	10	280
0.25	0.214	0.179	0.143	0.107	0.714	0.358	Sums to 1

4.4 Simple Additive Weighting Method

4.4.1 Description and Uses

This is a very straightforward and easily constructed process. Fisburn has referred this also as the weighted sum method (Fishburn, 1967). SAW is the simplest, and still one of the widest used MADM methods.

Its simplistic approach makes it easy to use. Depending on the type of the relational data used, we might either want the larger average or the smaller average.

4.4.2 Methodology

Here, each criterion (attribute) is given a weight, and the sum of all weights must be equal to 1. If criteria are equally weighted, then we merely need to sum the alternative values. Each alternative is assessed with regard to every criterion (attribute). The overall or composite performance score of an alternative is given simply by Equation 4.5 with m criteria.

$$P_i = \frac{\sum_{j=1}^{m} w_j m_{ij}}{m} \tag{4.5}$$

It was previously thought that all the units in the criteria must be identical units of measure such as dollars, pounds, and seconds. A normalization process can make the values unitless. So, we recommend normalizing the data as shown in Equation 4.6:

$$P_i = \frac{\sum_{j=1}^{m} w_j m_{ij\text{Normalized}}}{m} \tag{4.6}$$

where $(m_{ij\text{Normalized}})$ represents the normalized value of m_{ij}, and P_i is the overall or composite score of the alternative A_i. The alternative with the highest value of P_i is considered the best alternative.

4.4.3 Strengths and Limitations

The strengths are (1) the ease of use and (2) the normalized data allow for comparison across many differing criteria. Limitations include the following: larger is always better or smaller is always better. There is no flexibility in this method to state which criterion should be larger or smaller to achieve better performance. This makes gathering useful data of the same relational value scheme (larger or smaller) essential.

4.4.4 Sensitivity Analysis

Sensitivity analysis should be applied to the weighting scheme that is employed to determine how sensitive the model is to the weights. Weighting can be arbitrary for a decision-maker, or in order to obtain weights, you might choose to use a scheme to perform pairwise comparison as we show in AHP that we discuss in Section 4.5. Whenever subjectivity enters into the process for finding weights, then sensitivity analysis is recommended. Please see Sections 4.5.4 and 4.6.4 for our suggested scheme for dealing with sensitivity analysis for individual criterion weights.

4.4.5 Illustrative Examples: Simple Additive Weighting

Example 4.7: Car Selection by SAW (Data from Consumer's Report and U.S. News and World Report Online Data)

We are considering six cars: Ford Fusion, Toyota Prius, Toyota Camry, Nissan Leaf, Chevy Volt, and Hyundai Sonata. For each car, we have data on seven criteria that were extracted from Consumer's Report and U.S. News and World Report data sources. They are *cost, MPG city, MPG highway, performance, interior and style, safety*, and *reliability*. We provide the extracted information in Table 4.9.

Initially, we might assume that all weights are equal to obtain a baseline ranking. We substitute the rank orders (first to sixth) for the actual data. We compute the average rank that attempt to find the best ranking (smaller is better). We find that our rank ordering is Fusion, Sonata, Camry, Prius, Volt, and Leaf.

SAW USING RANK ORDERING OF THE DATA BY CRITERIA

Cars	Cost	MPG City	MPG Highway	Performance	Interior and Style	Safety	Reliability	Value	Rank
Prius	3	2	2	6	1	2	4	2.857	4
Fusion	4	1	1	1	3	1	3	2	1
Volt	6	6	2	2	6	1	4	3.857	6
Camry	1	3	3	4	5	2	1	2.714	1
Sonata	2	5	2	5	2	1	1	2.572	2
Leaf	5	4	2	2	4	2	4	3.285	5

TABLE 4.9

Raw Car Data

Cars	Cost ($000)	MPG City	MPG Highway	Performance	Interior and Style	Safety	Reliability
Prius	27.8	44	40	7.5	8.7	9.4	3
Fusion	28.5	47	47	8.4	8.1	9.6	4
Volt	38.668	35	40	8.2	6.3	9.6	3
Camry	25.5	43	39	7.8	7.5	9.4	5
Sonata	27.5	36	40	7.6	8.3	9.6	5
Leaf	36.2	40	40	8.1	8.0	9.4	3

Next, we apply a scheme to the weights and still use the ranks 1–6 as before. Perhaps we apply a technique similar to the pairwise comparison that we will discuss in Section 4.5. Using the pairwise comparison to obtain new weights, we obtain a new order: Camry, Sonata, Fusion, Prius, Leaf, and Volt. The changes in results of the rank ordering that differ from using equal weights show the sensitivity that the model has to the given criteria weights. We assume that the criteria in order of importance are cost, reliability, MPG city, safety, MPG highway, performance, and interior and style.

We use pairwise comparisons to obtain a new matrix:

	COST	Reliability	MPG City	Safety	MPG HW	Performance	Interior and Style
COST	1	2	3	4	5	6	7
Reliability	0.5	1	2	3	4	5	5
MPG City	0.333333	0.5	1	3	4	5	6
Safety	0.25	0.33333333	0.333333	1	2	3	4
MPG HW	0.2	0.25	0.25	0.5	1	3	4
Performance	0.166667	0.2	0.2	0.333333	0.333333	1	1
Interior and Style	0.142857	0.2	0.166667	0.25	0.25	1	1

The CR is 0.01862, and the new weights are as follows:

COST	0.38388
Reliability	0.22224
MPG City	0.15232
Safety	0.08777
MPG HW	0.06675
Performance	0.04612
Interior and Style	0.04092

Using these weights and applying to the previous ranking, we obtain values that we average and we select the smaller average. We find the rank ordering is Fusion, Sonata, Camry, Prius, Leaf, and Volt.

Prius	1.209897292	4
Fusion	0.801867414	1
Volt	1.470214753	6
Camry	1.15961718	3
Sonata	1.015172736	2
Leaf	1.343230626	5

SAW USING RAW DATA

We could also use the raw data directly from Table 4.5 except cost given that we use the ranks of the raw data. Now, only *cost* represents a value where smaller is better, so we can replace cost with its reciprocal. So *1/cost* represents a variable where larger is better. If we use the criteria weights from the previous results and if our raw data replace *cost* with 1/*cost*, we obtain a final ranking based on the criteria that larger values are better.

Cars	Value	Rank
Prius	0.16505	4
Fusion	0.17745	2
Volt	0.14177	6
Camry	0.1889	1
Sonata	0.1802	3
Leaf	0.14663	5

Our rank ordering is Camry, Fusion, Sonata, Prius, Leaf, and Volt.

SENSITIVITY ANALYSIS

We suggest employing sensitivity analysis on the criteria weights as described earlier. We modified the weights in a controlled manner and resolved the SAW values. These are displayed in Figure 4.5 where we see that the top ranked cars (Fusion, Camry, and Prius) do not change over our range of sensitivity analysis.

Example 4.8: Kite Network to Rank Nodes with SAW

We revisit the Kite Network described earlier. Here we present two methods that will work on the data from Example 4.2. Method I represents transforming the output data into rankings from first to last

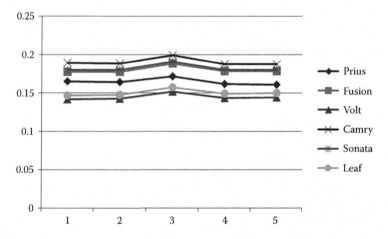

FIGURE 4.5
Sensitivity analysis of SAW values for cars.

place. Then we apply the weights and average all the values. We rank them smaller to larger to represent the alternative choices. We present only results using the pairwise comparison criteria to obtain the weighted criteria.

Weights	0.153209	0.144982	0.11944	0.067199	0.157688	0.357482
Susan	1	1	10	1	3	2
Steve	2	2	2	2	1	4
Sarah	3	2	1	10	1	7
Tom	4	4	2	2	5	5
Claire	4	4	2	2	5	5
Fred	6	7	2	2	7	8
David	6	7	2	2	7	8
Claudia	6	6	2	2	3	1
Ben	9	9	2	2	9	3
Jennifer	10	10	2	2	10	8

Susan	0.153209	0.144982	1.194396	0.067199	0.473064	0.714965		0.457969	Steve	0.426213
Steve	0.306418	0.289964	0.238879	0.134398	0.157688	1.42993		0.426213	Susan	0.457969
Sarah	0.459627	0.289964	0.11944	0.67199	0.157688	2.502377		0.700181	Claudia	0.498828
Tom	0.612835	0.579928	0.238879	0.134398	0.78844	1.787412		0.690316	Tom	0.690316
Claire	0.612835	0.579928	0.238879	0.134398	0.78844	1.787412		0.690316	Claire	0.690316
Fred	0.919253	1.014875	0.238879	0.134398	1.103816	2.859859		1.04518	Sarah	0.700181
David	0.919253	1.014875	0.238879	0.134398	1.103816	2.859859		1.04518	Ben	0.924772
Claudia	0.919253	0.869893	0.238879	0.134398	0.473064	0.357482		0.498828	Fred	1.04518
Ben	1.37888	1.304839	0.238879	0.134398	1.419192	1.072447		0.924772	David	1.04518
Jennifer	1.532089	1.449821	0.238879	0.134398	1.576879	2.859859		1.298654	Jennifer	1.298654

Method I rankings: Steve, Susan, Claudia. Tom, Claire, Sarah, Ben Fred, David, and Jennifer.

Method II uses the raw metric data and the weights we found earlier assuming that larger values imply better values.

Susan	0.027663	0.025384	0.010988	0.007262	0.017164	0.0723		0.026793	Claudia	0.029541
Steve	0.021279	0.019939	0.011903	0.006743	0.017833	0.055503		0.0222	Susan	0.026793
Sarah	0.019151	0.019939	0.013226	0.005994	0.017833	0.037245		0.018898	Steve	0.0222
Tom	0.017023	0.016586	0.011903	0.006743	0.015916	0.006938		0.012518	Sarah	0.018898
Claire	0.017023	0.016586	0.011903	0.006743	0.015916	0.006938		0.012518	Ben	0.018268
Fred	0.012767	0.013593	0.011903	0.006743	0.015381	0		0.010065	Tom	0.012518
David	0.012767	0.013593	0.011903	0.006743	0.015381	0		0.010065	Claire	0.012518
Claudia	0.012767	0.015108	0.011903	0.006743	0.017164	0.113562		0.029541	Fred	0.010065
Ben	0.008512	0.003497	0.011903	0.006743	0.013954	0.064997		0.018268	David	0.010065
Jennifer	0.004256	0.000757	0.011903	0.006743	0.011146	0		0.005801	Jennifer	0.005801

The results are Claudia, Susan, Steven, Sarah, Ben, Tom, Claire, Fred, David, and Jennifer. Although the top three are the same, their order is different. The model is sensitive both to the input format and the weights.

SENSITIVITY ANALYSIS

We apply sensitivity analysis to the weights, in controlled manner, and determine how each change impacts the final rankings. We recommend a controlled method to modify the weights. This is discussed later. You are asked in the Exercises set to perform sensitivity analysis to this problem.

Exercises 4.4

In each problem, use SAW to find the ranking under these weighted conditions:

1. All weights are equal.
2. Choose and state your weights.

  a. For a given hospital, rank order the procedure using the following data:

	Procedure			
	1	2	3	4
Profit	$200	$150	$100	$80
X-Ray times	6	5	4	3
Laboratory time	5	4	3	2

  b. For a given hospital, rank order the procedure using the following data:

	Procedure			
	1	2	3	4
Profit	$190	$150	$110	980
X-Ray times	6	5	5	3
Laboratory time	5	4	3	3

  c. Rank order the threats listed in Table 4.1:

Threat Alternatives/ Criterion	Reliability of Threat Assessment	Approximate Associated Deaths (000)	Cost to Fix Damages (in millions)	Location Density (in millions	Destructive Psychological Influence	Number of Intelligence Related Tips
Dirty bomb threat	0.40	10	150	4.5	9	3
Anthrax-bio terror threat	0.45	0.8	10	3.2	7.5	12
District of Columbia- Road and bridge network threat	0.35	0.005	300	0.85	6	8
New York subway threat	0.73	12	200	6.3	7	5
District of Columbia metro threat	0.69	11	200	2.5	7	5

(*Continued*)

Threat Alternatives/ Criterion	Reliability of Threat Assessment	Approximate Associated Deaths (000)	Cost to Fix Damages (in millions)	Location Density (in millions	Destructive Psychological Influence	Number of Intelligence Related Tips
Major bank robbery	0.81	0.0002	10	0.57	2	16
Federal Aviation Administration threat	0.70	0.001	5	0.15	4.5	15

d. Consider a scenario where we want to move and rank the cities.

City	Affordability of Housing (Average Home Cost in Hundreds of Thousands)	Cultural Opportunities— Events per Month	Crime Rate—Number of Reported # Crimes per Month (in Hundreds)	Quality of Schools on Average (Quality Rating between [0,1])
1	250	5	10	0.75
2	325	4	12	0.6
3	676	6	9	0.81
4	1,020	10	6	0.8
5	275	3	11	0.35
6	290	4	13	0.41
7	425	6	12	0.62
8	500	7	10	0.73
9	300	8	9	0.79

e. Perform sensitivity analysis to the node ranking for the Kite example.

f. Use the entropy weight method for the cars example and determine the rankings. Compare with our results shown in the text.

4.5 Analytical Hierarchy Process

4.5.1 Description and Uses

AHP is a multiobjective decision analysis tool that is first proposed by Saaty (1980). It is designed when either subjective and objective measures or just subjective measures are being evaluated in terms of a set of alternatives based on multiple criteria, organized in a hierarchical structure, see Figure 4.6.

At the top level is the goal. The next layer has the criteria to be evaluated or weighted, and at the bottom level the alternatives are measured against each criterion. The decision-maker assesses their evaluation by making pairwise comparisons in which every pair is subjectively or objectively compared. This subjective method involves a nine-point scale that we present later in Table 4.6.

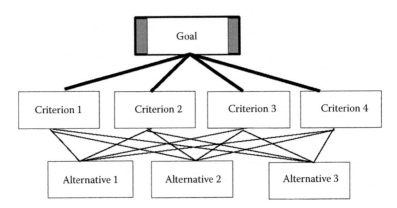

FIGURE 4.6
Generic AHP hierarchy.

We briefly discuss the elements in the framework of AHP. This process can be described as a method to decompose a problem into subproblems. In most decisions, the decision-maker has a choice among many alternatives. Each alternative has a set of attributes or characteristics that can be measured, either subjectively or objectively. We will call these as attributes or criteria. The attribute elements of the hierarchical process can relate to any aspect of the decision problem that is either tangible or intangible, carefully measured or roughly estimated, and well or poorly understood—anything that applies to the decision at hand.

We state simply that in order to perform AHP, we need a goal or an objective and a set of alternatives, each with criteria (attributes) to compare. Once the hierarchy is built, the decision-makers systematically evaluate the various elements pairwise (by comparing them to one another two at a time), with respect to their impact on an element above them in the hierarchy. In making the comparisons, the decision-makers can use concrete data about the elements or subjective judgments concerning the elements' relative meaning and importance. If we realize that humans can easily change their minds, then sensitivity analysis will be very important.

The AHP converts these subjective but numerical evaluations to numerical values that can be processed and compared over the entire range of the problem. A numerical weight or priority is derived for each element of the hierarchy, allowing diverse and often incommensurable elements to be compared to one another in a rational and consistent way.

In the final step of the process, numerical priorities are calculated for each of the decision alternatives. These numbers represent the alternatives' relative ability to achieve the decision goal, so they allow a straightforward consideration of the various courses of action (COA).

It can be used by individuals who are working on straightforward decision or teams that are working on complex problems. It has unique advantages when important elements of the decision are difficult to quantify or compare, or where communication among team members is impeded by their different specializations, terminologies, or perspectives. The techniques to do pairwise comparisons enable one to compare as will be shown in later examples.

4.5.2 Methodology of the Analytic Hierarchy Process

The procedure for using the AHP can be summarized as follows:

Step 1: Build the hierarchy for the decision

Goal	Select the best alternative
Criteria	$c_1, c_2, c_3, \ldots, c_m$
Alternatives	$a_1, a_2, a_3, \ldots, a_n$

Step 2: Judgments and Comparison

Build a numerical representation using a nine-point scale in a pairwise comparison for the attributes criterion and the alternatives. The goal, in AHP, is to obtain a set of eigenvectors of the system that measures the importance with respect to the criterion. We can put these values into a matrix or table based on the values from Saaty's nine-point scale, see Table 4.7.

We must ensure that this pairwise matrix is consistent according to Saaty's scheme to compute the CR. The value of CR must be less than or equal to 0.1 to be considered valid.

n	1	2	3	4	5	6	7	8	9	10
RI*	0	0	0.52	0.89	1.1	1.24	1.35	1.4	1.45	1.49

Next, we approximate the largest eigenvalue, λ, using the power method (Burden and Faires, 2013). We compute the consistency index (CI) using the formula:

$$CI = \frac{(\lambda - n)}{(n - 1)}$$

Then we compute the CR using:

$$CR = \frac{CI}{RI}$$

If CR < 0.1, then our pairwise comparison matrix is consistent, and we may continue the AHP process. If not, we must go back to our

---

* RI - ratio index.

pairwise comparison and must fix the inconsistencies until CR < 0.1. In general, the consistency ensures that if $A > B$, $B > C$ that we have $A > C$ for all A, B, and C. These A, B, C are the criteria.

Step 3: Finding all the eigenvectors combined in order to obtain a comparative ranking. Various methods are available for doing this.

Methods to solve for decision-maker weights: The use of technology is suggested to find the weights. We have found Excel as a useful technology to assist.

1. Power method to estimate the dominant eigenvectors

    We suggest the method from Burden and Faires (2013) using the power method as it is straightforward to implement using technology.

    Definition of a dominant eigenvalue and dominant eigenvector: Let $\lambda_1, \lambda_2, \ldots, \lambda_n$ be the eigenvalues of a $n \times n$ matrix A; λ_1 is called the dominant eigenvalue of A if $|\lambda_1| > |\lambda_i|$, for $i = 2, \ldots, n$. The eigenvectors corresponding to λ_1 are called the dominant eigenvectors of A. The power method to find these eigenvectors is iterative. First assume that the matrix A has a dominant eigenvalue with corresponding dominant eigenvectors. Then choose an initial nonzero vector in R^n as the approximation, x_0, of one of the dominant eigenvectors of A. Finally form the iterative sequence.

    $$x_1 = Ax_0$$
    $$x_2 = Ax_1 = A^2 x_0$$
    $$x_3 = Ax_2 = A^3 x_0$$
    $$\cdot$$
    $$\cdot$$
    $$\cdot$$
    $$x_k = Ax_{k-1} = A^k x_0$$

2. Discrete dynamical systems (DDS) approximation method (Fox, 2012).

Step 4: After the $m \times 1$ criterion weights and the $n \times m$ matrix for n alternatives by m criterion are found, we use matrix multiplication to obtain the $n \times 1$ final rankings.

Step 5. We order the final ranking.

4.5.3 Strengths and Limitations of Analytic Hierarchy Process

Similar to all modeling and MADM methods, the AHP has strengths and limitations.

The main strength of the AHP is its ability to rank choices in the order of their effectiveness in meeting objectives. If the judgments made about the relative importance of criteria and those about the alternatives' ability to satisfy those objectives have been made in good faith and effort, then the AHP calculations lead to the logical consequence of those judgments. It is quite hard, but not impossible, to manually change the pairwise judgments to get some predetermined result. A further strength of the AHP is its ability to detect inconsistent judgments in the pairwise comparisons using the CR value. If the CR value is greater than 0.1, then the judgments are deemed to be inconsistent.

The limitations of the AHP are that it only works because the matrices are all of the same mathematical form. This is known as a positive reciprocal matrix. The reasons for this are explained in Saaty's book (1990), so we will simply state that point in the form that is required. To create such a matrix requires that, if we use the number 9 to represent that *A is absolutely more important than B*, then we have to use 1/9 to define the relative importance of *B* with respect to *A*. Some people regard that as reasonable; others do not.

Another suggested limitation is in the possible scaling. However, understanding that the final values obtained simply say that one alternative is relatively better than another alternative. For example, if the AHP values for alternatives {*A*, *B*, *C*} found were (0.392, 0.406, 0.204), then they imply that only alternatives *A* and *B* are about equally good at approximately 0.4, whereas *C* is worse at 0.2. It does not mean that *A* and *B* are twice as good as *C*.

The AHP is a useful technique for discriminating between competing options in the light of a range of objectives to be met. The calculations are not complex, and although the AHP relies on what might be seen as a mathematical trick, you do not need to understand the mathematics to use the technique. Be aware that it only shows relative values.

Although AHP has been used in many applications in business, industry, and government as can be seen in literature searches of the procedure, Hartwich (1999) noted several limitations. First and foremost, AHP was criticized for not providing sufficient guidance about structuring the problem to be solved, forming the levels of the hierarchy for criteria and alternatives, and aggregating group opinions when team members are geographically dispersed or are subjected to time constraints. Team members may carry out rating items individually or as a group. As the levels of hierarchy increase, the difficulty and time it takes to synthesize weights will also increase. One simple fix involves having the decision-making participants (the analysts and decision-maker) review the basics of the AHP methodology and work through examples so that concepts are thoroughly and easily understood (Hartwich, 1999).

Another critique of AHP is the *rank reversal* problem. Rank reversal involves the change in the ordering of the alternatives when the procedure is changed, when more alternatives are added, or when the criteria are changed. This implies that changes in the importance ratings whenever criteria or alternatives are added to or deleted from the initial set of alternatives are compared. Several modifications to AHP have been proposed to cope with this and other related issues. Many of the enhancements involved ways of computing, synthesizing pairwise comparisons, and/or normalizing the priority and weighting vectors. We mention the importance of rank reversal now, because TOPSIS corrects this rank reversal issue.

4.5.4 Sensitivity Analysis

As AHP, at least in the pairwise comparisons, is based on subjective inputs using the nine-point scale, then sensitivity analysis is extremely important. Leonelli (2012) in his master's thesis outlines procedures for sensitivity analysis to enhance decision support tools, including numerical incremental analysis of a weight, probabilistic simulations, and mathematical models. How often do we change our minds about the relative importance of an object, place, or thing? Often enough that we should alter the pairwise comparison values to determine how robust our rankings are in the AHP process. We suggest doing enough sensitivity analysis to find the *break point* values, if they exist, of the decision-maker weights that change the rankings of our alternatives. As the pairwise comparisons are subjective matrices that are compiled using the Saaty method, we suggest a minimum *trial and error* sensitivity analysis using the numerical incremental analysis of the weights.

Chen and Kocaoglu (2008) grouped sensitivity analysis into three main groups that he called: numerical incremental analysis, probabilistic simulations, and mathematical models. The numerical incremental analysis, also known as one-at-a-time (OAT) or trial and error works by incrementally changing one parameter at a time, finds the new solution and shows graphically how the ranks change. There exist several variations of this method (Hurly, 2001; Barker, 2011). Probabilistic simulations employ the use of Monte Carlo simulation (Butler, 1997) that allows random changes in the weights and simultaneously explores the effect on the ranks. Modeling may be used when it is possible to express the relationship between the input data and the solution results.

We used Equation 4.7 (Alinezhad and Amini, 2011) for adjusting weights that fall under the incremental analysis:

$$w'_j = \frac{1-w'_p}{1-w_p} w_j \qquad (4.7)$$

where w'_j is the new weight and w_p is the original weight of the criterion to be adjusted and w'_p is the value after the criterion was adjusted. We found this to be an easy method to adjust weights to reenter back into our model.

4.5.5 Illustrative Examples with Analytic Hierarchy Process

Example 4.9: Car Selection Revisited

We revisit car selection with our raw data presented in Table 4.5 to illustrate AHP in selecting the best alternative based on pairwise comparisons of the decision criteria.

Step 1: Build the hierarchy and prioritize the criterion from your highest to lower priority.

Goal	Select the best car
Criteria	$c_1, c_2, c_3, \ldots, c_m$
Alternatives	$a_1, a_2, a_3, \ldots, a_n$

For our cars example, we choose the priority as follows: cost, MPG city, safety, reliability, MPG highway, performance, and interior and style. Putting these criteria in a priority order allows for an easier assessment of the pairwise comparisons. We used an Excel template that is prepared for these pairwise comparisons.

Step 2: Perform the pairwise comparisons using Saaty's nine-point scale. We used an Excel template that is created to organize the pairwise comparisons and obtained the pairwise comparison matrix.

This yields the following decision criterion matrix:

	Cost	MPG City	MPG Highway	Safety	Reliability	Performance	Interior and Style
Cost	1	2	2	3	4	5	6
MPG city	0.5	1	2	3	4	5	5
MPG Highway	0.5	0.5	1	2	2	3	3
Safety	0.3333	0.333	0.5	1	1	2	3
Reliability	0.25	0.25	0.5		1	2	3
Performance	0.2	0.2	0.333	0.5	1	1	2
Interior and style	0.166	0.2	0.333	0.333	0.333	0.5	1

We check the CR to ensure that it is less than 0.1. For our pairwise decision matrix, the CR is 0.00695. As the CR < 0.1, we continue.

We find the *eigenvector* as the decision weights:

Cost	0.342407554
City	0.230887543
Highway	0.151297361
Safety	0.094091851
Reliability	0.080127732
Performance	0.055515667
Interior and Style	0.045672293

Step 3: For the alternatives, we either have the data as we obtained it for each car under each decision criterion or we can use pairwise comparisons by criteria for how each car fares versus its competitors. In this example, we take the raw data from Table 4.6 for cost and instead of using the *cost* we use $1/cost$. We then normalize the columns.

We have other options for dealing with a criteria and variable such as *cost*. Thus, we have three COA: (1) use 1/*cost* to replace *cost*; (2) use a pairwise comparison using the nine-point scale; or (3) remove *cost* from a criteria and a variable, run the analysis, and then do a *benefit/cost* ratio to rerank the results.

Step 4: We multiply the matrix of the normalized raw data from Consumer Reports and the matrix of weights to obtain the rankings. Using COA (1) from Step 3, we obtain the following results:

Cars	Values AHP	Rank
Prius	0.170857046	4
Fusion	0.180776107	2
Volt	0.143888039	6
Camry	0.181037124	1
Sonata	0.171051618	3
Leaf	0.152825065	5

Camry is our first choice, followed by Fusin, Sonata, Leaf, and Volt.

If we use method COA (2) from Step 3, then within the final matrix, we replace the actual costs with these pairwise results (CR = 0.0576):

	Cost
Prius	0.139595
Fusion	0.121844
Volt	0.041493
Camry	0.43029
Sonata	0.217129
Leaf	0.049648

Then we obtain the ranked results as follows:

Cars	Values AHP	Rank
Prius	0.14708107	4
Fusion	0.152831274	3
Volt	0.106011611	6
Camry	0.252350537	1
Sonata	0.173520854	2
Leaf	0.113089654	5

If we do COA (3) from Step 3, then this method requires us to redo the pairwise criterion matrix without the cost criteria. These weights are as follows:

MPG city	0.363386
MPG Highway	0.241683
Safety	0.159679
Reliability	0.097
Performance	0.081418
Interior and style	0.056834

We normalize the original costs from Table 4.6 and divide these ranked values by the normalized *cost* to obtain a *cost/benefit* value. These are shown in ranked order:

Camry	1.211261
Fusion	1.178748
Prius	1.10449
Sonata	1.06931
Leaf	0.821187
Volt	0.759482

SENSITIVITY ANALYSIS

We alter our decision pairwise values to obtain a new set of decision weights to use in COA (1) from Step 3 to obtain new results: Camry, Fusion, Sonata, Prius, Leaf, and Volt. The new weights and model's results are as follows:

Cost	0.311155922	
MPG City	0.133614062	
MPG Highway	0.095786226	
Performance	0.055068606	
Interior	0.049997069	
Safety	0.129371535	
Reliability	0.225006578	
Alternatives	Values	
Prius	0.10882648	4
Fusion	0.11927995	2
Volt	0.04816882	5
Camry	0.18399172	1
Sonata	0.11816156	3
Leaf	0.04357927	6

The resulting values have changed but not the relative rankings of the cars. Again, we recommend using sensitivity analysis to find a *break point*, if one exists.

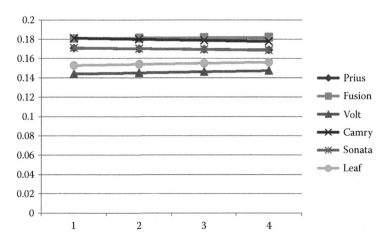

FIGURE 4.7
Camry overtakes Fusion as the top alternative as we change the weight of *Cost*.

We systemically varied the cost weights using Equation 4.5 with increments of (±) 0.05. We plotted the results to show the approximate break point of the criteria cost as weight of cost + 0.1 as shown in Figure 4.7.

Prius	0.170857	0.170181	0.169505	0.16883
Fusion	0.180776	0.18119	0.181604	0.182018
Volt	0.143888	0.145003	0.146118	0.147232
Camry	0.181037	0.179903	0.178768	0.177634
Sonata	0.171052	0.170242	0.169431	0.168621
Leaf	0.152825	0.15395	0.155074	0.156198

We see that as weight for the criteria cost decreases and other criteria weights proportionally increase that Fusion overtakes Camry as the number 1 ranked car.

Example 4.10: Kite Network Revisited with AHP

Assume that all we have are the outputs from ORA, which we do not show here due to the volume of output produced. We take the metrics from ORA and normalize each column. The columns for each criterion are placed in a matrix X with entries, x_{ij}. We define w_j as the weights for each criterion.

Next, we assume that we can obtain pairwise comparison matrix from the decision-maker concerning the criterion. We use the output from ORA and normalize the results for AHP to rate the alternatives within each criterion. We provide a sample pairwise comparison matrix for weighting the criterion from the Kite example using Saaty's nine-point scale. The CR is 0.0828, which is less than 0.1, so our pairwise matrix is consistent and we continue.

PAIRWISE COMPARISON MATRIX

	Central	Eigenvector	In-degree	Out-degree	Information Centrality	Betweenness
Central	1	3	2	2	1/2	1/3
Eigenvector	1/3	1	1/3	1	2	1/2
In-degree	1/2	3	1	1/2	1/2	1/4
Out-degree	1/2	1/2	1	1	1/4	1/4
Information Centrality	2	2	4	4	1	1/3
Betweenness	3	2	4	4	3	1

We obtain the steady-state values that will be our criterion weights, where the sum of the weights equals 1.0. There exist many methods to obtain these weights. The methods used here are the power method from numerical analysis (Burden and Faires, 2013) and DDS (Fox, 2012; Giordano et al., 2014).

0.1532	0.1532	0.1532	0.1532	0.1532	0.1532
0.1450	0.1450	0.1450	0.1450	0.1450	0.1450
0.1194	0.1195	0.1194	0.1194	0.1194	0.1194
0.0672	0.0672	0.0672	0.0672	0.0672	0.0672
0.1577	0.1577	0.1577	0.1577	0.1577	0.1577
0.3575	0.3575	0.3575	0.3575	0.3575	0.3575

These values provide the weights for each criterion: *centrality* = 0.1532, *eigenvectors* = 0.1450, *in-centrality* = 0.1194, *out-centrality* = 0.0672, *information centrality* = 0.1577, and *betweenness* = 0.3575.

We multiply the matrix of the weights and the normalized matrix of metrics from ORA to obtain our output and ranking:

Node	AHP Value	Rank
Susan	0.160762473	2
Steven	0.133201647	3
Sarah	0.113388361	4
Tom	0.075107843	6
Claire	0.075107843	6
Fred	0.060386019	8
David	0.060386019	8
Claudia	0.177251415	1
Ben	0.109606727	5
Jennifer	0.034801653	10

For this example with AHP Claudia, *cl* is the key node. However, the bias of the decision-maker is important in the analysis of the criterion weights. The criterion, *betweenness*, is two to three times more important than the other criterion.

SENSITIVITY ANALYSIS

Changes in the pairwise decision criterion cause fluctuations in the key nodes. We change our pairwise comparison so that *betweenness* is not a dominant criterion.

With these slight pairwise changes, we now find that Susan is now ranked first, followed by Steven and then Claudia. The AHP process is sensitive to changes in the criterion weights. We vary *betweenness* in increments of 0.05 to find the break point.

	Centrality	IN	OUT	Eigen	EIGENC	Close	IN-Close	Betw	INFO Cen.
t	0.111111	0.111111	0.111111	0.114399	0.114507	0.100734	0.099804	0.019408	0.110889
c	0.111111	0.111111	0.111111	0.114399	0.114507	0.100734	0.099804	0.019408	0.108891
f	0.083333	0.083333	0.083333	0.093758	0.094004	0.097348	0.09645	0	0.097902
s	0.125	0.138889	0.111111	0.137528	0.137331	0.100734	0.111826	0.104188	0.112887
su	0.180556	0.166667	0.194444	0.175081	0.174855	0.122743	0.107632	0.202247	0.132867
st	0.138889	0.138889	0.138889	0.137528	0.137331	0.112867	0.111826	0.15526	0.123876
d	0.083333	0.083333	0.083333	0.093758	0.094004	0.097348	0.107632	0	0.100899
cl	0.083333	0.083333	0.083333	0.104203	0.104062	0.108634	0.107632	0.317671	0.110889
b	0.055556	0.055556	0.055556	0.024123	0.023985	0.088318	0.087503	0.181818	0.061938
j	0.027778	0.027778	0.027778	0.005223	0.005416	0.070542	0.069891	0	0.038961

10 alternatives and 9 attirbutes or criterion

Criterion weights

w_1	0.034486	
w_2	0.037178	
w_3	0.045778	
w_4	0.398079	
w_5	0.055033	
w_6	0.086323	
w_7	0.135133	
w_8	0.207991	

Tom	0.098628		Susan	0.161609
Claire	0.098212		Steven	0.133528
Fred	0.081731		Claudia	0.133428
Sarah	0.12264		Sarah	0.12264
Susan	0.161609		Tom	0.098628
Steven	0.133528		Claire	0.098212
David	0.083319		David	0.083319
Claudia	0.133428		Fred	0.081731
Ben	0.0645		Ben	0.0645
Jennifer	0.022405		Jennifer	0.022405

Further, sensitivity analysis of the nodes is provided in Figure 4.8.

We varied the weight of the criterion *betweenness* by lowering it by 0.05 in each iteration and increasing the other weights using Equation 4.1. We see that Claudia and Susan change as the top node when we reduce *betweenness* by 0.1.

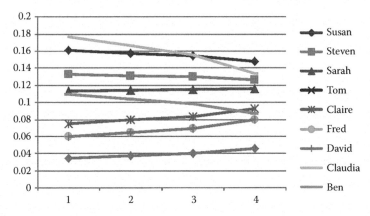

FIGURE 4.8
Sensitivity analysis for nodes varying only *betweenness*.

Exercises 4.5

1. For the problems in Section 4.3, solve by AHP. Compare your results to your results using SAW.
2. Perform sensitivity analysis by changing the weight of your highest criteria weight until it is no longer the highest weighted criteria. Did it change the rankings?

Projects 4.5

Construct a computer program to find the weights using AHP by using the power method.

4.6 Technique of Order Preference by Similarity to the Ideal Solution

4.6.1 Description and Uses

The technique for order of preference by similarity to ideal solution (TOPSIS) is a multicriteria decision analysis method, which was originally developed in a dissertation from Kansas State University (Hwang and Yoon, 1981). It has been further developed by others (Yoon, 1987; Hwang et al., 1993). TOPSIS is based on the concept that the chosen alternative should have the shortest geometric distance from the positive ideal solution and the longest geometric distance from the negative ideal solution. It is a method of compensatory aggregation that compares a set of alternatives by identifying weights for each criterion, normalizing the scores for each criterion, and calculating the geometric distance between each alternative and the ideal alternative, which is the best score in each criterion. An assumption of TOPSIS is that the criteria are monotonically increasing or decreasing. Normalization is usually required as the parameters or criteria that are often with incompatible dimensions in multicriteria problems. Compensatory methods such as TOPSIS allow trade-offs between criteria, where a poor result in one criterion can be negated by a good result in another criterion. This provides a more realistic form of modeling than noncompensatory methods, which include or exclude alternative solutions based on hard cutoffs.

We only desire to briefly discuss the elements in the framework of TOPSIS. TOPSIS can be described as a method to decompose a problem into subproblems. In most decisions, the decision-maker has a choice among many alternatives. Each alternative has a set of attributes or characteristics that can be measured, either subjectively or objectively. The attribute elements of the hierarchical process can relate to any aspect of the decision problem whether tangible or intangible, carefully measured or roughly estimated, well or poorly understood information. Basically, anything at all that applies to the decision at hand can be used in the TOPSIS process.

4.6.2 Methodology

The TOPSIS process is carried out as follows:

Step 1: Create an evaluation matrix consisting of m alternatives and n criteria, with the intersection of each alternative and criterion given as x_{ij}, giving us a matrix $(X_{ij})_{m \times n}$.

$$D = \begin{array}{c} \\ A_1 \\ A_2 \\ A_3 \\ \cdot \\ \cdot \\ \cdot \\ A_m \end{array} \begin{array}{cccccc} x_1 & x_2 & x_3 & & & x_n \\ \begin{bmatrix} x_{11} & x_{12} & x_{13} & \cdot & \cdot & x_{1n} \\ x_{21} & x_{22} & x_{23} & \cdot & \cdot & x_{2n} \\ x_{31} & x_{32} & x_{33} & \cdot & \cdot & x_{3n} \\ \cdot & \cdot & \cdot & \cdot & & \cdot \\ \cdot & \cdot & \cdot & \cdot & & \cdot \\ \cdot & \cdot & \cdot & \cdot & & \cdot \\ x_{m1} & x_{m2} & x_{m3} & \cdot & \cdot & x_{mn} \end{bmatrix} \end{array}$$

Step 2: The matrix shown as D above then is normalized to form the matrix $R = (r_{ij})_{m \times n}$ as shown using the normalization method:

$$r_{ij} = \frac{x_{ij}}{\sqrt{\sum x_{ij}^2}}$$

for $i = 1,2\dots,m; j = 1,2,\dots,n$

Step 3: Calculate the weighted normalized decision matrix. First, we need the weights. Weights can come from either the decision-maker or by computation.

Step 3a: Use either the decision-maker's weights for the attributes $x_1, x_2,...,x_n$ or compute the weights through the use of Saaty's (1980) AHP decision-maker weights method to obtain the weights as the eigenvector to the attributes versus attribute pairwise comparison matrix.

$$\sum_{j=1}^{n} w_j = 1$$

The sum of the weights over all attributes must equal 1 regardless of the method used. Use the methods described in Section 4.4.2 to find these weights.

Step 3b: Multiply the weights to each of the column entries in the matrix from *Step 2* to obtain the matrix, T.

$$T = (t_{ij})_{m \times n} = (w_j r_{ij})_{m \times n}, i = 1,2,...,m$$

Step 4: Determine the worst alternative (A_w) and the best alternative (A_b): Examine each attribute's column and select the largest and smallest values appropriately. If the values imply that larger is better (profit), then the best alternatives are the largest values, and if the values imply that smaller is better (such as cost), then the best alternative is the smallest value.

$$A_w = \{\max(t_{ij}| i = 1,2,...,m | j \in J_-, \min(t_{ij}| i = 1,2,...,m) | j \in J_+\} \equiv$$

$$\{t_{wj} | j = 1,2,...,n\},$$

$$A_{wb} = \{\min(t_{ij}| i = 1,2,...,m | j \in J_-, \max(t_{ij}| i = 1,2,...,m) | j \in J_+\} \equiv$$

$$\{t_{bj} | j = 1,2,...,n\},$$

where:

$J_+ = \{j = 1,2,...n | j)$ associated with the criteria having a positive impact

$J_- = \{j = 1,2,...n | j)$ associated with the criteria having a negative impact

We suggest that if possible, make all entry values in terms of positive impacts.

Step 5: Calculate the L2-distance between the target alternative *i* and the worst condition A_w.

$$d_{iw} = \sqrt{\sum_{j=1}^{n}(t_{ij}-t_{wj})^2}, i = 1,2,\ldots,m$$

and then calculate the distance between the alternative i and the best condition A_b

$$d_{ib} = \sqrt{\sum_{j=1}^{n}(t_{ij}-t_{bj})^2}, i = 1,2,\ldots m$$

where d_{iw} and d_{ib} are L2-norm distances from the target alternative i to the worst and best conditions, respectively.

Step 6: Calculate the similarity to the worst condition:

$$s_{iw} = \frac{d_{iw}}{(d_{iw}+d_{ib})}, 0 \le s_{iw} \le 1, i = 1,2,\ldots,m$$

where:

  S_{iw} = 1 if and only if the alternative solution has the worst condition

  S_{iw} = 0 if and only if the alternative solution has the best condition

Step 7: Rank the alternatives according to their value from S_{iw} ($i = 1,2, \ldots, m$).

4.6.2.1 Normalization

Two methods of normalization that have been used to deal with incongruous criteria dimensions are linear normalization and vector normalization.

Normalization can be calculated as in Step 2 of the TOPSIS process mentioned earlier. Vector normalization was incorporated with the original development of the TOPSIS method (Yoon, 1987) and is calculated using the following formula:

$$r_{ij} = \frac{x_{ij}}{\sqrt{\sum x_{ij}^2}} \text{ for } i = 1,2\ldots,m; \ j = 1,2,\ldots,n$$

In using vector normalization, the nonlinear distances between single dimension scores and ratios should produce smoother trade-offs (Hwang and Yoon, 1981).

Let us suggest two options for the weights in Step 3. First, the decision-maker might actually have a weighting scheme that he or she wants the analyst to use. If not, we suggest using Saaty's nine-point pairwise method that is developed for the AHP (Saaty, 1980) to obtain the criteria weights as described in the previous section.

4.6.3 Strengths and Limitations

TOPSIS is based on the concept that the chosen alternative should have the shortest geometric distance from the positive ideal solution and the longest geometric distance from the negative ideal solution.

TOPSIS is a method of many steps that compares a set of alternatives by identifying weights for each criterion, normalizing scores for each criterion, and calculating the geometric distance between each alternative and the ideal alternative, which is the best score in each criterion.

4.6.4 Sensitivity Analysis

The decision weights are subject to sensitivity analysis to determine how they affect the final ranking. The same procedures discussed in Section 4.4 are valid here. Sensitivity analysis is essential for good analysis. In addition, Alinezhad and Amini (2011) suggest sensitivity analysis for TOPSIS for changing an attribute weight. We will again use Equation 4.6 in our sensitivity analysis.

4.6.5 Illustrate Examples with Technique of Order Preference by Similarity to the Ideal Solution

Example 4.11: Car Selection Revisited with TOPSIS (Table 4.5)

We might assume that our decision-maker weights from the AHP section are still valid for our use. These criteria weights are:

Cost	0.38960838
MPG city	0.11759671
MPG Highway	0.04836533
Performance	0.0698967
Interior	0.05785692
Safety	0.10540328
Reliability	0.21127268

	Cost	MPG_city	MPG_HW	Perf.	Interior	Safety	Reliability	N/A
Cost	1	4	6	5	6	4	2	0
MPG_city	0.25	1	6	3	5	1	0.333333333	0
MPG_HW	0.166667	0.166667	1	0.5	0.5	0.333333	0.25	0
Perf.	0.2	0.333333	2	1	2	0.5	0.333333333	0
Interior	0.166667	0.2	2	0.5	1	0.5	0.333333333	0
Safety	0.25	1	3	2	2	1	0.5	0
Reliability	0.5	3	4	3	3	2	1	0
N/A	0	0	0	0	0	0	0	1

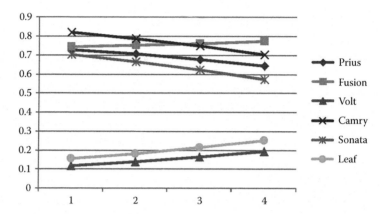

FIGURE 4.9
TOPSIS values of the cars by varying the weight for cost incrementally by −0.05 each of four increments along the *x-axis*.

We use the identical data from the car example from AHP, but we apply Steps 3–7 from TOPSIS to our data. We are able to keep the cost data and just inform TOPSIS that a smaller cost is better. We obtained the following rank ordering of the cars: Camry, Fusion, Prius, Sonata, Volt, and Leaf.

Car	TOPSIS Value	Rank
Camry	0.8215	1
Fusion	0.74623	2
Prius	0.7289	3
Sonata	0.70182	4
Leaf	0.15581	5
Volt	0.11772	6

It is critical to perform sensitivity analysis on the weights to see how they affect the final ranking. This time we work toward finding the break point where the order of cars actually changes. As cost is the largest criterion weight, we vary it using Equation 4.5 in increments of 0.05. We see from Figure 4.9 that Fusion overtakes Camry when cost is decreased by about 0.1, which allows reliability to overtake cost as the dominate-weighted decision criterion.

Example 4.12: Social Networks with TOPSIS

We revisit the Kite Network with TOPSIS to find influences in the network. We present the extended output from ORA that we used in Table 4.10.

We use the decision weights from AHP (unless a decision-maker gives us their own weights) and find the eigenvectors for our eight metrics as follows:

TABLE 4.10

Summary of Extended ORA's Output for Kite Network

	In	Out	Eigen	EigenL	Close	In-Close	Betweenness	INF Centrality
Tom	0.4	0.4	0.46	0.296	0.357	0.357	0.019	0.111
Claire	0.4	0.4	0.46	0.296	0.357	0.357	0.019	0.109
Fred	0.3	0.3	0.377	0.243	0.345	0.345	0	0.098
Sarah	0.5	0.4	0.553	0.355	0.357	0.4	0.102	0.113
Susan	0.6	0.7	0.704	0.452	0.435	0.385	0.198	0.133
Steven	0.5	0.5	0.553	0.355	0.4	0.4	0.152	0.124
David	0.3	0.3	0.377	0.243	0.345	0.385	0	0.101
Claudia	0.3	0.3	0.419	0.269	0.385	0.385	0.311	0.111
Ben	0.2	0.2	0.097	0.062	0.313	0.313	0.178	0.062
Jennifer	0.1	0.1	0.021	0.014	0.25	0.25	0	0.039

w_1	0.034486
w_2	0.037178
w_3	0.045778
w_4	0.398079
w_5	0.055033
w_6	0.086323
w_7	0.135133
w_8	0.207991

We take the metrics from ORA and perform Steps 2–7 of TOPSIS to obtain the results:

S+	S–	C	
0.0273861	0.181270536	0.86875041	Susan
0.0497878	0.148965362	0.749499497	Steven
0.0565358	0.14154449	0.714581437	Sarah
0.0801011	0.134445151	0.626648721	Tom
0.0803318	0.133785196	0.624822765	Claire
0.10599	0.138108941	0.565790826	Claudla
0.1112243	0.12987004	0.538668909	David
0.1115873	0.128942016	0.536076177	Fred
0.1714404	0.113580988	0.398499927	Ben
0.2042871	0.130399883	0.389617444	Jennifer

We rank order the final output from TOPSIS as shown in the last column earlier. We interpret the results as follows: The key node is *Susan* followed by *Steven, Sarah, Tom,* and *Claire.*

SENSITIVITY ANALYSIS

We used Equation 4.7 and systemically altered the value of the largest criteria weight, *EigenL* and depict this in Figure 4.10.

We note that Susan remains the most influential node.

COMPARISON OF RESULTS FOR THE KITE NETWORK

We have also used the other two MADM methods to rank order our nodes in previous work in social network analysis (SNA) (Fox and Everton 2013). When we applied DEA and AHP to compare to TOPSIS, we obtained the results displayed in Table 4.11 for the Kite Network.

It might be useful to use this table as input for another round of one of these presented methods and then use sensitivity analysis.

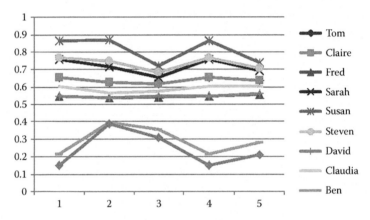

FIGURE 4.10

Sensitivity analysis plot as a function of varying *EigenL* weight in increments of −0.05 units.

TABLE 4.11

MADM Applied to Kite Network

Node	SAW	TOPSIS Value (Rank)	DEA Efficiency Value (Rank)	AHP Value (Rank)
Susan	0.046 (1)	0.862 (1)	1 (1)	0.159 (2)
Sarah	0.021 (4)	0.675 (3)	0.786 (2)	0.113 (4)
Steven	0.026 (3)	0.721 (2)	0.786 (2)	0.133 (3)
Claire	0.0115 (7)	0.649 (4)	0.653 (4)	0.076 (6)
Fred	0.0115 (7)	0.446 (8)	0.653 (4)	0.061 (8)
David	0.031 (2)	0.449 (7)	0.536 (8)	0.061 (8)
Claudia	0.012 (8)	0.540 (6)	0.595 (6)	0.176 (1)
Ben	0.018 (5)	0.246 (9)	0.138 (9)	0.109 (5)
Jennifer	0.005 (10)	0. (10)	0.030 (10)	0.036 (10)
Tom	0.0143 (6)	0.542 (5)	0.553 (7)	0.076 (6)

Exercises 4.6

1. For the problems in Section 4.3, solve by TOPSIS. Compare your results using both SAW and AHP.
2. Perform sensitivity analysis by changing the weight of your highest criteria weight until it is no longer the highest weighted criteria. Did it change the rankings?

Projects 4.6

1. Write a program using the technology of your choice to implement any of all of the following: (a) SAW, (b) AHP, and (C) TOPSIS.
2. Enable your program in (1) to perform sensitivity analysis.

References and Suggested Additional Readings

Alinezhad, A. and A. Amini. 2011. Sensitivity analysis of TOPSIS technique: The results of change in the weight of one attribute on the final ranking of alternatives. *Journal of Optimization in Industrial Engineering*, 7(2011), 23–28.

Baker, T. and Z. Zabinsky. 2011. A multicriteria decision making model for reverse logistics using analytical hierarchy process. *Omega*, 39, 558–573.

Burden, R. and D. Faires. 2013. *Numerical Analysis* (9th ed.). Boston, MA: Cengage Publishers.

Butler, J., J. Jia, and J. Dyer. 1997. Simulation techniques for the sensitivity analysis of multi-criteria decision models. *Eurpean Journal of Operations Research*, 103, 531–546.

Callen, J. 1991. Data envelopment analysis: Practical survey and managerial accounting applications. *Journal of Management Accounting Research*, 3(1991), 35–57.

Carley, K. M. 2011. *Organizational Risk Analyzer (ORA)*. Pittsburgh, PA: Center for Computational Analysis of Social and Organizational Systems (CASOS), Carnegie Mellon University.

Charnes, A., W. Cooper, and E. Rhodes. 1978. Measuring the efficiency of decision making units. *European Journal of Operations Research*, 2(1978), 429–444.

Chen, H. and D. Kocaoglu. 2008. A sensitivity analysis algorithm for hierarchical decision models. *European Journal of Operations Research*, 185(1), 266–288.

Consumer's Reports Car Guide. 2012. The Editors of Consumer Reports. http://e-libdigital.com/download/consumer-reports-car-guide-2012.pdf (accessed September 6, 2017).

Cooper, W., S. Li, L. Seiford, L. R. M. Thrall, and J. Zhu. 2001. Sensitivity and stability analysis in DEA: Some recent developments. *Journal of Productivity Analysis*, 15(3), 217–246.

Cooper, W., L. Seiford, and K. Tone, 2000. *Data Envelopment Analysis*. Norwell, MA: Kluwer Academic Press.

Figueroa, S. 2014. Improving recruiting in the 6th recruiting brigade through statistical analysis and efficiency measures. Master's Thesis, Naval Postgraduate School, Monterey, CA.

Fishburn, P. C. 1967. Additive utilities with incomplete product set: Applications to priorities and assignments. *Operations Research Society of America (ORSA)*, 15, 537–542.

Fox, W. P. 2012. Mathematical modeling of the analytical hierarchy process using discrete dynamical systems in decision analysis. *Computers in Education Journal*, 3, 27–34.

Fox, W. and S. Everton. 2013. Mathematical modeling in social network analysis: Using TOPSIS to find node influences in a social network. *Journal of Mathematics and System Science*, 3(10), 531–541.

Fox, W. and S. Everton. 2014. Mathematical modeling in social network analysis: Using data envelopment analysis and analytical hierarchy process to find node influences in a social network. *Journal of Defense Modeling and Simulation*, 2(2014), 1–9.

Giordano, F. R., W. Fox, and S. Horton. 2014. *A First Course in Mathematical Modeling* (5th ed.). Boston, MA: Brooks-Cole Publishers.

Hartwich, F. 1999. *Weighting of Agricultural Research Results: Strength and Limitations of the Analytic Hierarchy Process (AHP)*, Universitat Hohenheim. https://entwick-lungspolitik.uni-hohenheim.de/uploads/media/DP_09_1999_Hartwich_02.pdf (accessed September 6, 2017).

Hurly, W. J. 2001. The analytical hierarchy process: A note on an approach to sensitivity which preserves rank order. *Computers and Operations Research*, 28, 185–188.

Hwang, C. L., Y. Lai, and T. Y. Liu. 1993. A new approach for multiple objective decision making. *Computers and Operational Research*, 20, 889–899.

Hwang, C. L. and K. Yoon. 1981. *Multiple Attribute Decision Making: Methods and Applications*. New York: Springer-Verlag.

Krackhardt, D. 1990. Assessing the political landscape: Structure, cognition, and power in organizations. *AdministrativeScience Quarterly*, 35, 342–369.

Leonelli, R. 2012. Enhancing a decision support tool with sensitivity analysis. Thesis, University of Manchester.

Neralic, L. 1998. Sensitivity analysis in models of data envelopment analysis. *Mathematical Communications*, 3, 41–59.

Saaty, T. 1980. *The Analytical Hierarchy Process*. New York: McGraw Hill.

Thanassoulis, E. 2011. *Introduction to the Theory and Application of Data Envelopment Analysis-A Foundation Text with Integrated Software*. London, UK: Kluwer Academic Press.

Trick, M. A. 1996. Multiple Criteria Decision Making for Consultants. http://mat.gsia.cmu.edu/classes/mstc/multiple/multiple.html (accessed April 2014).

Trick, M. A. 2014. Data Envelopment Analysis, Chapter 12. http://mat.gsia.cmu.edu/classes/QUANT/NOTES/chap12.pdf (accessed April 2014).

Winston, W. 1995. *Introduction to Mathematical Programming*. Pacific Grove, CA: Duxbury Press, pp. 322–325.

Yoon, K. 1987. A reconciliation among discrete compromise situations. *Journal of Operational Research Society*, 38, 277–286.

Zhenhua, G. 2009. The application of DEA/AHP method to supplier selection. *2009 International Conference on Information Management, Innovation Management and Industrial Engineering*, Washington, DC, pp. 449–451.

5

Modeling with Game Theory

OBJECTIVES

1. Know the concept of formulating a two-person game.
2. Understand total and partial conflict games.
3. Understand solution methodologies for each type of game.
4. Understand and interpret the solutions.

The Battle of the Bismarck Sea is set in the South Pacific in 1943. The historical facts are that General Imamura had been ordered to transport Japanese troops across the Bismarck Sea to New Guinea, and General Kenney, the U.S. commander in the region, wanted to bomb the troop transports before their arrival at their destination. Imamura had two options to choose from as routes to New Guinea: (1) a shorter northern route or (2) a longer southern route. Kenney must decide where to send his search planes and bombers to find the Japanese fleet. If Kenney sends his planes to the wrong route he can recall them later, but the number of bombing days is reduced.

We assume that both commanders, Imamura and Kenney, are rational players, each trying to obtain their best outcome. Further, we assume that there is no communications or cooperation which may be inferred since the two are enemies engaging in war. Further, each is aware of the intelligence assets that are available to each side and are aware of what the intelligence assets are producing. We assume that the number of days the U.S. planes can bomb as well as the number of days to sail to New Guinea are accurate estimates. Can we build a model that shows the strategies which each player should employ?

5.1 Introduction

Conflict has been a central theme in human history. Conflict arises when two or more individuals with different views, goals, or aspirations compete to control the course of future events. Game theory studies competition. It uses mathematics and mathematical tools to study situations in which rational players are involved in conflict both with and without cooperation. According to Wiens (2003), game theory studies situations in which parties compete, and also possibly cooperate, to influence the outcome of the parties' interaction to each party's advantage. The situation involves conflict between the participants called players because some outcomes favor one player at the possible expense of the other players. What each player obtains from a particular outcome is called the player's payoff. Each player can choose among a number of strategies to influence his payoff. However, each player's payoff depends on the other players' choices. According to Straffin (2004), rational players desire to maximize their own payoffs. Game theory is a branch of applied mathematics that is used in the social sciences (most notably in economics), business, biology, decision sciences, engineering, political science, international relations, operations research, applied mathematics, computer science, and philosophy. Game theory mathematically captures behavior in strategic situations in which an individual's success in making choices depends on the choices of others. Although initially developed to analyze competitions in which one individual does better at another's expense, game theory has grown to treat a wide class of interactions among players in competition.

Games can have several features; a few of the most common are listed in the following:

- *Number of players*: Each person who makes a choice in a game or who receives a payoff from the outcome of those choices is a player. A two-person game has two players. A three or more person is referred to as an *N*-person game.

- *Strategies per player*: Each player chooses from a set of possible actions known as strategies. In a two-person game, we allow the row player to have up to *m* strategies and the column player to have up to *n* strategies. The choice of a particular strategy by each player determines the payoff to each player.

- *Pure-strategy solution*: If a player should always choose one strategy over all other strategies to obtain their best outcome in a game, then that strategy represents a pure-strategy solution. Otherwise if strategies should be played randomly, then the solution is a mixed-strategy solution.

- *Nash equilibrium*: A Nash equilibrium is a set of strategies, which represents mutual best responses to the other player's strategies. In other words, if every player is playing their part of Nash equilibrium, no player has an incentive to unilaterally change his or her strategy. Considering only situations where players play a single strategy without randomizing (a pure strategy), a game can have any number of Nash equilibrium.

- *Sequential game*: A game is sequential if one player performs his or her actions after another; otherwise the game is a simultaneous game.

- *Simultaneous game*: A game is simultaneous if each player chooses their strategy for the game and implements them at the same time.

- *Perfect information*: A game has perfect information if either in a sequential game, every player knows the strategies chosen by the players who preceded them or in a simultaneous game, each player knows the other players' strategies and outcomes in advance.

- *Constant sum or zero sum*: A game is constant sum if the sums of the payoffs are the same for every set of strategies and zero sum if the sum is always equal to zero. In these games, one player gains if and only if another player loses; otherwise we have variable sum game.

- Extensive form presents the game in a tree diagram, whereas normative form presents the game in a payoff matrix. In this chapter, we only present the normative form and its associated solution methodologies.

- *Outcomes*: An outcome is a set of strategies taken by the players, or it is their payoffs resulting from the actions or strategies taken by all players.

- *Total conflict* games are games between players where the sums of the outcomes for all strategy pairs are either the same constant or zero. Games whose outcome sums are variable are known as *partial conflict* games.

The study of game theory has provided many classical and standard games that provide insights into the playing of games. Table 5.1 provides a short

TABLE 5.1

Summary of Classical Games in Game Theory

Game	Players	Strategies per Player	Number of Pure Strategy Nash Equilibrium	Simultaneous	Perfect Information	Zero Sum
Battle of the sexes	2	2	2	Yes	No	No
Blotto games	2	Variable	Variable	Yes	No	Yes
Chicken (aka hawk-dove)	2	2	2	Yes	No	No
Matching pennies	2	2	0 (mixed equilibrium)	Yes	No	Yes
Nash bargaining game	2	Infinite	Infinite	Yes	No	No
Prisoner's dilemma	2	2	1	Yes	No	No
Rock, paper, scissors	2	3	0 (mixed equilibrium)	Yes	No	Yes
Stag hunt	2	2	2	Yes	No	No
Trust game	2	Infinite	1	No	Yes	No

Source: Adapted and modified from http://en.wikipedia.org/wiki/List_of_games_in_game_theory

summary of some of these games. A complete list maybe viewed at the following website: http://en.wikipedia.org/wiki/List_of_games_in_game_theory.

This chapter is primarily concerned with two-person games. The irreconcilable, conflicting interests between the two players in a game resemble parlor games and military encounters between enemy states. Giordano et al. (2014) explained the two-person game in a context of modeling. Players make moves and counter-moves, until the rules of engagement declare the game is ended. The rules of engagement determine what each player can or must do at each stage (the available and/or required moves given the circumstances of the game at this stage) as the game unfolds. For example, in the game rock, paper, scissors, both players simultaneously make one move, with rock beating scissors beating paper beating rock. Although this game consists of only one move, games like chess require many moves to resolve the conflict.

Outcomes or payoffs used in a game are determined as a result of playing a pair of strategies by the players. These outcomes may come from calculated values or expected values, ordinal ranking, cardinal values developed from a lottery system (Von Neumann and Morgenstern, 1944; Straffin, 2004), or cardinal values from pairwise comparisons; Fox, 2015). Here we will assume that we have the cardinal (interval or ratio data) outcomes or payoffs for our games since cardinal outcomes allow us to do mathematical calculations.

We will present only the movement diagram to look for pure-strategy solutions and the linear programming (LP) formulation for all solutions in a zero-sum game. There are other methods and shortcut methods available to solve many of these total conflict games (see additional readings).

For partial conflict games, we will present the movement diagram to look for pure-strategy solutions if they exist, LP formulations for two-person two-strategy games for equalizing strategies, nonlinear methods for more than two strategies each, and LP to find security levels for as all players now seek to maximize their preferred outcome. We use the concept that every partial conflict game has a Nash's equalizing mixed-strategy solution even if it has a pure-strategy solution (Gillman and Housman, 2009).

We do not cover concepts and solution methodologies of *N*-person games, such as three-person total and partial conflict games. We briefly discuss the Nash arbitration scheme and its nonlinear formulation.

5.2 Background of Game Theory

Game theory is the study of strategic decision-making used for "the study of mathematical models of conflict and cooperation between intelligent rational decision-makers" (Myerson, 1991). Game theory has applications in many areas of business, military, government, networks, and industry. For more information on applications of game theory in these areas, see Chatterjee and Samuelson (2001), Cantwell (2003), Mansbridge (2013), and Aiginger (1999). In addition, McCormick and Fritz (2006) discussed game theory in warlord politics that blends military and diplomatic decisions.

The study of game theory began with total conflict games, also known as zero-sum games, such that one person's gain exactly equals the net losses of the other player(s). Game theory continues to grow with application to a wide range of applied problems.

The Nash equilibrium for a two-player, zeros-sum game can be found by solving a LP problem and its dual solution (Danzig, 1951, 2002; Dorfman, 1951). In their work, they assume that every element of the payoff matrix containing outcomes or payoffs to the row player, M_{ij}, is positive. A more current approach (Fox, 2008, 2012) shows that the payoff matrix entries can be positive or negative.

5.2.1 Two-Person Total Conflict Games

We begin with characteristics of the two-person total conflict game, as follows:

There are two persons (called the row player who we will be referred to as Rose and the column player who we will be referred to as Colin).

Rose must choose from among her 1 to *m* strategies and Colin must choose from among his 1 to *n* strategies.

TABLE 5.2

General Payoff Matrix, M, of a Two-Person Total Conflict Game with Outcomes (M_{ij}, N_{ij}) When Players Play Strategies i and j, Respectively

Rose's Strategies	Colin's Strategies			
	Column 1	Column 2	...	Column n
Row 1	$M_{1,1}, N_{1,1}$	$M_{1,2}, N_{1,2}$	...	$M_{1,n}, N_{1,n}$
Row 2	$M_{2,1}, N_{2,1}$	$M_{2,2}, N_{2,2}$	...	$M_{2,n}, N_{2,n}$
.	.	.	.	.
.	.	.	.	.
.	.	.	.	.
Row m	$M_{m,1}, N_{m,1}$	$M_{m,2}, N_{m,2}$	...	$M_{m,n}, N_{m,n}$

If Rose chooses the ith strategy and Colin chooses the jth strategy, then Rose receives a payoff of a_{ij} and Colin loses an amount of a_{ij}. In Table 5.2, this is shown as a payoff pair where Rose receives a payoff of M_{ij} and Colin receives a payoff of N_{ij}.

5.2.2 Games Are Simultaneous and Repetitive

There are two types of possible solutions: (1) *Pure-strategy* solutions are solutions when each player achieves their best outcomes by always choosing the same strategy in repeated games. (2) *Mixed-strategy* solutions are solutions when players play a random selection of their strategies to obtain their best outcomes in simultaneous repeated games.

We do not address sequential games in this chapter, where in sequential games, we look ahead and reason back to obtain pure strategy solutions, if they exist. We do discuss simultaneous games, and we illustrate a payoff matrix as shown in Table 5.2.

The game is a total conflict game if and only if the sum of the pairs, $M_{i,j} + N_{i,j}$, always equals either 0 for all strategies i and j or always equals the same constant for all strategies i and j. If the sum equals zero, then we might only list only the row payoff, M_{ij}, that we call a_{ij}.

For example, if a player wins x when the other player loses x then their sum is zero or in business marketing strategy, based upon 100% of the market, if one player get $x\%$ of the market when the other player gets $y\%$ then $x + y = 100$. We might choose to only list $x\%$ as the outcomes because the row players get $x\%$ and the column players lose $x\%$.

Movement diagrams: We define a movement diagram as a procedure where we draw arrows in each row (vertical arrow) and column (horizontal arrow) from the smaller payoff to the larger payoff. If there exists one or more payoff where all arrows point in, then those payoffs constitute pure-strategy Nash equilibriums.

Example 5.1: Trader Joe's versus Whole Foods

Suppose that grocery chain stores want to enter into the market in a new area. They have choices to locate in more densely populated area or less densely populated town surrounded by other towns. Let us assume a smaller grocery store such as Trader Joe's will locate a franchise in either densely populated area or a less densely populated area. Further, a mega grocery store franchise such as Whole Foods is making the same decision—they will locate either in denser populated area or the less dense area. Analysts have estimated the market shares and we place both sets of payoffs in a single game matrix. Listing the row player's payoffs first, we have the payoff as shown in Table 5.3. The payoff matrix represents a constant sum total conflict game. We apply the movement diagram.

All of the arrows point to the payoff (65,35) in the densely populated areas for both stores strategies for both players and no arrow exits that outcome. This indicates that neither player can unilaterally improve their solution, known as a Nash equilibrium (Straffin, 2004).

LINEAR PROGRAMMING IN TOTAL CONFLICT GAMES

The minimax theorem (von Neumann, 1928) states that for every two-person, zero-sum game with finitely many strategies, there exists an outcome value V and a set of strategies for each player, such that

1. Given Player 2's strategy, the best payoff possible for Player 1 is V
2. Given Player 1's strategy, the best payoff possible for Player 2 is $-V$

Equivalently, Player 1's strategy guarantees him a payoff of V regardless of Player 2's strategy, and similarly Player 2 can guarantee himself a payoff of $-V$. The name minimax arises because each player minimizes the maximum payoff possible for the other; since the game is zero sum, he or she also minimizes his or her own maximum loss (i.e., maximize his or her minimum payoff).

Every total conflict game may be formulated as a LP problem (Danzig, 1951; Dorfman, 1951). Consider a total conflict two-person game in which

TABLE 5.3

Payoff Matrix and Movement Diagram of Example 1

			Trader Joe's	
		Densely populated		Less densely populated
	Densely populated	(65,35) ⭕ ⇑	⟸	(70,30)
Whole Foods				⇑
	Less densely populated	(55,45)	⟸	(60,40)

maximizing Player X has m strategies and minimizing Player Y has n strategies. The entry (M_{ij}, N_{ij}) from the ith row and jth column of the payoff matrix represents the payoff for those strategies. We present the following formulation using only the elements of M for the maximizing player who provides results for the value of the game and the probabilities x_i (Winston, 2003; Giordano et al. 2014).

We note that if there are negative values in the entries of the payoff matrix, then we need a slight modification to the formulation in LP since all variables need to be nonnegative for the Simplex method. In order to obtain a possible negative value solution for the game, we will use a method described by Winston (2003) and replace any variable that could take on negative values with the difference in two positive variables. Since only the value of game V might be positive or negative, then replace V with $V_j - V_j'$ where both $V_j - V_j'$ are positive variables. We only assume that the value of the game could be positive or negative. The other values we are looking for are probabilities that are always nonnegative. In these games the players want to maximize the value of the game that the player receives. We present Equation 5.1 as a formulation to find the optimal strategies and value of the game.

Maximize V

Subject to

$$M_{1,1}x_1 + M_{2,1}x_2 + \ldots + M_{m,1}x_m - V \geq 0$$

$$M_{1,2}x_1 + M_{2,2}x_2 + \ldots + M_{m,2}x_m - V \geq 0$$

$$\ldots$$

$$M_{1,m}x_1 + M_{2,m}x_2 + \ldots + M_{m,n}x_m - V \geq 0 \tag{5.1}$$

$$x_1 + x_2 + \ldots + x_m = 1$$

Nonnegativity

where the weights, x_i, yield Rose's strategy and the value of V is the value of the game to Rose. When the solution to this total conflict game is obtained, we also have the solution to Colin's game through what is known as the dual solution (Winston, 2003). As an alternative to the dual, we could formulate Colin's game and solve it directly as shown in Equation 5.1 using the original N_{ij} entries from the matrix M. We call the value of the game for Colin, v, to distinguish it from V.

Maximize v

Subject to

$$N_{1,1}y_1 + N_{2,1}y_2 + \ldots + N_{m,1}y_n - v \geq 0$$

$$N_{2,1}y_1 + N_{2,2}y_2 + \ldots + N_{2,m}y_n - v \geq 0$$

$$\ldots \tag{5.2}$$

$$N_{m,1}y_1 + N_{m,2}y_2 + \ldots + N_{m,n}y_n - v \geq 0$$

$$y_1 + y_2 + \ldots + y_n = 1$$

Nonnegativity

TABLE 5.4

Zero-Sum Game Payoffs for Example 1

		Trader Joe's	
		Densely populated	Less densely populated
Whole Foods	Densely populated	65	67
	Less densely populated	55	60

where the weights, y_i, yield Colin's strategy and the value of v is the value of the game to Colin.

If we put our Example 5.1 into our two formulations and solve, we obtain the solution $x_1 = 1$, $x_2 = 0$, $V = 65$, $y_1 = 1$, $y_2 = 0$, and $v = 35$. The overall solution is that the businesses each choose to place their business in a large city with a game solution value of (65,35) split of the market share.

The primal–dual simplex method only works in the zero-sum game format (Fox, 2010). We may convert this game to the zero-sum game format to obtain the solution. Since this is a constant sum game, whatever Rose gains, Colin loses. For example, out of 100% if Whole Foods gains 60%, then Trader Joe's loses 60% of the market as shown in Table 5.4.

For a zero-sum game, we only need a single formulation of the linear program. The row player maximizes and the column player minimizes with rows' values. This constitutes a primal and dual relationship. The linear program used in zero-sum games is equivalent to formulation (1) except we use a_{ij} to designate the zero-sum outcomes for Rose (let $M_{ij} = a_{ij}$), as shown in Equation 5.3:

Maximize V

Subject to

$$a_{1,1}x_1 + a_{1,2}x_2 + \ldots + a_{1,n}x_m - V \geq 0$$
$$a_{2,1}x_1 + a_{2,2}x_2 + \ldots + a_{2,n}x_n - V \geq 0$$
$$\vdots$$
$$\vdots \tag{5.3}$$
$$a_{m,1}x_1 + a_{m,2}x_2 + \ldots + a_{m,n}x_m - V \geq 0$$
$$x_1 + x_2 + \ldots + x_m = 1$$
$$V, x_i \geq 0$$

where V is the value of the game, $a_{m,n}$ are payoff-matrix entries, and x's are the weights (probabilities to play the strategies).

We place these payoffs into our formulation and we let V stand for the value of the game to row player.

In Section 5.2.1, we defined Rose to represent the row player with strategies 1, 2, …, m and Colin to represent the column player with strategies 1, 2, …, n as shown in Table 5.2. In this example, Rose would represent Whole Foods and Colin would represent Trader Joe's.

Maximize V
Subject to

$$65x_1 + 55x_2 - V \geq 0$$
$$70x_1 + 60x_2 - V \geq 0$$
$$x_1 + x_2 = 1$$
$$x_1, x_2, V \geq 0$$

The optimal solution strategies found are identical as before with both players choosing more densely populated area as their best strategy. The use of LP is very suitable for large games between two players each having many strategies (Fox, 2010, 2012; Giordano et al., 2014). We note that the solution to Trader Joe's is found as the dual solution that is found in different locations of outputs for different LP software packages (see Winston, 2003).

THE PARTIAL CONFLICT GAME

Partial conflict games are games in which one player wins but the other player does not have to lose. Both players could win something or both could lose something.

Solution methods for partial conflict games include looking for dominance, movement diagrams, and equalizing strategies. Here we present an extension as an application of LP from the total conflict game to the partial conflict game, see Fox (2010, 2012). Because of the nature of partial conflict games, where both players are trying to maximize their outcomes, we can model all players' strategies as their own maximizing linear programs. We treat each player as separate maximizing LP problems.

Again, we use the payoff matrix from Table 5.2. We now assume that $M_{ij} + N_{ij}$ is not always equal to zero or the same constant for all i and j. In noncooperative partial conflict games, we first look for a pure-strategy solution using the movement diagram.

Rose maximizes payoffs, so she would prefer the highest payoff in each column. Vertical arrows are in columns, but the values are Rose's. Similarly for Colin, he maximizes his payoffs, so he would prefer the highest payoff in each row. We draw an arrow to the highest payoff in that row. Horizontal arrows are in rows, but the values are Colin's. If all arrows point in from every direction, then that or those points will be pure Nash equilibrium.

If all the arrows do not point at a value or values, then we must use equalizing strategies to find the weights (probabilities) for each player for a game with two players and two strategies each. We would proceed as follows:

- *Rose's game*: Rose maximizing, Colin *equalizing* is a total conflict game that yields Colin's equalizing strategy.
- *Colin's game*: Colin maximizing, Rose *equalizing* is a total conflict game that yields Rose's equalizing strategy.

- *Note*: If either side plays its equalizing strategy, then the other side *unilaterally* cannot improve its own situation (it stymies the other player).

This translates into two maximizing LP formulations that are shown as Equations 5.4 and 5.5. Formulation (4) provides the Nash equalizing solution for Colin with strategies played by Rose, whereas formulation (5) provides the Nash equalizing solution for Rose and strategies played by Colin.

Maximize V
Subject to

$$N_{1,1}x_1 + N_{2,1}x_2 - V \geq 0$$
$$N_{1,2}x_1 + N_{2,2}x_2 - V \geq 0 \tag{5.4}$$
$$x_1 + x_2 = 1$$
$$\text{Nonnegativity}$$

Maximize v
Subject to

$$M_{1,1}y_1 + M_{1,2}y_2 - v \geq 0$$
$$M_{2,1}y_1 + M_{2,2}y_2 - v \geq 0 \tag{5.5}$$
$$y_1 + y_2 = 1$$
$$\text{Nonnegativity}$$

If there is a pure-strategy solution, you must find it through dominance or the movement diagrams. You will not find those pure-strategy results via these LP formulations that only provide the Nash equilibrium using equalizing strategies (Straffin, 2004).

For games with two players and more than two strategies each, we present the nonlinear optimization approach by Barron (2013). Consider a two-person game with a payoff matrix as before. Let us separate the payoff matrix into two matrices M and N for Players 1 and 2. We solve the following nonlinear optimization formulation in expanded form, in Equation 5.6.

$$\text{Maximize } \sum_{i=1}^{n}\sum_{j=1}^{m} x_i a_{ij} y_j + \sum_{i=1}^{n}\sum_{j=1}^{m} x_i b_{ij} y_j + -p - q$$

Subject to

$$\sum_{j=1}^{m} a_{ij} y_j \leq p, \quad i = 1, 2, \ldots, n$$

$$\sum_{i=1}^{n} x_i b_{ij} \leq q, \quad j = 1, 2, \ldots, m \qquad (5.6)$$

$$\sum_{i=1}^{n} x_i = \sum_{j=1}^{m} y_j = 1$$

$$x_i \geq 0, y_j \geq 0$$

Example 5.2: Partial Conflict Equalizing Strategy (Game Solution with Linear Programming with Equations 5.4 and 5.5)

The following partial conflict game has revised payoff estimates of marketing shares that have no pure-strategy solution as indicated by the arrows in the movement diagram.

		Colin	
		More dense	Less dense
Rose	More dense	(20,40) ⟸	(10,0) ⇑
	Less dense	(30,10) ⟹	(0,40)

We use Equations 5.4 and 5.5 to formulate and solve these partial conflict games for the Nash equalizing strategies.

Maximize Vc
Subject to

$$40x_1 + 10x_2 - Vc \geq 0$$
$$0x_1 + 40x_2 - Vc \geq 0$$
$$x_1 + x_2 = 1$$
Nonnegativity

and
Maximize vr
Subject to

$$20y_1 + 10y_2 - vr \geq 0$$
$$30y_1 + 0y_2 - vr \geq 0$$
$$y_1 + y_2 = 1$$
Nonnegativity

The solutions to this partial conflict game are (a) $Vc = 22.85714286$ when $x_1 = 0.5714285714$ and $x_2 = 0.4285714286$ and (b) $Vr = 15.000$ when $y_1 = 0.5000$ and $y_2 = 0.5000$. This game results in Colin playing Large City and Small City half the time and ensuring a value of the game of 15.00 for Rose. Rose plays a mixed strategy of 4/7 Large City and 3/7 Small City, which yields a value of the game of 22.85714286 for Colin.

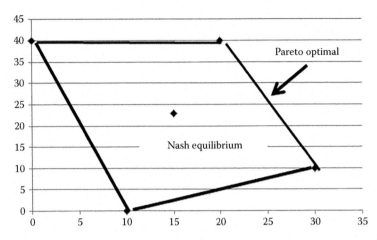

FIGURE 5.1
Payoff polygon for Example 5.2.

To be a solution the Nash equilibrium must be Pareto optimal (north-east region) as defined by Straffin (2004). Visually, we can plot the coordinates of the outcomes and connect them into a convex set. The Nash equilibrium (15, 22.85) is an interior point and is not Pareto optimal (Figure 5.1).

When the results are mixed strategies the implication is that the game is repeated many times to achieve that outcome.

Example 5.3: 3 × 3 Nonzero-Sum Game Using Equation 5.6

Assume for example that both Rose and Colin have three strategies each with payoffs as follows:

		Colin		
		C1	C2	C3
	R1	(−1,1)	(0,2)	(0,2)
Rose	R2	(2,1)	(1,−1)	(0,0)
	R3	(0,0)	(1,1)	(1,2)

First, if we use a movement diagram, we find two Nash points. They are R2C1 (2,1) and R3C3 (1,2). These pure-strategy solutions are not equivalent and trying to achieve them might lead to other results. We might employ the nonlinear method described earlier to look for other equilibrium solutions, if they exist. We find another solution does exist when $p = q = 0.667$ when $x_1 = 0$, $x_2 = 0.667$, $x_3 = 0.333$, $y_1 = 0.333$, $y_2 = 0$, $y_3 = 0.667$.

We provide the Maple commands from our template that we used to obtain the solution.

> *with(LinearAlgebra)* : *with(Optimazation)* :

> *A* :− Matrix([[−1, 0, 0], [2, 1, 0], [0, 1, 1]]);

$$A := \begin{bmatrix} -1 & 0 & 0 \\ 2 & 1 & 0 \\ 0 & 1 & 1 \end{bmatrix}$$

> *B* := Matrix([[1, 2, 2], [1, −1, 0], [0, 1, 2]]);

$$B := \begin{bmatrix} 1 & 2 & 2 \\ 1 & -1 & 0 \\ 0 & 1 & 2 \end{bmatrix}$$

> *PartialConflictGame*(3, 3, *A*, *B*, 0, 0);

$$[3.77713416099823 \ 10^{-9}, [p = 0.666666666322467,$$
$$q = 0.666666664413520, x_1 = 0., x_2 = 0.666666666437955,$$
$$x_3 = 0.333333334037583, y_1 = 0.333333333333333, y_2 = 0.,$$
$$y_3 = 0.666666666666667]]$$

COMMUNICATIONS AND COOPERATION IN PARTIAL CONFLICT GAMES

Allowing for communication might change the game's solution. First moves, threats, promises, and combinations of threats and promises could be considered in an attempt to obtain better outcomes. The strategy of moves can be examined in the additional and reference readings.

NASH ARBITRATION METHOD

When we have not achieved a solution by other methods that are acceptable by the players, then a game might move to arbitration. The Nash arbitration theorem has 4 axioms and a theorem to be used with a status quo point, that can be either a security level or a threat level. We illustrate using the security levels from the Prudential Strategies. We find the security levels first.

PRUDENTIAL STRATEGY (SECURITY LEVELS)

The security levels are the payoffs to the players in a partial conflict game where each players attempt to maximize their own payoff. We can solve for these payoff using a separate linear programs for each security level.

We have Rose in Rose's game and Colin in Colin's game. So the LP formulations are in Equations 5.7 and 5.8.

Maximize V
Subject to

$$N_{1,1}y_1 + N_{1,2}y_2 + ... + N_{1,m}y_n - V \geq 0$$

$$N_{2,1}y_1 + N_{2,2}y_2 + ... + N_{2,m}y_n - V \geq 0$$

...

$$N_{m,1}x_1 + N_{m,2}x_2 + ... + N_{m,n}x_n - V \geq 0$$ (5.7)

$$y_1 + y_2 + ... + y_n = 1$$

$$y_j \leq 1 \quad \text{for} \quad j = 1,...,n$$

Nonnegativity

where the weights y_i yield Colin's prudential strategy and the value of V is the security level for Colin.

Maximize v
Subject to

$$M_{1,1}x_1 + M_{2,1}x_2 + ... + M_{n,1}x_n - v \geq 0$$

$$M_{1,2}x_1 + M_{2,2}x_2 + ... + M_{n,2}x_n - v \geq 0$$

...

$$M_{1,m}y_1 + M_{2,m}y_2 + ... + M_{m,n}y_n - v \geq 0$$ (5.8)

$$x_1 + x_2 + ... + x_m = 1$$

$$x_i \leq 1 \quad \text{for} \quad i = 1,...,m$$

Nonnegativity

Let us return to Example 5.2 to illustrate finding the security levels. Let SLR and SLC represent the security levels for Rose and Colin, respectively. We use linear programming to find these values using formulations (5.7) and (5.8).

Max SLR
$20x_1 + 30x_2 - \text{SLR} \geq 0$
$10x_1 + 0x_2 - \text{SLR} \geq 0$
$x_1 + x_2 = 1$
Nonnegativity

Max SLC
$40y_1 - \text{SLC} \geq 0$
$10y_1 + 40y_2 - \text{SLC} \geq 0$
$y_1 + y_2 = 1$
Nonnegativity

The solution yields both how the game is played and the security levels. Player 1 always plays R1 and Player 2 plays 4/7 C1 and 3/7 C2. The security level is (10, 160/7) = (10, 22.86).

We use this security level as our status quo (SQ) point and we can now formulate the Nash arbitration scheme. There are four axioms that are met using the arbitration scheme.

> *Axiom* 1: Rationality. The solution should be in the negotiation set.
> *Axiom* 2: Linear invariance. If either Rose's or Colin's utilities are transformed by a positive linear function, the solution point should be transformed by the same function.
> *Axiom* 3: Symmetry. If the polygon happens to be symmetric about the line of slope + 1 through a SQ point, then the solution should be on this line.
> *Axiom* 4: Independence of irrelevant alternatives. Suppose N is the solution point for a polygon, P with status quo point SQ. Suppose Q is another polygon that contains both SQ and N and is totally contained in P. Then N should also be the solution point to Q with status quo point SQ.

We can state Nash theorem as:

Nash's Theorem (1950). "there is one and only arbitration scheme which satisfies rationality, linear invariance, symmetry, and independence of irrelevant alternatives. It is this: if the status quo (SQ) point is (x_0, y_0), then the arbitrated solution point N is the point (x, y) in the polygon with $x \geq x_0$, $y \geq y_0$ which maximizes the product of $(x-x_0)(y-y_0)$."

We apply this theorem in a nonlinear optimization model framework. The formulation is

Maximize $(x-10)(y-22.86)$
Subject to

$$3x + y = 100$$
$$x \geq 10$$
$$y \geq 22.86$$
Nonnegativity

We find the solution. In this example, that our Nash arbitration point is (20, 40) for our given status quo point of SQ point (10, 22.86).

5.3 Illustrative Modeling Examples of Zero-Sum Games

In this section, we present some illustrative examples of game theory. We present the scenario, discuss the outcomes used in the payoff matrix, and present a possible solution for the game. In most game theory problems the solution suggests insights into how to play the game rather than a definite methodology to *winning* the game. We present only total conflict game as illustrative examples in this section.

Example 5.4: The Battle of the Bismarck Sea

The Battle of the Bismarck Sea is set in the South Pacific in 1943, see Figure 5.2. The historical facts are that General Imamura had been ordered to transport Japanese troops across the Bismarck Sea to New Guinea, and General Kenney—the United States commander in the region—wanted to bomb the troop transports before their arrival at their destination. Imamura had two options to choose from as routes to New Guinea: (1) a shorter northern route or (2) a longer southern route. Kenney must decide where to send his search planes and bombers to find the Japanese fleet. If Kenney sends his planes to the wrong route he can recall them later, but the number of bombing days is reduced. A good background can be found at the following site: (http://policonomics. com/video-d4-battle-of-the-bismarck-sea/).

We assume that both commanders, Imamura and Kenney, are rational players, each trying to obtain their best outcome. Further, we assume that there is no communication or cooperation, which may be inferred since the two are enemies engaging in war. Further, each is aware of the intelligence assets that are available to each side and is aware of what the intelligence assets are producing. We assume that the number of days that U.S. planes can bomb and the number of days to sail to New Guinea are accurate estimates.

The players, Kenney and Imamura, both have the same set of strategies for routes: {*North, South*} and their payoffs given as the numbers of exposed days for bombing are shown in Table 5.5. Imamura loses exactly what Kenney gains. This is a total conflict as seen in Figure 5.3.

As a total conflict game, we might only list the outcomes to Kenney to find our solution. The payoffs are provided in Table 5.6.

There is a dominant column strategy for Imamura to sail *North* since the values in the column are correspondingly less than or equal to the values for sailing *South*. This would eliminate the column for *South*. Seeing that as an option, Kenney would search *North* as this option

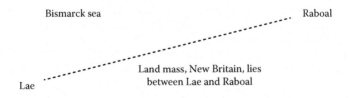

Northern route

Bismarck sea   Raboal

                        Land mass, New Britain, lies
Lae   between Lae and Raboal

Soloman sea, the southern route

FIGURE 5.2
The battle of the bismarck sea region. Japanese troops were being taken from Rabaul to Lae.

TABLE 5.5

The Battle of the Bismarck Sea with Payoffs to
(Kenney, Imamura)

		Imamura	
		North	South
Kenney	North	(2,–2)	(2,–2)
	South	(1,–1)	(3,–3)

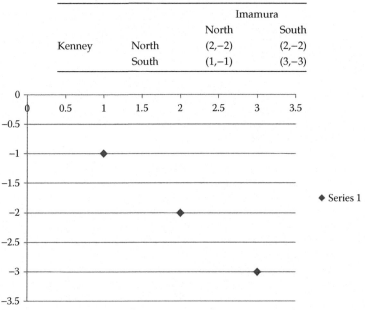

FIGURE 5.3
Plot of the game payoffs for the battle of the bismarck sea.

TABLE 5.6

The Battle of the Bismarck Sea as a Zero-Sum Game

		Imamura	
		North	South
Kenney	North	2	2
	South	1	3

provides a greater outcome than searching *South*, 2 > 1. We could also apply the *minimax* theorem from Section 5.3 (saddle point method) to find a plausible Nash equilibrium as Kenney searches North and Imamura takes the Northern route as shown in Table 5.7.

Applied to the Battle of the Bismarck Sea, this Nash equilibrium of (*North, North*) implies that no player can do unilaterally better by changing their strategy. The solution is for the Japanese to sail *North* and for Kenney to search *North* yielding two bombing days. This result (*North, North*) was indeed the real outcome in 1943.

Next, let us assume that communications are allowed. We will consider first moves by each player. If Kenney moved first, (*North, North*) would remain the outcome. However, (*North, South*) also becomes a valid response also with value 2.

TABLE 5.7

Minimax Method (Saddle Point Method)

			Imamura		
		North	South	Row min	Max
Kenney	North	2	2	2	②
	South	1	3	1	
	Column max	2	3		
	Min	②			

If Imamura moved first, (*North, North*) would be the outcome. What is important about moving first in a zero-sum game is that although it gives more information, neither player can be better than the Nash equilibrium in the original zero-sum game. We conclude from our brief analysis that moving first does not alter the equilibrium of the game. It is true that in zero-sum games that moving first does not alter the equilibrium strategies of the game.

Example 5.5: Penalty Kicks in Soccer

This example is adapted from an article by Chiappori et al. (2002) (see additional readings). Let us consider a penalty kick in soccer between two players: the kicker and the opponent's goalie. The kicker basically has two alternatives of strategies that we will consider: he or she might kick the ball left or right. The goalie will also have two strategies: he or she can prepare to dive left or right to block the kick. We will start very simply in the payoff matrix by awarding a 1 to the player who is successful and a −1 to the player who is unsuccessful. The payoff matrix would simplify be as follows:

		Goalie	
		Dive left	Dive right
Kicker	Kick left	(−1,1)	(1,−1)
	Kick right	(1,−1)	(−1,1)

or as just from the kicker's prospective.

		Goalie	
		Dive left	Dive right
Kicker	Kick left	−1	1
	Kick right	1	−1

There is no pure strategy. We find a mixed strategy to the zero-sum game using either LP or the method of oddments. The mixed-strategy results are that the kicker randomly kicks 50% left and 50% right, whereas the goalie randomly dives 50% left and 50% right. The value of the game to each player is 0.

Let us refine the game using real data. A study was done in the Italian Football league in 2002 by Ignacio Palacios-Huerta. As he observed the kicker can aim the ball to the left or to the right of the goalie, and the goalie can dive either left or right as well. The ball is kicked with enough speed that the decisions of the kicker and goalie are effectively made simultaneously. Based on these decisions the kicker is likely to score or not score. The structure of the game is remarkably similar to our simplified game. If the goalie dives in the direction that the ball is kicked then he or she has a good chance of stopping the goal, and if he or she dives in the wrong direction then the kicker is likely to get a goal.

Based upon an analysis of roughly 1400 penalty kicks, Palacios-Huerta determined the empirical probabilities of scoring for each of the four outcomes: the kicker kicks left or right and the goalie dives left or right. His results led to the following payoff matrix:

		Goalie	
		Dive left	Dive right
Kicker	Kick left	0.58,−0.58	0.95,−0.95
	Kick right	0.93,−0.93	0.70,−0.70

There is no pure-strategy equilibrium as expected. The kicker and the goalie must use a mixture of strategies since the game is played over and over. Neither player wants to reveal a pattern to their decision. We apply the method of oddments to determine the mixed strategies for each player based on these data.

		Goalie		Oddments	Probabilities
		Dive left	Dive right		
Kicker	Kick left	0.58	0.95	0.37	0.23/0.60 = 0.383
	Kick right	0.93	0.70	0.23	0.37/0.60 = 0.6166
Oddments		0.35	0.25		
Probabilities		0.25/0.60 = 0.416	0.35/0.60 = 0.5833		

We find that the mixed strategy for the kicker is 38.3% kicking left and 61.7% kicking right, whereas the goalie dove right 58.3% and dives left 41.7%. If we merely count percentages from the data that were collected by Palacios-Huerta in his study from 459 penalty kicks over 5 years of data, we find that the kicker did 40% kicking left and 60% kicking right, whereas the goalie dove left 42% and right 58%. Since our model closely approximates our data, our game theory approach adequately models the penalty kick.

Example 5.6: Batter–Pitcher Duel

In view of the use of technology in sports today, we present an example of the hitter–pitcher duel. First in this example, we extend the strategies for each player in our model. We consider a batter–pitcher duel between Buster Posey of the San Francisco Giants and various pitchers in the national league where the pitcher throws a fastball, a split-finger fastball, a curve ball, or a change-up. The batter, aware of these pitches, must prepare appropriately for the pitch. Data are available from many websites that we might use. We consider only Posey facing a right-handed pitcher (RHP) in this analysis.

For a National League RHP versus Buster Posey, assume we have compiled the following data. Let FB=fastball, CB=curveball, CH=change-up, SF=split-fingered fastball.

Posey/RHP	FB	CB	CH	SF
FB	0.337	0.246	0.220	0.200
CB	0.283	0.571	0.339	0.303
CH	0.188	0.347	0.714	0.227
SF	0.200	0.227	0.154	0.500

Both the batter and pitcher want the best possible result. We set this up as a LP problem.

Our decision variables are x_1, x_2, x_3, x_4 as the percentages to guess FB, CB, CH, SF, respectively and V represents Posey's batting average.

Max V
Subject to

$$0.337x_1 + 0.283x_2 + 0.188x_3 + 0.200x_4 - V \geq 0$$
$$0.246x_1 + 0.571x_2 + 0.347x_3 + 0.227x_4 - V \geq 0$$
$$0.220x_1 + 0.339x_2 + 0.714x_3 + 0.154x_4 - V \geq 0$$
$$0.200x_1 + 0.303x_2 + 0.227x_3 + 0.500x_4 - V \geq 0$$
$$x_1 + x_2 + x_3 + x_4 = 1$$
$$x_1, x_2, x_3, x_4, V \geq 0$$

We solved this LP problem and found that the optimal solution (strategy) is to guess the fastball (FB) 27.49%, guess the curveball (CB) 64.23%, never guess change-up (CH), and guess split-finger fastball (SF) 8.27% of the time to obtain a 0.291 batting average.

The pitcher then wants to keep the batting average also as low as possible. We can set up the linear program for the pitcher as follows:

Our decision variables are y_1, y_2, y_3, y_4 as the percentages to throw the FB, CB, CH, SF, respectively, and V represents Posey's batting average.

Min V
Subject to

$$0.337y_1 + 0.246y_2 + 0.220y_3 + 0.200y_4 - V \leq 0$$
$$0.283y_1 + 0.571y_2 + 0.339y_3 + 0.303y_4 - V \leq 0$$

$$0.188y_1 + 0.347y_2 + 0.714y_3 + 0.227y_4 - V \leq 0$$
$$0.200y_1 + 0.227y_2 + 0.154y_3 + 0.500y_4 - V \leq 0$$
$$y_1 + y_2 + y_3 + y_4 = 1$$
$$x_1, y_2, y_3, y_4, V \geq 0$$

We found that the RHP should randomly throw 65.93% fastballs, no curveballs, 3.25% change-ups, and 30.82% split-finger fastballs for Posey to keep only that 0.291 batting average.

Now, assume we also have statistics for Posey versus a LHP.

Example 5.7: Operation Overlord

Operation Overlord from World War II can be viewed from the context of game theory. In *Thinking Strategically* (Dixit and Nalebuff, 1991, Case Study #7, pp. 195–198). In 1994, the Allies were planning an operation for the liberation of Europe and the Germans were planning their defense against it. There were two known possibilities for an initial amphibious landing: the beaches at (1) Normandy and (2) the Calais. Any landing would succeed against a weak defense, so the Germans did want a weak defense at the potential landing site. Calais was more difficult for a landing but it was closer to the Allies' targets for success.

Suppose the probabilities of an allied success are based as follows:

		German defense	
		Normandy	Calais
Allied landing	Normandy	75%	100%
	Calais	100%	20%

Similar to the other example, we decide to use a 100 point scale. The Allies' successful landing at Calais would earn 100 points, a successful landing at Normandy would earn 80 points, and failure at either landing would earn 0 points. What decisions should be made?

We compute the expected values and place in a payoff matrix.

		German defense	
		Normandy	Calais
Allied landing	Normandy	0.75 * 80 = 60	1 * 80 = 80
	Calais	1 * 100 = 100	0.20 * 100 = 20

There are no pure-strategy solutions in this example. We used mixed strategies and determined the game's outcome.

The Allies would employ a mixed strategy of 80% Normandy and 20% Calais to achieve an outcome of 68 points. At the same time, the Germans should employ a strategy of 6320% Normandy and 40% Calais for their defenses to keep the Allies at 68 points.

Implementation of the landing was certainly not two pronged. So, what do the mixed strategies imply in the strategic thinking? Most likely a strong feint at Calais and lots of information leak about Calais happened, whereas the real landing was a secret at Normandy. The Germans had a choice as to believe the information about Calais or somewhat equally divide their defenses. Although the true results were in doubt for a while, the Allies prevailed.

Example 5.8: Choosing the right Course of action (Cantwell, 2003; Fox, 2016)

The U.S. Army Command and General Staff College presented this approach for choosing the best course of action (COA) for a mission. For a possible battle between two forces, we compute the optimal COAs by the two opponents by game theory as follows:

> **Steps 1 and 2** List the friendly COAs and rank order them as best to worst.
> COA 1–Decisive victory
> COA 4–Attrition-based victory
> COA 2–Failure by culmination
> COA 3–Defeat in place
> **Step 3** COA for the enemy how they rank best to worst in each row, where the row represents the friendly COA. In this example, the enemy is thought to have six distinct possible COAs. The enemy COA #1 is best against the friendly COAs #1 and #4.
> **Step 4** Decide in each case if we think we will win, lose, draw

Friendly						
COA 1	Best	Win	Win	Draw	Loss	Loss
COA 4	Win	Win	Win	Win	Loss	Loss
COA 2	Win	Win	Loss	Loss	Loss	Worst
COA 3	Draw	Draw	Draw	Loss	Loss	Loss

> **Step 5 and 6** Provide scores. Since there are 4 friendly COAs and 6 enemy COAs, we use scores from 24 (Best) to 1 (Worst).

Friendly						
COA 1	Best 24	Win 23	Win 22	Draw	Loss	Loss
COA 4	Win 21	Win 20	Win 19	Win 18	Loss	Loss
COA 2	Win 17	Win 16	Loss	Loss	Loss	Worst
COA 3	Draw	Draw	Draw	Loss	Loss	Loss

Step 7 and 8 Put into numerical order for loss.

Friendly						
COA 1	Best 24	Win 23	Win 22	Draw	Loss 3	Loss 2
COA 4	Win 21	Win 20	Win 19	Win 18	Loss 10	Loss 9
COA 2	Win 17	Win 16	Loss 11	Loss 7	Loss 8	Worst 1
COA 3	Draw	Draw	Draw	Loss 6	Loss 5	Loss 4

Step 9 Fill in the scores for the draw

Friendly						
COA 1	Best 24	Win 23	Win 22	Draw 15	Loss 3	Loss 2
COA 4	Win 21	Win 20	Win 19	Win 18	Loss 10	Loss 9
COA 2	Win 17	Win 16	Loss 11	Loss 7	Loss 8	Worst 1
COA 3	Draw 14	Draw 13	Draw 12	Loss 6	Loss 5	Loss 4

There is no pure-strategy solution. Because of the size of the payoff matrix, we did not use the movement diagram instead we used the MINIMAX theorem. Basically, we find the minimum in each row and then the maximum of these minimums. Then we find the maximums in each column and then the minimum of those maximums. If the maximum of the row minimums equals to minimum of the column maximums then we have a pure-strategy solution. If not, we have to find the mixed-strategy solution. In Step 10 above-mentioned, the maximum of the minimums is {9}, whereas the minimum of the maximums is {10}. They are not equal.

LP may be used in zero-sum games to find the solutions whether they are pure-strategy or mixed-strategy solutions.

So we solve using LP. We let V be the value of the game, x_1–x_4 be the probabilities in which to play strategies 1 through 4 for the friendly side. The values y_1–y_6 represent the probabilities that the enemy should employ to obtain their best results.

Maximize V
$$24x_1 + 16x_2 + 13x_3 + 21x_4 - V \geq 0$$
$$23x_1 + 17x_2 + 12x_3 + 20x_4 - V \geq 0$$
$$22x_1 + 11x_2 + 6x_3 + 19x_4 - V \geq 0$$
$$3x_1 + 7x_2 + 5x_3 + 10x_4 - V \geq 0$$
$$15x_1 + 8x_2 + 4x_3 + 9x_4 - V \geq 0$$
$$2x_1 + 1x_2 + 14x_3 + 18x_4 - V \geq 0$$
$$x_1 + x_2 + x_3 + x_4 = 1$$
Nonnegativity

Solution: $V = 9.462$ when *friendly* chooses $x_1 = 7.7\%$, $x_2 = 0$, $x_3 = 0$, $x_4 = 92.3\%$, whereas the *enemy* best results come when $y_1 = 0$, $y_2 = 0$, $y_3 = 0$, $y_4 = 46.2\%$, and $y_5 = 53.8\%$.

Interpretation: At 92.3% we see that we should defend along the Vistula River almost all the time. The value of the game, 9.462, is greater than the pure-strategy solution of 9 for always picking to defend the Vistula River. This implies that we benefit from secrecy and employing deception. We can benefit *selling the enemy* on our attacking north and fixing in the south. A negative 9.462 for the enemy does not mean that they lose. We need to further consider the significance of the values and mission analysis.

This modeling procedure has direct applications to business and industry as well.

Exercises 5.3

Solve problems using any method

1. Given the payoff table

		Colonel. blotto	
		Defend City I C_1	Defend City II C_2
	Attack City I R_1	15	15
Colonel sotto	Attack City II R_2	10	0

What should each player do?

2. Given the payoff table

		Colin	
		C1	C2
Rose	R1	A	b
	R2	C	d

   a. What assumptions have to be true for *a* at R1C1 to be the pure-strategy solution?

   b. What assumptions have to be true for *b* at R1C2 to be the pure-strategy solution?

   c. What assumptions have to be true for *c* at R2C1 to be the pure-strategy solution?

   What assumptions have to be true for *d* at R2C2 to be the pure-strategy solution?

3. Given the payoff table

		Colin	
		C1	C2
Rose	R1	1/2	1/2
	R2	1	0

What is the pure-strategy solution for this game?

4. Consider the following batter–pitcher duel. All the entries in the payoff matrix reflect the percent of hits off the pitcher, the batting average. What strategies are optimal for each player?

		Pitcher	
		C1	C2
Batter	R1	0.350	0.150
	R2	0.300	0.250

5. Consider the following game, find the solution.

Rose/Colin	X	Y	Z
#1	80	40	75
#2	70	35	30

6. Consider the following game, find the solution.

Rose/Colin	W	X	Y	Z
#1	40	80	35	60
#2	65	90	55	70
#3	55	40	45	75
#4	45	25	50	50

7. The following represents a game between a professional athlete (Rose) and management (Colin) for contract decisions. The athlete has two strategies and management has three strategies. The values are in (1000s). What decision should each make?

Rose/Colin	X	Y	Z
#1	490	220	195
#2	425	350	150

8. Solve the following game:

		Colin	
		C	D
Rose	A	(2,−2)	(1,−1)
	B	(3,−3)	(4,−4)

9. The predator has two strategies for catching the prey (ambush or pursuit). The prey has two strategies for escaping (hide or run). The following game matrix shows the values to the players:

		Predator	
Prey		Ambush	Pursue
	Hide	0.25	0.30
	Run	0.70	0.550

10. A professional football team has collected data for certain plays against certain defenses. In the payoff matrix is the yards gained or lost for a particular play against a particular defense.

Team A/Team B	1	2	3
1	0	−1	5
2	7	5	10
3	15	−4	−5
4	5	0	10
5	−5	−10	10

11. Given the payoff table

		Colin	
		C1	C2
Rose	R1	a	b
	R2	c	d

What assumptions have to be true for there not to be a saddle point solution? Show that the two largest entries must be diagonally opposite to each other.

12. Given the payoff table

		Colin	
		C1	C2
Rose	R1	a	b
	R2	c	d

where $a > d > b > c$. Show that Colin plays C1 and C2 with probabilities x and $(1-x)$ that

$$x = \frac{d-b}{(a-c)+(d-b)}$$

13. In the game in exercise 12 earlier, show that the value of the game is

$$v = \frac{ad-bc}{(a-c)+(d-b)}$$

Solve the following games

14.

	Colin	
Rose	C1	C2
R1	5	3
R2	3	2

15.

	Colin	
Rose	C1	C2
R1	−2	4
R2	3	−2

16. Solve the hitter–pitcher duel for the following players:

		RR	
		Fast ball	Split-finger
Payoff tableau		C1	C2
Derek jeter	Guess fast ball R1	0.333	0.257
	Guess split finger R2	0.185	0.416

17. Solve the following game where each player has three strategies each.

		Colin		
		C1	C2	C3
Rose	R1	3	7	2
	R2	8	5	1
	R3	6	9	4

18. Solve the following game where each player has three strategies each.

		Colin		
		C1	C2	C3
Rose	R1	0.5	0.9	0.9
	R2	0.1	0	0.1
	R3	0.9	0.9	0.5

19. Solve the following game where Rose has three strategies and Colin has only two strategies.

		Colin	
		C1	C2
Rose	R1	6	5
	R2	1	4
	R3	8	5

20. Solve the following game where Rose has three strategies and Colin has only two strategies.

		Colin	
		C1	C2
Rose	R1	2	−1
	R2	1	4
	R3	6	2

21.

Rose		Colin			
	C1	C2	C3	C4
R1	1	−1	2	3
R2	2	4	0	5

22.

		Pitcher	
		Fastball	Curve
Batter	Fastball	0.300	0.200
	Curve	0.100	0.500

23. Given the payoff table

		Colin	
		C1	C2
Rose	R1	2	−3
	R2	0	2
	R3	−5	10

24. Given the payoff table

Rose		Colin		
	C1	C2	C3
R1	1	1	10
R2	2	3	−4

25.

		Colin		
		A	B	C
Rose	A	1	2	2
	B	2	1	2
	C	2	2	0

26. Determine the solution to the following zero-sum games by any method or methods but show/state work. State the value of the game for both Rose and Colin and what strategies each player chooses.

   a.

		Colin		
		C1	C2	C3
Rose	R1	10	20	14
	R2	5	21	8
	R3	8	22	0

   b.

		Colin	
		C1	C2
Rose	R1	-8	12
	R2	2	6
	R3	0	-2

   c.

		Colin	
		C1	C2
Rose	R1	8	2
	R2	5	16

   d.

		Colin		
		C1	C2	C3
Rose	R1	15	12	11
	R2	14	16	17

   e.

		Colin			
		C1	C2	C3	C4
Rose	R1	3	2	4	1
	R2	-9	1	-1	0
	R3	6	4	7	3

27. For the following game between Rose and Colin, write out the LP formulation for Rose. Using the SIMPLEX LP of the Solver or any other technology including the template, find and state the *complete solution* to the game in context of the game.

		Colin			
		C1	C2	C3	C4
	R1	0	1	2	6
Rose	R2	2	4	1	2
	R3	1	-1	4	-1
	R4	-1	1	-1	3
	R5	-2	-2	2	2

Projects 5.3

1. Research the solution methodologies for the three-person games. Staffin (1989), is a good reference. Consider the following three-person zero-sum game between Rose, Colin, and Larry.

Larry D1

		Colin	
		C1	C2
Rose	R1	(4,4,−8)	(−2,4,−2)
	R2	(4,−5,1)	(3,−3,0)

Larry D2

		Colin	
		C1	C2
Rose	R1	(−2,0,2)	(−2,−1,3)
	R2	(−4,5,−1)	(1,2,−3)

a. Draw the movement diagram for this zero-sum game. Find any and all equilibrium. If this game were played without any possible coalitions, what would you expect to happen?

b. If Colin and Rose were to form a coalition against, Larry sets up and solves the resulting 2 × 4 game. What are the strategies to be played and the payoffs for each of the three players? Solve by hand (show work) and then check using the three-person template.

Given the results of the coalitions of Colin versus Rose–Larry and Rose versus Colin–Larry as follows:

Colin versus Rose–Larry (3, −3, 0)

Rose versus Colin–Larry (−2, 0, 2)

c. If no *side payments* were allowed, would any player be worse off joining a coalition than playing alone? Briefly explain (include the values to justify decisions).

d. What is (are) the preferred coalition(s), if any?

e. Briefly explain how *side payments* could work in this game.

5.4 Partial Conflict Games Illustrative Examples

In this section, we present some example of partial conflict games and their solution techniques.

Example 5.9: Cuban Missile Crisis (Classic Game of Chicken)

"We're eyeball to eyeball, and I think the other fellow just blinked" were the eerie words of Secretary of State Dean Rusk at the height of the Cuban missile crisis in October 1962. He was referring to signals by the Soviet Union that it desired to defuse the most dangerous nuclear confrontation ever to occur between the superpowers, which many analysts have interpreted as a classic instance of nuclear *Chicken*.

We will highlight the scenario from 1962. The Cuban missile crisis was precipitated by a Soviet attempt in October 1962 to install medium-range and intermediate-range nuclear-armed ballistic missiles in Cuba that were capable of hitting a large portion of the United States. The goal of the United States was immediate removal of the Soviet missiles, and U.S. policy-makers seriously considered two strategies to achieve this end: (1) naval blockade or (2) airstrikes.

The proximity of Cuba to Florida made the situation dire. Further the range of the missiles from Cuba allowed for major political, population, and economic centers to become targets.

In Kennedy's speech to the nation, he explained the situation as well as the goals of the United States. He set several initial steps. First, to halt the offensive buildup, a strict quarantine on all offensive military equipment under shipment to Cuba is being initiated. He went on to say that any launch of missiles from Cuba at anyone would be considered an act of war by the Soviet Union against the United States resulting in a full retaliatory nuclear strike against the Soviet Union. He called upon Krushchev to end this threat to the world and restore world peace.

According to Bram's analysis of the Cuban missile crisis (1994; 2001), Bram discussed the use of the theory of strategic moves. Indeed, Theodore Sorensen, special counsel to President John Kennedy, used the language of *moves* to describe the deliberations of EXCOM, the Executive Committee of key advisors to Kennedy during the Cuban missile crisis:

> We discussed what the Soviet reaction would be to any possible move by the United States, what our reaction with them would have to be to that Soviet action, and so on, trying to follow each of those roads to their ultimate conclusion.

Problem statement: Build a mathematical model that allows for consideration of alternative decisions by the two opponents.

Assumptions: We assume the two opponents are rational players.

MODEL BUILDING

The goal of the United States was the immediate removal of the Soviet missiles, and U.S. policy-makers seriously considered the following two strategies to achieve this end:

1. A *naval blockade* (B), or *quarantine* as it was euphemistically called, to prevent shipment of more missiles, possibly followed by stronger action to induce the Soviet Union to withdraw the missiles already installed.
2. A *surgical air strike* (A) to wipe out the missiles already installed, insofar as possible, perhaps followed by an invasion of the island.

The alternatives open to Soviet policy-makers were

1. *Withdrawal* (W) of their missiles.
2. *Maintenance* (M) of their missiles.

We list the payoffs (x, y) as payoffs to the United States, payoffs to the Soviet Union where 4 is the best result, 3 is next best, 2 is next worst, and 1 is the worst.

		Soviet union	
		Withdraw missiles (W)	Maintain missiles (M)
United States	Blockade (B)	(3,3)	(2,4)
	Air strike (A)	(4,2)	(1,1)

The arrow diagram is shown in Figure 5.4 where we have two equilibriums at (4,2) and (2,4).

The Nash equilibrium is circled. Note that there are two equilibria (4,2) and (2,4) that can be found by our arrow diagram.

As in Chicken, as both players attempt to get to their equilibrium, the outcome of the games ends up at (1,1). This is disastrous for both countries and their leaders. The best solution is the (3,3) compromise position.

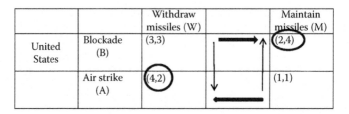

		Withdraw missiles (W)		Maintain missiles (M)
United States	Blockade (B)	(3,3)		(2,4)
	Air strike (A)	(4,2)		(1,1)

FIGURE 5.4
Cuban missile crisis as a game of chicken.

However, (3,3) is not stable. This will eventually put us back at (1,1). In this situation, one way to avoid this chicken dilemma is to try strategic moves.

Both sides did not choose their strategies simultaneously or independently. Soviets responded to our blockade after it was imposed. The United States held out the chance of an air strike as a viable choice even after the blockade. If the Union of Soviet Socialist Republics (USSR) would agree to remove the weapons from Cuba, the United States would agree to (a) remove the quarantine and (b) agree not to invade Cuba. If the Soviets maintain their missiles, the United States would prefer the airstrike to the blockade. Attorney General Robert Kennedy said, "if they do not remove the missiles, then we will." The United States used a combination of promises and threats. The Soviets know that our credibility in both areas was high (strong resolve). Therefore, they withdrew the missiles and the crisis ended. Khrushchev and Kennedy were wise.

Needless to say, the strategy choices, probable outcomes, and associated payoffs shown in Figure 5.1 provide only a skeletal picture of the crisis as it developed over a period of thirteen days. Both sides considered more than the two alternatives listed and several variations on each. The Soviets, for example, demanded withdrawal of American missiles from Turkey as a *quid pro quo* for withdrawal of their own missiles from Cuba, a demand publicly ignored by the United States.

Nevertheless, most observers of this crisis believe that the two superpowers were on a collision course, which is actually the title of one book describing this nuclear confrontation. They also agree that neither side was eager to take any irreversible step, such as one of the drivers in Chicken might do by defiantly ripping off the steering wheel in full view of the other driver, thereby foreclosing the option of swerving.

Although in one sense the United States *won* by getting the Soviets to withdraw their missiles; Premier Nikita Khrushchev of the Soviet Union at the same time extracted from President Kennedy a promise not to invade Cuba, which seems to indicate that the eventual outcome was a compromise of sorts. But this is not game theory's prediction for Chicken, because the strategies associated with compromise do not constitute a Nash equilibrium.

To see this, assume play is at the compromise position (3,3), that is, the U.S. blockades Cuba and the S.U. withdraw its missiles. This strategy is not stable, because both players would have an incentive to defect to their more belligerent strategy. If the United States were to defect by changing its strategy to airstrike, play would move to (4,2), improving the payoff the United States received; if the Soviet Union were to defect by changing its strategy to maintenance, play would move to (2,4), giving the Soviet Union a payoff of 4. (This classic game theory setup gives us no information about which outcome would be chosen, because the table of payoffs is symmetric for the two players. This is a frequent problem in interpreting the results of a game theoretic analysis, where more than one equilibrium position can arise.) Finally, if the players be at the mutually worst outcome of (1,1), that is, nuclear war, both would obviously desire to move away from it, making the strategies associated with it, like those with (3,3), unstable.

Example 5.10: Writers Guild Strike 2007–2008 (Fox, 2008)

The 2007–2008 Writers Guild of America strike was a strike by the Writers Guild of America, East (WGAE) and the Writers Guild of America, West (WGAW) that started on November 5, 2007. The WGAE and WGAW were two labor unions representing film, television, and radio writers working in the United States. Over 12,000 writers joined the strike. These entities will be referred to in the model as the Writers Guild.

The strike was against the Alliance of Motion Picture and Television Producers (AMPTP), a trade organization representing the interests of 397 American film and television producers. The most influential of these are eight corporations: CBS Corporation, Metro-Goldwyn-Mayer, NBC Universal, News Corp/Fox, Paramount Pictures, Sony Pictures Entertainment, the Walt Disney Company, and Warner Brothers. We will refer to this group as Management.

The Writers Guild has indicated their industrial action would be a *marathon*. AMPTP negotiator Nick Counter has indicated that negotiations would not resume as long as strike action continues, stating, "We're not going to negotiate with a gun to our heads—that's just stupid."

The last such strike in 1988 lasted 21 weeks and 6 days, costing the American entertainment industry an estimated $500 million ($870 million in 2007 dollars). According to a report on the January 13, 2008 edition of *NBC Nightly News*, if one takes into account everyone affected by the current strike, the strike has cost the industry $1 billion so far; this is a combination of lost wages to cast and crew members of television and film productions and payments for services provided by janitorial services, caterers, prop and costume rental companies, and so on.

The TV and movie companies stockpiled *output* so that they could possibly outlast the strike rather than working to meet the demands of the writer's and to avoid the strike.

Build a model that presents how each side should progress.

GAME THEORY APPROACH

Let us begin by stating strategies for each side. Our two rational players will be the Writers Guild and the Management. We develop strategies for each player.

Strategies:

- *Writers Guild*: Their strategies are to strike (S) or not to strike (NS).
- *Management*: Salary increase and revenue sharing (In) or SQ.

First, we rank order the outcomes for each side in order of preference. (These rank orderings are ordinal utilities.)

WRITER'S ALTERNATIVES AND RANKINGS

- *Strike*: Status Quo S SQ—writer's worst case (1)
- *No strike*: Status Quo NS SQ—writer's next to worst case (2)
- *Strike*: Salary increase and revenue sharing S IN—writer's next to best case (3)
- *No strike*: Salary increase and revenue sharing NS IN—writer's best case (4)

MANAGEMENT'S ALTERNATIVES AND RANKINGS

- *Strike*: Status Quo—managements next to best case (3)
- *No strike*: Status Quo—management best case (4)
- *Strike*: Salary increase and revenue sharing—management next to worst case (2)
- *No strike*: Salary increase and revenue sharing—management worst case (1)

This provides us with a payoff matrix consisting of ordinal values, see Figure 5.5. We will refer to the Writers as Rose and the Management as Colin.

Payoff Matrix for Writers and Management

We use the movement diagram, see Figure 5.6, to find (2,4) as the likely outcome.

We notice that the movement arrows point toward (2,4) as the pure Nash equilibrium. We also note that this result is not satisfying to the Writers Guild and that they would like to have a better outcome. Both

		Management (Colin)	
		SQ	IN
Writer's (Rose)	S	(1,3)	(3,2)
	NS	(2,4)	(4,1)

FIGURE 5.5
Payoff matrix for Writers Guild strike.

		Management	
		SQ	IN
Writer's	S	(1,3)	(3,2)
	NS	(2,4)	(4,1)

FIGURE 5.6
Diagram for Writers Guild strike.

(3,2) and (4,1) within the payoff matrix provide better outcomes to the Writers.

We can employ several options to try to secure a better outcome for the Writer's. We can first try strategic moves and if that fails to produce a better outcome, then we can move on to Nash arbitration. Both of these methods employ communications in the game. In strategic moves, we examine the game to see if *moving first* changes the outcome, if threatening our opponent changes the outcome, or if making promises to our opponent changes our outcome, or a combination of threats and promises changes the outcome.

We examine strategic moves. The writers move first and their best result is again (2,4). If management moves first the best result is (2,4). First moves keep us at the Nash equilibrium. The writers consider a threat and tell management that if they choose SQ, they will strike putting us at (1,3). This result is indeed a threat as it is worse for both the writers and management. However the options for management under IN are both worse than (1,3), so they do not accept the threat. The writers do not have a promise. At this point we might involve an arbiter using the method as suggested earlier.

Writers and management security levels are found from prudential strategies and our Equations 5.7 and 5.8. The security levels are (2, 3). We show this in Figure 5.7.

The Nash arbitration formulation is
  Maximize $(x-2)(y-3)$
  Subject to

$$3/2\,x + y = 7$$
$$x > 2$$
$$y > 3$$

The Nash equilibrium value, (2,4), lies along the Pareto optimal line segment. But the writers can do better by going on strike and forcing arbitration, which is what they did.

In this example, we consider *binding arbitration* where the players have a third party to work out the outcomes that best meets their desires and is acceptable to all players. We apply the Nash arbitration method to the Writers Guild problem and we found the following results.

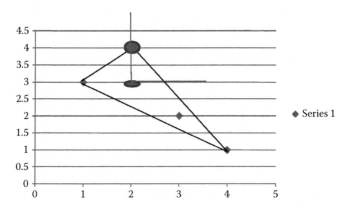

FIGURE 5.7
Payoff polygon for Writers Guild strike.

The SQ point is the security level of each side. We find these values using prudential strategies as (2,3). The function for the Nash Arbitration scheme is *Maximize* $(x–2)(y–3)$.

Using technology, we find the desired solution to our nonlinear programming (NLP) as

$$x = 2.3333$$
$$y = 3.5$$

We have the x and y coordinates (2.3333, 3.5) as our arbitrated solution. We can also determine how the arbiters should proceed. We solve the following simultaneous equations:

$$2p_1 + 4p_2 = 2.3333$$
$$4p_1 + p_2 = 3.5$$

We find that the probabilities to be played are 5/6 and 1/6. Further we see that Player 1, the writers, always plays R_2 so management arbiter plays $5/6\ C_1$ and $1/6\ C_2$ during the arbitration.

**Example 5.11: Game Theory Applied to a
Dark Money Network (Modified from Couch et al. 2016)**

Discussing the strategies for defeating the dark money network (DMN) leads well into the game theory analysis of the strategies for the DMN and the State trying to defeat them. When conducting game theory analysis, we originally limited the analysis by using ordinal scaling and ranking each of the four strategic options one through four. The game was set up in the following. Strategy R1 is for the state to pursue a nonkinetic strategy and R2 is a kinetic strategy. Strategy C1 is for the DMN to maintain its organization and C2 is for it to decentralize.

		Dark money network	
As		C1	C2
State	R1	2, 1	4, 2
	R2	1, 4	3, 3

This ordinal scaling worked when allowing communications and strategic moves, however, without a way of determining interval scaling it was impossible to conduct analysis of prudential strategies, Nash arbitration, or Nash's equalizing strategies. Here we show an application of analytic hierarchy process (AHP) (Chapter 4) to determine the interval scaled payoffs of each strategy for both the DMN and the State. Again, we will the use Saaty's standard nine-point preference in the pairwise comparison of combined strategies. For the State, the evaluation criteria we chose for the four possible outcomes were how well it degraded the DMN, how well it maintained the states' own ability to fund raise, how well the strategy would rally their base, and finally how well it removed nodes from the DMN. The evaluation criteria we chose for the DMN's four possible outcomes were how anonymity was maintained, how much money the outcome would raise, and finally how well a DMN could maintain control of the network.

After conducting AHP analysis, we obtained a new payoff matrix with cardinal utility values.

		Dark money network	
		C1	C2
State	R1	0.168, 0.089, 1	0.366, 0.099
	R2	0.140, 0.462	0.324, 0.348

With these cardinal scaling it is now possible to conduct a proper analysis that might include mixed strategies or arbitrated results such as to find prudential strategies, Nash's equalizing strategies, and Nash arbitration. Using a series of game theory solvers developed by Feix (2007), we obtain the following results:

- *Nash equilibrium*: A pure-strategy Nash equilibrium was found at R1C2 (0.366, 0.099) using strategies of nonkinetic and decentralization.
- Mixed Nash equalizing strategies is for the State to play nonkinetic 91.9% of the time and kinetic 8.1% of the time, and DMN to play maintain organization.
- Prudential strategies (security levels) (0.168, 0.099) when the State plays C1 and the DMN to play A.

Since there is no equalizing strategy for the DMN, the State should attempt to equalize the DMN. The result is as follows:

This is a significant departure from our original analysis before including the AHP pairwise comparisons in our analysis. The recommendations for the state were to use a kinetic strategy 50% of the time and a nonkinetic strategy 50% of the time. However, it is obvious that with proper scaling the recommendation should have been to execute a nonkinetic strategy for the vast majority (92%) of the time, and only occasionally (8%) to conduct kinetic targeting of network nodes. This greatly reinforces the recommendation to execute a nonkinetic strategy to defeat the DMN.

Finally, if the State and the DMN could enter into arbitration, the result would be at R2C2, which was the same prediction as before the proper scaling.

Example 5.12: Course of Actions Revisited

Let us return to Example 5.4 of Section 5.3, which was a zero-sum game approach. In many real-world analysis, if a player wins it does not necessarily mean that another player loses. Both might win or both might lose based upon the mission and courses of action played. Again let us assume the Player 1 has four strategies and Player 2 has six strategies that they can play. We use an AHP approach (see Fox, 2015) to obtain cardinal values of the payoff to each player. Although this example is based on combat courses of action, this methodology can be sued when both competing players have courses of action that they could employ. As in Example 5.11, we use AHP to convert the ordinal rankings of the combined course of actions (COAs) in order to obtain the cardinal values used in our example.

> $with(LinearAlgebra) : with(Optimazation) :$

> $M := Matrix([[6, 5.75, 5.5, 0.75, 3.75, 0.5], [4, 4.25, 2.75, 1.75, 2, 0.25], [3.25, 3, 1.5, 1.25, 1., 3.5], [5.25, 5, 4.75, 2.5, 2.25, 4.5]]);$

$$M := \begin{bmatrix} 6 & 5.75 & 5.5 & 0.75 & 3.75 & 0.5 \\ 4 & 4.25 & 2.75 & 1.75 & 2 & 0.25 \\ 3.25 & 3 & 1.5 & 1.25 & 1. & 3.5 \\ 5.25 & 5 & 4.75 & 2.5 & 2.25 & 4.5 \end{bmatrix}$$

> $N := Matrix\left(\left[\left[\frac{1}{6}, \frac{1}{3}, 0.5, .6336, 3.667, 3.8333\right], [1.5, 1.3333, 2.333, .8554, 2.833, 4], [2.1667, 2, 3.1667, .3574, 3.5, 1.83333], \left[\frac{2}{3}, 0.88333, 1, 5.95, 2.667, 1.1667\right]\right]\right)$

$$N := \begin{bmatrix} \dfrac{1}{6} & \dfrac{1}{3} & 0.5 & 0.6336 & 3.667 & 3.8333 \\[2ex] 1.5 & 1.3333 & 2.333 & 0.8554 & 2.833 & 4 \\[2ex] 2.1667 & 2 & 3.1667 & 0.3574 & 3.5 & 1.83333 \\[2ex] \dfrac{2}{3} & 0.8333 & 1 & 5.95 & 2.667 & 1.1667 \end{bmatrix}$$

> $X := \ `<,>`\ (x[1], x[2], x[3], x[4]);$

$$X := \begin{bmatrix} x_1 \\ x_2 \\ x_3 \\ x_4 \end{bmatrix}$$

> $Y := \ `<,>`\ (y[1], y[2], y[3], y[4], y[5], y[6]);$

$$Y := \begin{bmatrix} y_1 \\ y_2 \\ y_3 \\ y_4 \\ y_5 \\ y_6 \end{bmatrix}$$

> $Cnst := \{seq((M.Y)\ [i] \le p, i = 1\ ..\ 4), seq((Transpose(X).N)\ [i] \le q,$
> $i = 1\ ..\ 4), add(x[i], i = 1\ ..\ 4) = 1, add(y[i], i = 1\ ..\ 6) = 1\};$

$$Cnst := \left\{ x_1 + x_2 + x_3 + x_4 = 1, y_1 + y_2 + y_3 + y_4 + y_6 = 1, \frac{1}{3} x_1 \right.$$
$$+ 1.3333\, x_2 + 2\, x_3 + 0.8333\, x_4 \le q, \frac{1}{6} x_1 + 1.5\, x_2 + 2.1667\, x_3$$
$$+ \frac{2}{3} x_4 \le q, 0.5\, x_1 + 2.333\, x_2 + 3.1667\, x_3 + x_4 \le q, 0.6336\, x_1$$
$$+ 0.8554\, x_2 + 0.3574\, x_3 + 5.95\, x_4 \le q, 4\, y_1 + 4.25\, y_2 + 2.75\, y_3$$
$$+ 1.75\, y_4 + 2\, y_5 + 0.25\, y_6 \le p, 6\, y_1 + 5.75\, y_2 + 5.5\, y_3 + 0.75\, y_4$$
$$+ 3.75\, y_5 + 0.5\, y_6 \le p, 3.25\, y_1 + 3\, y_2 + 1.5\, y_3 + 1.25\, y_4 + 1\, y_5$$
$$+ 3.5\, y_6 \le p, 5.25\, y_1 + 5\, y_2 + 4.75\, y_3 + 2.5\, y_4 + 2.25\, y_5 + 4.5\, y_6$$
$$\left. \le p \right\}$$

> *with(LinearAlgebra) : with(Optimization) :*

> *objective := expand(Transpose(X).M.Y + Transpose(X).N.Y − p*
  − q);

$$objective := -q - p + \frac{37}{6} y_1 x_1 + 5.5 y_1 x_2 + 5.4167 y_1 x_3$$
$$+ 5.916666667 y_1 x_4 + 6.083333333 y_2 x_1 + 5.5833 y_2 x_2 + 5 y_2 x_3$$
$$+ 5.8333 y_2 x_4 + 6.03 y_3 x_1 + 5.083 y_3 x_2 + 4.6667 y_3 x_3$$
$$+ 5.75 y_3 x_4 + 1.3836 y_4 x_1 + 2.6054 y_4 x_2 + 1.6074 y_4 x_3$$
$$+ 8.45 y_4 x_4 + 7.417 y_5 x_1 + 4.833 y_5 x_2 + 4.5 y_5 x_3 + 4.917 y_5 x_4$$
$$+ 4.3333 y_6 x_1 + 4.25 y_6 x_2 + 5.33333 y_6 x_3 + 5.6667 y_6 x_4$$

> *QPSolve(objective, Cnst, assume = nonnegative, maximize);*

$$[3.07875455071633, [p = 2.86363636268048,$$
$$q = 0.633599997658894, x_1 = 1., x_2 = 0., x_3 = 0., x_4 = 0., y_1 = 0.,$$
$$y_2 = 0., y_3 = 0., y_4 = 0., y_5 = 0.727272727631539,$$
$$y_6 = 0.272727272566538]]$$

The solution depending on the starting conditions for P and Q are found using Maple. We just found

> *QPSolve(objective, Cnst, assume = nonnegative, maximize);*

$$\Big[3.07875455071633, \Big[p = 2.86363636268048,$$
$$q = 0.633599997658894, x_1 = 1., x_2 = 0., x_3 = 0., x_4 = 0., y_1 = 0.,$$
$$y_2 = 0., y_3 = 0., y_4 = 0., y_5 = 0.727272727631539,$$
$$y_6 = 0.272727272566538 \Big] \Big]$$

If we vary P and Q, we find additional solutions:

> *NLPSolve(objective, Cnst, assume = nonnegative, maximize, initialpoint*
  = {p = 3, q = 6});

$$\Big[3.6891878441025483410^{-15}, \Big[p = 2.50000000000000,$$
$$q = 5.95000000000000, x_1 = 0., x_2 = 0., x_3 = 0., x_4 = 0., y_1 = 0.,$$
$$y_2 = 0., y_3 = 0., y_4 = 1.00000000000000.,$$
$$y_5 = 2.77555756156289 \ 10^{-17}., y_6 = 0. \Big] \Big]$$

> QPSolve(objective, Cnst, assume = nonnegative, maximize, initialpoint
> = {p = 2, q = 1});

$$\left[3.07875454546495, \left[p = 2.86363636343297,\right.\right.$$
$$q = 0.633600000000000, x_1 = 1., x_2 = 0., x_3 = 0., x_4 = 0., y_1 = 0.,$$
$$y_2 = 0., y_3 = 0., y_4 = 0., y_5 = 0.727272727210145,$$
$$\left.\left.y_6 = 0.272727272789854\right]\right]$$

We found that Player 1 should play a pure strategy of COA 4, whereas Player 2 should play either a mixed strategy of $y_5 = 0.7272$ and $y_6 = 0.2727$ or a pure strategy of $y_4 = 1$. We also employed sensitivity analysis by varying the criteria weights that represent the cardinal values. We found that not much of a change in the results from this solution analysis is presented here.

Exercises 5.4

1. Find all the solutions:

	State sponsor	
	Sponsor Terrorism	Stop sponsoring Terrorism
The United States Strike militarily	(2, 4)	(1.5, 0)
Do not strike Militarily	(1, 1.5)	(0, 4)

2. Find all the solutions:

	State sponsor	
	Sponsor Terrorism	Stop sponsoring Terrorism
The United States Strike militarily	(3,5)	(2,1)
Do not strike	(4,2.5) Militarily	(1,2)

3. Find all the solutions:

		Colin	
		Arm	Disarm
Rose	Arm	(2,2)	(1,1)
	Disarm	(4,1)	(3,3)

4. Consider the following classical game (chicken). Find the solution:

		Colin	
		C	D
Rose	A	(3,3)	(2,4)
	B	(4,2)	(1,1)

5. Chicken

		Colin	
		C1	C2
Rose	R1	3,3	2,4
	R2	4,2	1,1

6.

		Colin	
		C1	C2
Rose	R1	2,3	4,1
	R2	1,2	3,4

7.

		Colin	
		C1	C2
Rose	R1	4,3	3,4
	R2	2,1	1,2

8. Prisoner's Dilemma

		Colin	
		C1	C2
Rose	R1	3,3	−1,5
	R2	5,−1	0,0

9.

		Colin	
		C1	C2
Rose	R1	3,3	1,5
	R2	4,0	0,2

Projects 5.4

1. Corporation XYZ consists of Companies Rose and Colin. Company Rose can make Products R1 and R2. Company Colin can make Products C1 and C2. These products are not in strict competition with one another, but there is an interactive effect depending on which products are on the market at the same time as reflected in the following table. The table reflects profits in millions of dollars per year. For example, if products R2 and C1 are produced and marketed simultaneously, Rose's profits are 4 million and Colin's profits are 5 million annually. Rose can make any mix of R1 and R2, and Colin can make any mix of C1 and C2. Assume the following information is known to each company.

   Note: The chief executive officer (CEO) is *not satisfied* with just summing the total profits. He might want the *Nash Arbitration Point* to award each company proportionately based on their strategic positions, if other options fail to produce the results he or she desires. Further, he or she does not believe that a dollar to Rose has the same importance to the corporation as a dollar to Colin.

		Colin	
		C1	C2
Rose	R1	(3,7)	(8,5)
	R2	(4,5)	(5,6)

   a. Suppose the companies have perfect knowledge and implement market strategies independently without communicating with one another. What are the likely outcomes? Justify your choice.

   b. Suppose each company has the opportunity to exercise a strategic move. Try *first moves* for each player and determine if a first move improves the results of the game.

   c. In the event things turn *hostile* between Rose and Colin, find, state, and then interpret the following:

   i. Rose's security level and prudential strategy?

   ii. Colin's security level and prudential strategy?

   Now suppose that the CEO is disappointed with the lack of spontaneous cooperation between Rose and Colin and decides to intervene and dictate the *best* solution for the corporation and employs an arbiter to determine an *optimal production and marketing schedule* for the corporation.

   d. Explain the concept of *Pareto optimal* from the CEO's point of view. Is the *likely outcome* you found in question (1) at or above Pareto optimal? Briefly explain and provide a payoff polygon plot.

e. Find and state the Nash arbitration point using the security levels found in question (3).

f. Briefly discuss how you would implement the Nash point. In particular, what mix of the products R1 and R2 should Rose produce and market, and what mix of the products C1 and C2 should Colin do? Must their efforts be coordinated, or do they simply need to produce the *optimal mix*? Explain briefly.

g. How much annual profit will Rose and Colin each make when the CEO's dictated solution is implemented?

5.5 Summary and Conclusions

We have presented some basic material concerning applied game theory and its uses in business, government, and industry. We presented some solution methodologies to solve these simultaneous total and partial conflict games. For analysis of sequential games, games with cooperation, and *N*-person games, please see the additional readings.

References and Suggested Readings

Aiginger, K. 1999. *The Use of the Game Theoretic Models for Empirical Industrial Organizations.* The Netherlands: Klewer Academic Press, pp. 253–277.

Aumann, R. 1987. *Game theory: The New Palgrave Dictionary of Economics.* London: Palgrave Macmillan.

Barron, E. N. 2013. *Game Theory: An Introduction.* Hoboken, NJ: John Wiley & Sons.

Bazarra, M., H. Sherali, and C. Shetty. 2006. *Nonlinear Programming* (3rd ed.). New York, NY: John Wiley.

Brams, S. 1994. Theory of moves. *American Scientist*, 81, 562–570.

Brams, S. 2001. Game theory and the Cuban missile crisis, © 1997–2004, Millennium Mathematics Project, University of Cambridge.

Cantwell, G. 2003. *Can Two Person Zero Sum Game Theory Improve Military Decision-Making Course of Action Selection?* Monograph. School of Advanced Military Studies, United States Army Command and General Staff College, Fort Leavenworth, KS.

Chatterjee, K. and W. Samuelson. 2001. *Game Theory and Business Applications.* Boston, MA: Klewer Academic Press.

Chiappori, A., S. Levitt, and T. Groseclose. 2002. Testing mixed-strategy equilibria when players are heterogeneous: The case of penalty kicks in soccer. *American Economic Review*, 92(4), 1138–1151.

Couch, C., W. Fox, and S. Everton. 2016. Mathematical modeling of a dark money network. *Journal of Defense Modeling and Simulation*, 1–12, doi:10.1177/1548512915625337.

Danzig, G. 1951. Maximization of a linear function of variables subject to linear inequalities. In T. Koopman (Ed.). *Activity Analysis of Production and Allocation Conference Proceeding*. New York, NY: John Wiley, pp. 339–347.

Danzig, G. 2002. Linear programming. *Operations Research*, 50(1), 42–47.

Dixit, A. and B. Nalebuff. 1991. *Thinking Strategically: The Competitive Edge in Business, Politics, and Everyday Life*. New York, NY: W.W. Norton Publisher.

Dorfman, R. 1951. Application of the simplex method to a game theory problem. In T. Koopman (Ed.). *Activity Analysis of Production and Allocation Conference Proceeding*. New York, NY: John Wiley, pp. 348–358.

Feix, M. 2007. Game Theory: Toolkit and Workbook for Defense Analysis Students, M.S. Thesis, Naval Postgraduate School, Monterey, CA.

Fox, W. 2008. Mathematical modeling of conflict and decision making: The writers' guild strike 2007–2008. *Computers in Education Journal*, 18(3), 2–11.

Fox, W. 2010. Teaching the applications of optimization in game theory's zero-sum and non-zero sum games. *International Journal of Data Analysis Techniques and Strategies*, 2(3), 258–284.

Fox, W. 2012. *Mathematical Modeling with Maple*. Boston, MA: Cengage Publishers.

Fox, W. 2015. An alternative approach to the lottery method in utility theory for game theory. *American Journal of Operations Research*, 5, 199–208. doi:10.4236/ajor.2015.53016.

Fox, W. 2016. Applied game theory to improve strategic and tactical military decision theory. *Journal of Defense Management*, 6, 2. doi:10.4172/2167-0374.1000147.

Game theory. http://en.wikipedia.org/wiki/List_of_games_in_game_theory (accessed April 26, 2013).

Gillman, R. and D. Housman. 2009. *Models of Conflict and Cooperation*. Providence, RI: American Mathematical Society.

Giordano, F., W. Fox, and S. Horton. 2014. *A First Course in Mathematical Modeling* (5th ed.). Chapter 10. Boston, MA: Brooks-Cole.

Klarrich, E. 2009. The mathematics of strategy: Classics of the Scientific Literature. www.pnas.org/site/misc/classics5.shtml (accessed October 2009).

Kleinberg, J. and D. Easly. 2010. *Networks, Clouds, and Markets: Reasoning about a Highly Connected World*. Cambridge, UK: Cambridge Press.

Kuhn, H. and A. Tucker. 1951. Nonlinear programming. *Proceedings of the Second Berkley Symposium on Mathematical Statistics and Probability*, J. Newman (Ed.). Berkeley, CA: University of California Press.

Mansbridge, J. 2013. Game theory and government. www.hks.harvard.edu/news-events/publications/insights/democratic/jane-mansbridge (accessed March 7, 2013).

McCormick, G. and L. Fritz. 2009. The logic of warlord politics. *Third World Quarterly*, 30(1), 81–112.

Myerson, R. 1991. *Game Theory: Analysis of Conflict*. Cambridge, MA: Harvard University Press.

Nash, J. 1950. The bargaining problem. *Econometrica*, 18, 155–162.

Nash, J. 1951. Non-cooperative games. *Annals of Mathematics*, 54, 289–295.

Nash, J. 2009. Lecture at NPS. February 19, 2009.

Straffin, P. 1980. The prisoner's dilemma. *UMAP Journal*, 1, 101–113.

Straffin, P. 1989. Game theory and nuclear deterrence. *UMAP Journal*, 10, 87–92.

Straffin, P. 2004. Game Theory and Strategy. Washington, DC: Mathematical Association of America.

Von Neumann, J. 1928. Zur theorie der gesellschaftsspiele. *Mathematische Annalen*, 100, 295–320. doi:10.1007/BF01448847.

Von Neumann, J. and O. Morgenstern. 1944. *Theory of Games and Economic Behavior*. Princeton, NJ: Princeton University Press.

Wiens, E. 2003. Game theory: Introduction, battle of the sexes, prisoner's dilemma, free rider problem, game of chicken, online two person zero sum game. http://www.egwald.ca/operationsresearch/gameintroduction.php (accessed April 26, 2013).

Williams, J. 1986. *The Compleat Strategyst*. New York, NY: Dover Press.

Winston, W. 2003. *Introduction to Mathematical Programming*, (4th ed.). Pacific Grove, CA: Duxbury Press.

6

Regression and Advanced Regression Models

OBJECTIVES

1. Understand the concept of correlation and linearity.
2. Build and interpret nonlinear regression models.
3. Build and interpret logistics regression models.
4. Build and interpret Poisson regression models.

The government of the Philippines has collected data on violence acts, committed by terrorists and insurgents over the past decade. Over the same period they have collected data on the population such as education levels (literacy), employment, government satisfaction, ethnicity, and such. The government is looking to see what it can do as improvements might reduce the number of violent acts committed. What should the government do to improve the situation? We will revisit this scenario later in this chapter. In this chapter, we will briefly discuss some regression techniques as background information and point keys to finding adequate models. We do not try to cover all the regression topics that exist but we do illustrate some real examples and the techniques used to gain insights, predict, explain, and answer scenario-related questions. We confine ourselves to simple linear regression, multiple regression, nonlinear regression (exponential and sine), binary logistics regression, and Poisson regression.

6.1 Introduction to Regression

In a simple linear regression, we want to build a model to minimize the sum of squared error. Equation 6.1 illustrates this.

$$\text{Minimize SSE} = \sum_{i=1}^{n} (y_i - f(x_i))^2 \qquad (6.1)$$

Most technology will do this for the user. EXCEL, SPSS, SAS, MINITAB, R, and JMP are among the most commonly used.

In the ordinary least squares method, linear regression, we are more concerned with the mathematical modeling and the use of the model for *explaining* or *predicting* the phenomena being studied or analyzed. We provide only a few diagnostic measures at this point such as percent relative error and residual plots. Percent relative error (Giordano et al., 2014) is calculated by Equation 6.2:

$$\%\text{RE} = \frac{100(y_i - f(x_i))}{y_i} \qquad (6.2)$$

We usually provide a rule of thumb when looking at the magnitude of the percent relative errors and we want most of them to be less than 20% and those near where we need to predict less than 10%.

Second, we provide the visualization of the plot of residuals versus the model. We examine the plot for trends or patterns (Affi and Azen, 1979, pp. 143–144; Giordano et al., 2014, pp. 636–638), such as shown in Figure 6.1. If a pattern is seen, we deem the model that is not adequate even though we might have to use it. If there is no visual pattern, then we may conclude the model is adequate.

When should we use simple linear regression and when should we use something more advanced? Let us begin with a concept called *correlation*.

Decision-makers have several misconceptions about *correlation*, many of which are supported by a poor definition. A good definition states that *correlation is a measure of the* linear relationship *between variables*. The key term is *linear*. Some definitions for correlation actually state it is a measure of the relationship between two variables and they do not even mention linear. This is true in Excel. The following Excel definition from the help menu is found in the Excel help menu (www.office.com) that we believe helps fuel the misconception:

> Returns the correlation coefficient of the array1 and array2 cell ranges. Use the correlation coefficient to determine the relationship between two properties. For example, you can examine the relationship between a location's average temperature and the use of air conditioners.

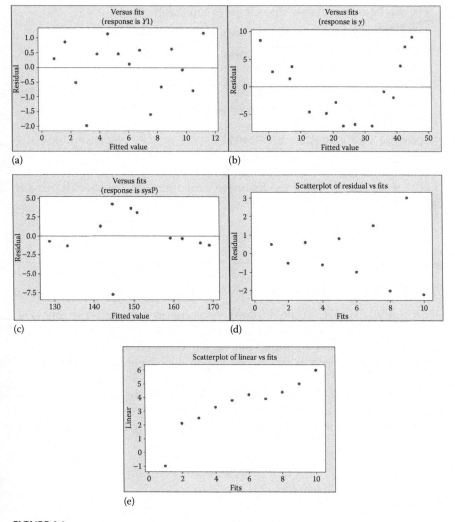

FIGURE 6.1
Patterns for residuals: (a) no pattern, (b) curved pattern, (c) outliers, (d) fanning pattern, and (e) linear trend.

It is no wonder decision-makers have misconceptions. Decision-makers, similar to the latter definition, often lose the term linear and state or think that correlation does measure the relationship between variables. As we will show that might be false thinking.

Some of the misconceptions that decision-makers have expressed are listed as follows:

- Correlation does everything.
- Correlation measures relationship.

- Correlation or relationships implies causation.
- If it looks linear (visual) and the correlation value is large, then the model found will be useful.
- If the computer lets me do it, it must be right.
- Model diagnostics are not useful if the correlation value is large.
- I have to use the *regression* package taught from class.

We provide two examples to illustrate some of the misconceptions surrounding the correlation and possible corrections that we use in our mathematical modeling courses. One diagnostic test that we now cover is the *common sense* test, does the model answer the question and does it provide realistic results?

6.2 Modeling, Correlation, and Regression

Let us initially return to our Philippines scenario introduced earlier. The researcher actually began with linear regression. For example, in 2008 an analysis of the literacy index per region versus the number of sigacts (Violence acts by terrorists or insurgents) produced the following plot and a linear model.

Within all the data, Figure 6.2 represents one of the *best* result overall model and we note that R^2 is only 0.14104, which represents a really poor fit. The regression formula is sigacts = −1.5525*literacy index + 147.01 with

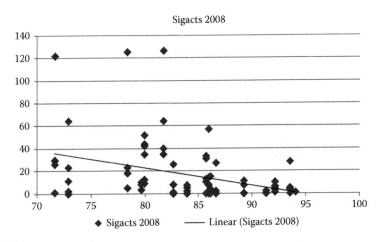

FIGURE 6.2
Plot of literacy versus sigacts for 2008.

a correlation of −0.3755. We will return to this scenario again later in this chapter but suffice it to say, linear regression is not useful to explain, predict, or answer the questions the government has imposed.

6.2.1 Linear, Multiple, and Nonlinear Regression

Example 6.1: Exponential Decay

We desire to build a mathematical model to predict the degree of recovery after discharge for orthopedic surgical patients. We have two variables: (1) time in days in the hospital, t and (2) a medical prognostic index for recovery, y, where a large value of this index indicates a good prognosis. Data taken from Neter et al. (1996).

t	2	5	7	10	14	19	26	31	34	38	45	52	53	60	65
y	54	50	45	37	35	25	20	16	18	13	8	11	8	4	6

We provide the scatterplot in Figure 6.3 showing the negative trend.

We obtain the correlation coefficient, ρ. Using Excel, =correl(array 1, array 2), we obtain the correlation coefficient of −0.94105. We present two rules of thumb for correlation from the literature. First, from Devore (2012), for mathematics, science, and engineering data we have the following:

$0.8 < \lvert \rho \rvert \leq 1.0$	Strong linear relationship
$0.5 < \lvert \rho \rvert \leq 0.8$	Moderate linear relationship
$\lvert \rho \rvert \leq 0.5$	Weak linear relationship

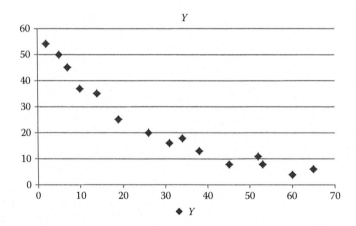

FIGURE 6.3
Scatterplot of the data with negative trend.

According to Johnson (2012) for nonmath, nonscience, and nonengineering data we find a more liberal interpretation of ρ:

$0.5 < \lvert\rho\rvert \le 1.0$	Strong linear relationship
$0.5 < \lvert\rho\rvert \le 0.3$	Moderate linear relationship
$0.1 < \lvert\rho\rvert \le 0.3$	Weak linear relationship
$\lvert\rho\rvert \le 0.1$	No linear relationship

Further, in our modeling efforts we emphasize the interpretation of $\lvert\rho\rvert \approx 0$. This can be interpreted as either no linear relationship or the existence of a nonlinear relationship. Most students fail to pick up on the importance of the nonlinear relationship aspect of the interpretation.

Using either rule of thumb the correlation coefficient, $\lvert\rho\rvert = 0.94105$, indicates a strong linear relationship. We obtain this value, look at Figure 6.3, and think they will have an excellent regression model.

Using Excel we obtain the following model, $y = 49.4601 - 0.75251 * t$. The sum of squared error is 451.1945. The correlation is, as we stated, -0.94105, and R^2, the coefficient of determination, is 0.88558. These are all indicators of a *good* model. Figure 6.4 gives this output.

Next, we examine both the percent relative error and the residual plot. The percent relative errors are shown as

	Residual output			
Observation	Predicted Y	Residuals	Percent relative error	
1	44.95539652	9.044603	16.7492657	
2	42.69787252	7.302127	14.60425495	
3	41.19285653	3.807143	8.460318833	
4	38.93533253	−1.93533	5.230628449	
5	35.92530053	−0.9253	2.643715793	
6	32.16276053	−7.16276	28.65104212	
7	26.89520453	−6.8952	34.47602266	
8	23.13266453	−7.13266	44.57915333	
9	20.87514053	−2.87514	15.97300297	
10	17.86510854	−4.86511	37.42391181	
11	12.59755254	−4.59755	57.46940673	
12	7.329996541	3.670003	33.36366781	
13	6.577488541	1.422511	17.78139324	
14	1.309932543	2.690067	67.25168641	
15	−2.452607455	8.452607	140.8767909	

Although some are small, others are quite large with 8 of the 15 over 20% in error. The last two are over 67% and over 140%. How much confidence would you have in predicting? The residual plot in Figure 6.5 clearly shows a curved pattern.

	v	w	x	Y	Z	AA	AB	AC	AD
SUMMARY OUTPUT									
Regression Statistics									
Multiple R	0.941052825								
R Square	0.88558042								
Adjusted R Square	0.876778914								
Standard Error	5.89128785								
Observations	15								
ANOVA									
	df	*SS*	*MS*	*F*	*Significance F*				
Regression	1	3492.13879	3492.139	100.6169	1.73616E-07				
Residual	13	451.1945429	34.70727						
Total	14	3943.333333							
	Coefficients	*Standard Error*	*t Stat*	*P-value*	*Lower 95%*	*Upper 95%*	*Lower 95.0%*	*Upper 95.0%*	
Intercept	46.46041252	2.762180628	16.82019	3.33E-10	40.49308407	52.42774	40.4930841	52.42774097	
T	-0.752508	0.075019751	-10.0308	1.74E-07	-0.914578318	-0.59044	-0.9145783	-0.590437682	

FIGURE 6.4
Excel's regression output.

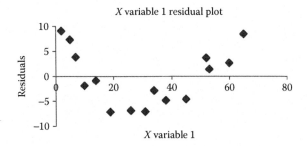

X variable 1 residual plot

FIGURE 6.5
Residual plot showing a curved pattern.

Furthermore, assume we need to predict the index when time was 100 days. Using our regression model, we would predict the index as −29.7906. A negative value is clearly unacceptable and makes no common sense since we expect the index, expressed by the dependent variable y, is always be positive. The model does not pass the common sense test. So, with a strong correlation of −0.94105 what went wrong? The residual plot diagnostic shows a curved pattern. In regression analysis (Affi and Azen, 1979, pp. 143–144), this suggests that a nonlinear term is missing from the model.

6.2.2 Multiple Linear Regression

If we try a parabolic model, $y = b_0 + b_1 x + b_2 x^2$ we get similar results to the linear model. We run the regression of y versus x and x^2 with an intercept. The model is $y = 55.82213 − 1.71026 * x + 0.014806 * x^2$. The correlation is now 0.990785 and the R^2 is 0.981654. The sum of squared error is now 72.3472. The output is seen in Figure 6.6. The residual plot appears to have removed the curved pattern as shown in Figure 6.7.

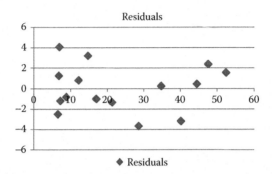

FIGURE 6.6
Excel's regression output for the parabolic model.

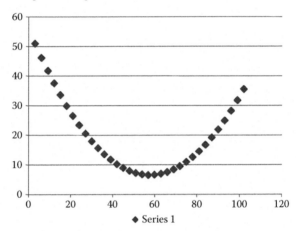

FIGURE 6.7
Residual plot of quadratic model.

If we use the model to predict at $x = 100$ days we find $y = 32.8321$. The answer is now positive but again does not pass the common sense test. The quadratic is now curving upward toward positive infinity as shown in Figure 6.8. This is an unexpected and an unacceptable outcome. We certainly cannot use the model to predict future outcomes.

6.2.3 Nonlinear Regression (Exponential Decay)

Let us try a nonlinear regression model (Fox, 2011; 2012b). We desire the model, $y = ae^{bx}$. Excel does not have a nonlinear regression package, so we use the solver to obtain a solution as shown by Fox (2011). We minimize the function, $\sum_{i=1}^{n}(y_i - a(\exp(bx_i)))^2$. We obtain the model, $y = 58.60663\ e^{-0.03959t}$. Care must be taken with the selection of the initial values for the unknown parameters (Fox, 2011; 2012b). We find that both Excel and Maple yield good models based on *good* input parameters. As a nonlinear model, correlation has no meaning nor does R^2. We computed the sum of squared error as that

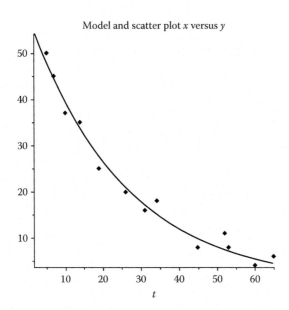

Model and scatter plot *x* versus *y*

FIGURE 6.8
Quadratic model, $y = 55.82213 - 1.71026*x + 0.014806*x^2$.

was our objective function. Our *SSE* was 45.4953. This value is substantially smaller than 451.1945 obtained in the linear model. The model and the residual plot showing no pattern are shown in Figures 6.9 and 6.10.

The percent relative errors are also much improved with only 4 being greater than 20% and none larger than 36.25122%.

Predicted	Residuals	Percent Rel. Error
54.14512	0.145119	0.268739
48.08152	−1.91848	3.836967
44.42124	−0.57876	1.286124
39.4466	2.446599	6.612429
33.66935	−1.33065	3.801867
27.6227	2.622698	10.49079
20.93679	0.936789	4.683943
17.17677	1.176769	7.354806
15.25318	−2.74682	15.26013
13.01923	0.019234	0.147957
9.868006	1.868006	23.35008
7.479514	−3.52049	32.00442
7.189185	−0.81081	10.13519
5.449086	1.449086	36.22714
4.470489	−1.52951	25.49185

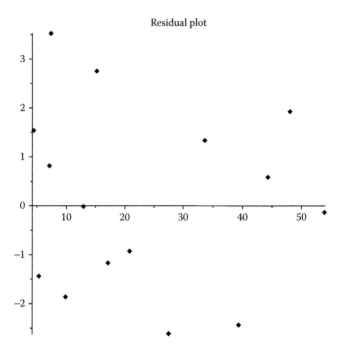

FIGURE 6.9
Plot of the model, $y = 56.60663*e^{(-0.039059*x)}$.

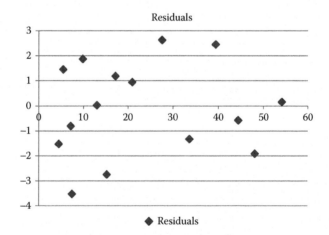

FIGURE 6.10
Residual plot from nonlinear model with no pattern.

If we use this model to predict at $x = 100$ we find by substitution that $y = 1.118$. This new result passes the common sense test. This model is our recommended model for these data.

Exercises 6.2

Compute the correlation and then fit the following data with the models using linear regression:

1.

x	1	2	3	4	5
y	1	1	2	2	4

a. $y = b + ax$
b. $y = ax^2$

2. Stretch of a spring data:

x ($\times 10^{-3}$)	5	10	20	30	40	50	60	70	80	90	100
y ($\times 10^5$)	0	19	57	94	134	173	216	256	297	343	390

a. $y = ax$
b. $y = b + ax$
c. $y = ax^2$

3. Data for the ponderosa pine:

x	17	19	20	22	23	25	28	31	32	33	36	37	39	42
y	19	25	32	51	57	71	113	140	153	187	192	205	250	260

a. $y = ax + b$
b. $y = ax^2$
c. $y = ax^3$
d. $y = ax^3 + bx^2 + c$

4. Given Kepler's data:

Body	Period (s)	Distance from Sun (m)
Mercury	7.60×10^6	5.79×10^{10}
Venus	1.94×10^7	1.08×10^{11}
Earth	3.16×10^7	1.5×10^{11}
Mars	5.94×10^7	2.28×10^{11}
Jupiter	3.74×10^8	7.79×10^{11}
Saturn	9.35×10^8	1.43×10^{12}
Uranus	2.64×10^9	2.87×10^{12}
Neptune	5.22×10^9	4.5×10^{12}

Fit the model $y = ax^{3/2}$.

Project 6.2

Write a program using nonstatistical technology to (a) compute the correlation between the data and (b) find the least squares for the general proportionality model: $y = ax^n$.

Write a program using nonstatistical technology to (a) compute the correlation between the data and (b) find the least squares for any of the general model: $y = a_0 + a_1x + a_2x^2 + \ldots + a_nx^n$.

6.3 Advanced Regression Techniques with Examples

Let us begin by looking at shipping data over time. Management is asking for a model that explains the behavior over time so that predictions might be made concerning future decisions.

6.3.1 Data

The following data represent logistical supply train information over 20 months.

Month	Usage (tons)	Month	Usage (tons)
1	20	11	19
2	15	12	25
3	10	13	32
4	18	14	26
5	28	15	21
6	18	16	29
7	13	17	35
8	21	18	28
9	28	19	22
10	22	20	32

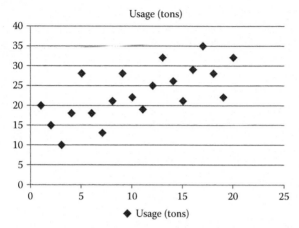

FIGURE 6.11
Scatterplot of usage in metric tons over 20 months.

First, we find the correlation coefficient. In this case, it is 0.712. According to our measures, this is a moderate to strong value. So, is the model to use linear? We plot the data to observe the trends and patterns as shown in Figure 6.11. Although we have seen linear regression used here, using linear regression does not allow the modeler to capture the seasonal trends. Thus, we see what appears to be an oscillating pattern with a slight linear trend. As this oscillating pattern is repeating five times within the data, it is more likely for the trend to be oscillating rather than linear.

We connect the data to help see the oscillating pattern as shown in Figure 6.12.

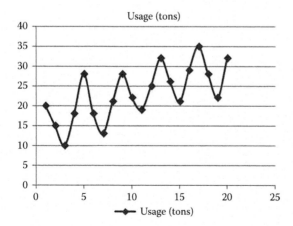

FIGURE 6.12
Scatterplot of usage in metric tons over 20 months with dots connected.

For illustrative purposes a linear model is found, Usage = 14.58 + 0.792*months. The sum of squared error is 406.16 with an $R^2 = 0.506944$. Now, let us assume the fit for the data needs to represent oscillations with a slight line trend. We move directly to the sine model with a linear trend. The model formulation is

$$\text{Minimize } \sum(y_i - (a * \sin(b * x_i + c) + d * x_i + e))^2$$

We fit the model,

$$\text{Usage} = a * \sin(b * \text{time} + c) + d * \text{time} + e$$

And find we have a good fit. Using the Excel Solver, see Fox (2011), we obtain the model,

$$\text{Usage} = 6.316 * \sin(1.574 * \text{months} + 0.0835) + 0.876 * \text{months} + 13.6929$$

The sum of squared error is only 11.58. The new SSE is quite a bit smaller than with the linear model. Our nonlinear (oscillating) model is overlaid with the data in Figure 6.13, visually representing a good model.

Clearly, the sine regression does a much better job in predicting the trends than just using a simple linear regression. The largest percent relative error is slightly larger than 10% (10.224%) and the average percent relative error is about 3.03%.

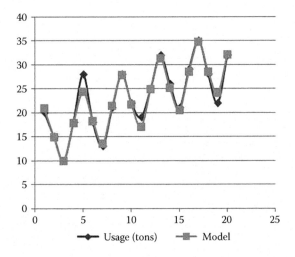

FIGURE 6.13
Overlay of model and data.

Example 6.2: Modeling Casualties in Afghanistan

In a recent report issued in GEN(Ret) Barry McCaffrey stated that the situation in Afghanistan is going to get much worse. As a matter of fact, he said casualties would double over the next year,

> **Newser**—America should prepare for thousands of casualties when the beefed-up US force in Afghanistan begins its spring offensive against the Taliban, warns a retired general and military expert. Gen. Barry McCaffrey, an analyst who has visited the war zone several times, predicts US casualties will rise to as many as 500 killed and wounded per month, matching or exceeding the deadliest months of last year's fighting.
>
> "What I want to do is signal that this thing is going to be up to $10 billion a month and 300 to 500 killed and wounded a month by next summer," McCaffrey tells the *Army Times*. "That's what we probably should expect. And that's light casualties." McCaffrey predicts that building a viable Afghan state able to take care of its own security will take as long as 10 years.
>
> Read more: http://www.newser.com/story/77563/general-brace-for-thousands-of-gi-casualties.html#ixzz0v5jSGM3W.

The data that Gen McCaffrey was analyzing were the 2001–2009 data. Table 6.1 provides the initial data table. In this analysis, we used 36 months of data from 2006 to 2008.

We need to analyze the data, produce a model and use the model to prove or disprove Gen (Ret) McCaffrey's prediction concerning the casualties in Afghanistan doubling over the next year. The data for 36 months are provided in Figure 6.14. We note that it is oscillating and linearly increasing over time.

Using the techniques described in the previous nonlinear supply example, we build a nonlinear sine with linear trend model. We estimated the

TABLE 6.1

Casualty Data from Afghanistan

Month	2001	2002	2003	2004	2005	2006	2007	2008	2009	2010	2011
1		12	10	25	6	7	21	19	83	199	308
2		13	9	17	5	17	39	18	52	247	245
3		53	14	12	16	7	26	53	78	346	345
4		8	13	11	29	13	61	37	60	307	411
5		2	8	31	34	39	87	117	156	443	
6		3	4	34	60	68	100	167	213	583	
7		6	10	25	38	59	100	151	394	667	
8		2	13	22	72	56	103	167	493	631	
9		5	19	34	47	70	88	122	390	674	
10	5	8	5	38	27	68	131	90	348	631	
11	10	4	27	18	12	51	75	37	214	605	
12	28	6	12	9	20	23	46	50	168	359	

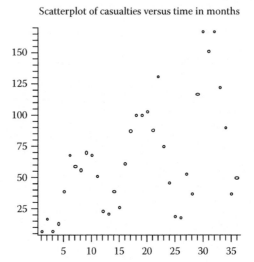

FIGURE 6.14
Casualties per month versus months.

five parameters as [75,0.5735−0.5,5.74,80]. Using a nonlinear application model such as the Leven–Marquard method (Fox, 2012b), we obtained the following model:

$$\text{Casualties} = -44.640 \sin(0.54705x + 0.27954) + 1.8494x + 33.535$$

The fit shown in Figure 6.15 appears to be a good fit. The residual plot, Figure 6.16, also shows no clear pattern suggesting the model appears adequate. The *SSE* is 14415.2125 with an R^2 of 0.93654. The sum of casualties in year three is 1028 casualties. Our predictions for year four sum to only 1321 casualties. This is not a doubling effect. Thus, the model does not support the doubling hypothesis of General McCaffrey.

LOGISTICS AND POISSON REGRESSION

Often our dependent variable has special characteristics. Here is examined two such special cases: (1) the dependent variables are binary {0,1} and (2) the dependent variables are counts that follow a Poisson distribution.

Example 6.3: Logistics Regression Example

Dehumanization and the Outcome of Conflict with Logistics Regression
  Dehumanization is not a new phenomenon in interhuman conflict. Man has arguably *dehumanized* his human adversaries to allow man to coerce, maim, or ultimately kill while avoiding the pain of conscience for committing the extreme, violent action. By taking away the human traits of his opponents, man has made his adversaries to be objects deserving wrath and self-actualizing his justice of the action. Dehumanization still occurs today in both developed and underdeveloped societies within

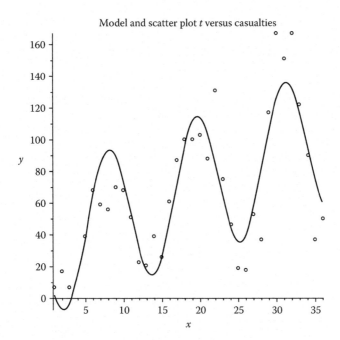

FIGURE 6.15
Casualty data with nonlinear regression model.

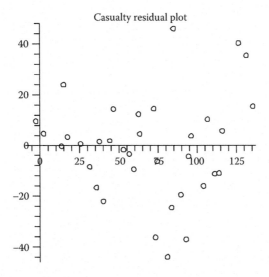

FIGURE 6.16
Residual plot for casualty model.

the interstate system. This case analyzes the impact that dehumanization has, in its various manifested forms, on the outcome of a state's ability to win a conflict.

DATA SPECIFICS

To examine at dehumanization as a quantitative statistic, this case amalgamated data from a series of 25 conflicts and a previous study of civilian casualties from the respective conflicts. The conflict casualty data set derived from Erik Melander, Magnus Oberg, and Jonathan Hall's Uppsala Peace and the conflict research paper, "The 'New Wars' Debate Revisited: An Empirical Evaluation of the Atrociousness of 'New Wars'," is shown in Figure 6.17.

Country	Year	Civilian	Military	Total
India	1946–1948	800,000	0	800,000
Columbia	1949–1962	200,000	100,000	300,000
China	1950–1951	1,000,000	*	1,000,000
Korea	1950–1953	1,000,000	1,899,000	2,899,000
Algeria	1954–1962	82,000	18,000	100,000
Tibet	1956–1959	60,000	40,000	100,000
Rwanda	1956–1965	102,000	3,000	105,000
Iraq	1961–1970	100,000	5,000	105,000
Sudan	1963–1972	250,000	250,000	500,000
Indonesia	1965–1966	500,000	*	500,000
Vietnam	1965–1975	1,000,000	1,058,000	2,058,000
Guatemala	1966–1987	100,000	38,000	138,000
Nigeria	1967–1970	1,000,000	1,000,000	2,000,000
Egypt	1967–1970	50,000	25,000	75,000
Bangladesh	1971–1971	1,000,000	500,000	1,500,000
Uganda	1971–1978	300,000	0	300,000
Burundi	1972–1972	80,000	20,000	100,000
Ethiopia	1974–1987	500,000	46,000	546,000
Lebanon	1975–1976	76,000	25,000	100,000
Cambodia	1975–1978	1,500,000	500,000	2,000,000
Angola	1975–1987	200,000	13,000	213,000
Afghanistan	1978–1987	50,000	50,000	100,000
El Salvador	1979–1987	50,000	15,000	65,000
Uganda	1981–1987	100,000	2,000	102,000
Mozambique	1981–1987	350,000	51,000	401,000

* denotes missing values.

FIGURE 6.17
Civilian and military casualties resultant from high- and low-intensity conflicts. (Adapted from Sivard, R. L., *World Military and Social Expenditures 1987–1988*, World Priorities, Washington, DC, 1989; Melander, E. et al., The "new wars" debate revisited: An empirical evaluation of the atrociousness of "new wars," Uppsala Peace Research Papers No. 9, Department of Peace and Conflict, 2006.)

As stated earlier, the above-mentioned conflicts represent the high- and low-intensity spectrum of conflict and include both inter- and intrastate conflicts. Thus, the data are a fair representation of conflict in general. However, the above-mentioned data table was used in support of a study that focused on the casualty output of conflict and not on the interrelation of civilian casualties that we define as an indicator of dehumanization to the outcome of the conflict for the state. Typically, there is no unambiguous victor or vanquished in conflict, but to allow us to analyze the relationship of civilian casualty ratios and the outcome of the conflict it is necessary to utilize a definitive binary assessment of each of the above-mentioned conflicts' winners and losers. To this end, we utilized an additional data set that codified conflicts in terms of two sides with the determination of which side *won* each respective conflict. The implications of this case study vary broadly, but we were singularly focused on civilian deaths in a conflict as an indicator of dehumanization's occurrence and subsequently dehumanization's effect on the state's ability to win the conflict.

By taking a ratio of the civilian casualties in relationship to the total casualties we were able to determine what percentages of casualties in each conflict were civilian, as shown in Figure 6.18. This provided us a quantifiable independent variable to analyze. In addition, we made the inference that the conflicts with higher civilian casualty percentages likely incurred a higher amount of *value targeting*, a previously discussed symptom of dehumanization. By using the civilian casualty percentage independent variable and comparing it to the assessed binary outcome of either a win or a loss as the dependent variable, we were able to synthesize the data into a binary logistical regression model to assess the significance of the civilian casualty percentages on the outcome of the state's (*Side A*) ability to win the conflict. For more information, see Kreutz (2010).

A BINARY LOGISTICAL REGRESSION ANALYSIS OF DEHUMANIZATION

Binary logistical regression analysis is an ideal method to analyze the interrelation of dehumanization's effects (shown through higher percentages of civilian casualties) on the outcome of conflict (shown to be a win "1" or a loss "0"). Binary logistical regression model statistics will allow us to explain whether or not the civilian casualties' percentage independent variable has a significant level on the outcome. Using the data shown in Figure 6.18, we assessed the civilian casualty percentages to be the independent variable X and *Side A*'s win/loss outcome from the conflict to be the dependent variable Y. From these data we were able to develop a binary logistical regression model. Using statistical analysis software package, we derived the logistics regression statistics from the model as shown in Figure 6.19 from Minitab©.

From the statistics, we can use the listed hypothesis testing of *Ho* for both the model's chi-square goodness of fit and the independent variable coefficient to measure each value's significance. In the case of the model itself, we notice a chi-square *P-value* of 0.468. Being that our

Conflict's Country Location	Side A	Side A Win (1) or Loss (0)	Side B	Side A Win (1) or Loss (0)	Year	Civilian Casualties	Military Casualties	Total Casualties	Civilian Deaths Percentage
India	India	1	CPI	0	1946-1948	800,000	0	800,000	1.0000
Columbia	Columbia	1	Military Junta	0	1949-1962	200,000	100,000	300,000	0.6667
China	China	1	Taiwan	0	1950-1951	1,000,000	*	1,000,000	1.0000
Korea	North Korea	0	South Korea	1	1950-1953	1,000,000	1,899,000	2,899,000	0.3460
Algeria	France	0	FLN	1	1954-1962	82,000	18,000	100,000	0.8200
Tibet	China	1	Tibet	0	1956-1959	60,000	40,000	100,000	0.6000
Rwanda	Tutsi	0	Hutu	1	1956-1965	102,000	3,000	105,000	0.9714
Iraq	Iraq	1	KDP	0	1961-1970	100,000	5,000	105,000	0.9524
Sudan	Sudan	1	Anya Nya	0	1963-1972	250,000	250,000	500,000	0.5000
Indonesia	Indonesia	1	OPM	0	1965-1966	500,000	*	500,000	1.0000
Vietnam	North Vietnam	1	South Vietnam	0	1965-1975	1,000,000	1,058,000	2,058,000	0.4859
Guatemala	Guatemala	1	FAR	0	1966-1987	100,000	38,000	138,000	0.7246
Nigeria	Nigeria	1	Republic of Biafra	0	1967-1970	1,000,000	1,000,000	2,000,000	0.5000
Egypt	Egypt	0	Israel	1	1967-1970	50,000	25,000	75,000	0.6667
Bangladesh	Bangladesh	1	JSS/SB	0	1971-1971	1,000,000	500,000	1,500,000	0.6667
Uganda	Uganda	1	Military Faction	0	1971-1978	300,000	0	300,000	1.0000
Burundi	Burundi	1	Military Faction	0	1972-1972	80,000	20,000	100,000	0.8000
Ethiopia	Ethiopia	1	OLF	0	1974-1987	500,000	46,000	546,000	0.9158
Lebanon	Lebanon	1	LNM	0	1975-1976	76,000	25,000	100,000	0.7600
Cambodia	Cambodia	0	Khmer Rouge	1	1975-1978	1,500,000	500,000	2,000,000	0.7500
Angola	Angola	1	FNLA	0	1975-1987	200,000	13,000	213,000	0.9390
Afghanistan	Afghanistan	1	USSR	0	1978-1987	50,000	50,000	100,000	0.5000
El Salvador	El Salvador	1	FMLN	0	1979-1987	50,000	15,000	65,000	0.7692
Mozambique	Mozambique	1	Renamo	0	1981-1987	350,000	51,000	401,000	0.8728
Uganda	Uganda	1	Kikosi Maalum et al.	0	1981-1987	100,000	2,000	102,000	0.9804

FIGURE 6.18
Conflict outcomes and civilian casualty percentages data set. (Kreutz, J., *J. Peace Res.*, 47, 243–250, 2010.)

Binary Logistic Regression: Side A versus Civ_DeathP

```
Link Function: Logit


Response Information

Variable  Value  Count
Side A    1        20   (Event)
          0         5
          Total    25


Logistic Regression Table

                                            Odds      95% CI
Predictor       Coef   SE Coef     Z      P  Ratio  Lower   Upper
Constant    0.0044187  1.92497  0.00  0.998
Civ_DeathP  1.85015    2.55605  0.72  0.469   6.36   0.04  953.29


Log-Likelihood = -12.247
Test that all slopes are zero: G = 0.526, DF = 1, P-Value = 0.468


Goodness-of-Fit Tests

Method            Chi-Square  DF      P
Pearson             22.1302   17  0.180
Deviance            21.7211   17  0.196
Hosmer-Lemeshow      5.8228    7  0.561


Table of Observed and Expected Frequencies:
(See Hosmer-Lemeshow Test for the Pearson Chi-Square Statistic)

                            Group
Value   1    2    3    4    5    6    7    8    9   Total
1
  Obs   1    3    2    2    1    2    2    2    5     20
  Exp 1.4  2.2  2.3  1.6  1.6  2.4  1.7  2.6  4.3
0
  Obs   1    0    1    0    1    1    0    1    0      5
  Exp 0.6  0.8  0.7  0.4  0.4  0.6  0.3  0.4  0.7
Total   2    3    3    2    2    3    2    3    5     25


Measures of Association:
(Between the Response Variable and Predicted Probabilities)

Pairs        Number  Percent  Summary Measures
Concordant      59     59.0   Somers' D               0.20
Discordant      39     39.0   Goodman-Kruskal Gamma   0.20
Ties             2      2.0   Kendall's Tau-a         0.07
Total          100    100.0
```

FIGURE 6.19
Civilian casualty–conflict outcome binary logistical regression model descriptive statistics.

ascribed alpha value for this test was 0.05, the model's *P-value* is significantly greater than the 0.05 value. This leads us to fail to reject the null hypothesis, Ho, and therefore accept the logistics model as adequate. We can therefore conclude that the model's chi-square value is significant. However, when we examine the independent variable's coefficient of 1.84505 and its corresponding *P-value* of 0.469, we come to a different conclusion. Using the same alpha value of 0.05 we see that the independent variable's *P-value* is significantly greater than the alpha value in leading us to fail to reject the null hypothesis. This leads us to conclude that although the model itself is adequate, the independent variable is not significant in determining the outcome of the dependent variable.

Interpretation: As a conclusion from our case study analysis, we find that the civilian casualty percentages actually are not significant in determining the outcome of a conflict for Side A as it is defined as either a win or a loss. Therefore, from this limited study, we can loosely conclude that dehumanization does not have a significant effect on the outcome of a state's ability to win or lose a conflict.

Example 6.4: Poisson Regression for SIGACTS (Fox and Durante, 2015)

We return to our Philippines example briefly described earlier in this chapter. Since the number of violence acts, SIGACTS, are counts. We examine the histogram in Figure 6.20 noticing that it appears to follow a Poisson distribution. A goodness of fit test for the Poisson distribution confirms that it follows a Poisson distribution ($\chi^2 = 933.11, p = 0.000$).

We ran a Poisson regression yielding the following model:

$$y = e^{(7.828 - 0.034 \times 1 - 0.01799 \times 2 + 0.00400 \times 3)}$$

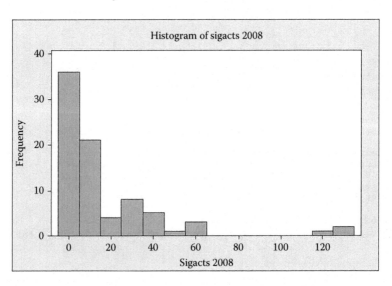

FIGURE 6.20
Histogram of SIGACTS 2008.

where:
y is the counts of violent activities
x_1 is the government satisfaction level
x_2 is the literacy level
x_3 is the poverty level

The regression output is displayed.

	Table 1				
	Coefficients	Estimates	se	t*	P-Value
	bo	7.8281417	0.535	14.63204056	0.002319158
	b1	−0.0341572	0.002	−17.07862043	0.001705446
	b2	−0.0179907	0.006	−2.998447403	0.047775567
	b3	0.00400015	0.003	1.333383431	0.156996007
	Table 2				
		df	Deviance	Mean Dev	Ratio
	Regression	3	2214.064499	738.0214997	738.02
	Residual	77	169.8891451	2.206352534	
	Total	80	2383.953644	29.79942055	

We interpret the odds ratio for the coefficients to help explain our results.

- Exp(−0.0342) = 0.96647. This means for one more value of government satisfaction the violence goes down slightly about 3.4%.
- Exp(−0.01799) = 0.9812. This means for every 1 unit increase in literacy the violence goes down slightly, 1.2%.
- Exp(0.004) = 1.004. This means that as poverty increases, violence also increases slightly, 0.400%.

Interpretation: Based on the findings of this research poverty, good governance and literacy are the ones that are most strongly related to violent conflict in the Philippines. For governance, the poor delivery of basic social services, corruption and inefficiency in the bureaucracy, and poor implementation of laws instigates frustration that could eventually lead to aggression. A calibrated level of authority or political control is required to attain social and political stability. For literacy, the findings correspond to the claim that conflict and aggression are influenced by literacy.

Consistency in quantitative analysis among separate data sets indicates that good governance and literacy may be the causal factors of conflict. Although the presence of correlations or relationship is present with the variables, this does not necessarily prove a causal link. Conflict is rarely caused by a single factor. It is usually caused by the interplay of long-term structural conditions with short-term proximate issues.

Insurgency has been an enduring problem in the Philippines. The government's Whole of the Nation Approach should be appropriately implemented with the full cooperation of the stakeholders. Insurgency is mainly driven by structural problems in the Philippine society that is beyond the scope of the military (Armed Forces of the Philippines, 2011; Sriram et al., 2009). The local government units should take the lead in resolving the conflict with the military and police as support. The military and police should only handle security concerns, whereas the local government units address the socioeconomic factors of the conflict.

Among the variables, good governance is considered to be the primordial factor in the resolution of conflict. Failure in governance can lead to the escalation of conflict that could further result in the breakdown in the delivery of critical political goods such as security, rule of law, and social services. Good governance would eventually have a causal effect that leads to the eradication of poverty through the creation of jobs and improved social benefits; enhancement of literacy through the establishment of an efficient education program; and recognition and appreciation of ethnic diversity through formulation and implementation of laws that would protect culture, identity, and ancestral domain of ethnic groups.

This research was mainly dependent on the availability of data from the National Statistical Office (NSO) and National Statistical Coordination Board (NSCB). Census is not being conducted annually and is mostly limited to the regional or provincial level. More significant analysis would have been made with the availability of an annual census data down to the municipal level. In addition, further research is recommended to determine the effects on conflict with all variables integrated in a given time period.

Exercises 6.3

For the following data, (a) plot the data and (b) state the type of regression that should be used to model the data.

1. Tire tread

Number	Hours	Tread (cm)
1	2	5.4
2	5	5.0
3	7	4.5
4	10	3.7
5	14	3.5
6	19	2.5
7	26	2.0
8	31	1.6
9	34	1.8
10	38	1.3
11	45	0.8
12	52	1.1
13	53	0.8
14	60	0.4
15	65	0.6

2. Let us assume our suspected nonlinear model form is $Z = a(x^b/y^c)$ for the following data. If we use our ln – ln transformation, we obtain: $\ln Z = \ln a + b \ln x - c \ln y$. Use regression techniques to estimate the parameters a, b, and c.

ROW	x	y	Z
1	101	15	0.788
2	73	3	304.149
3	122	5	98.245
4	56	20	0.051
5	107	20	0.270
6	77	5	30.485
7	140	15	1.653
8	66	16	0.192
9	109	5	159.918
10	103	14	1.109
11	93	3	699.447
12	98	4	281.184
13	76	14	0.476
14	83	5	54.468
15	113	12	2.810
16	167	6	144.923
17	82	5	79.733
18	85	6	21.821
19	103	20	0.223
20	86	11	1.899
21	67	8	5.180
22	104	13	1.334
23	114	5	110.378
24	118	21	0.274
25	94	5	81.304

3. Using the basic linear model, $y_i = \beta_0 + \beta_1 x_i$ fit the following data sets. Provide the model, the analysis of variance information, the value of R^2, and a residual plot.

a.

x	y
100	150
125	140
125	180
150	210
150	190
200	320
200	280
250	400
250	430
300	440
300	390
350	600
400	610
400	670

b. The following data represent changes in growth where x = body weight and y = normalized metabolic rate for 13 animals.

x	y
110	198
115	173
120	174
230	149
235	124
240	115
360	130
362	102
363	95
500	122
505	112
510	98
515	96

4. Ten observations of college acceptances to graduate school.

ADMIT	GRE	TOPNOTCH	GPA
1	380	0	3.61
0	660	1	3.67
0	800	1	4
0	640	0	3.19
1	520	0	2.93
0	760	0	3
0	560	0	2.98
1	400	0	3.08
0	540	0	3.39
1	700	1	3.92

5. Data set for lung cancer from Frome (1983). The number of person years is in parenthesis broken down by age and daily cigarette consumption.

Age	Nonsmokers	Smokes 1–9 per day	Smokes 10–14 per day	Smokes 15–19 per day	Smokes 20–24 per day	Smokes 25–34 per day	Smokes >35 per day
15–20	1 (10366)	0 (3121)	0 (3577)	0 (4319)	0 (5683)	0 (3042)	0 (670)
20–25	0 (8162)	0 (2397)	1 (3286)	0 (4214)	1 (6385)	1 (4050)	0 (1166)
25–30	0 (5969)	0 (2288)	1 (2546)	0 (3185)	1 (5483)	4 (4290)	0 (1482)
30–35	0 (4496)	0 (2015)	2 (2219)	4 (2560)	6 (4687)	9 (4268)	4 (1580)
35–40	0 (3152)	1 (1648)	0 (1826)	0 (1893)	5 (3646)	9 (3529)	6 (1136)
40–45	0 (2201)	2 (1310)	1 (1386)	2 (1334)	12 (2411)	11 (2424)	10 (924)
45–50	0 (1421)	0 (927)	2 (988)	2 (849)	9 (1567)	10 (1409)	7 (556)
50–55	0 (1121)	3 (710)	4 (684)	2 (470)	7 (857)	5 (663)	4 (255)
>55	2 (826)	0 (606)	3 (449)	5 (280)	7 (416	3 (284)	1 (104)

6. Modeling absences from class in which

Gender—1-female 2-males
Ethnicity—6 categories
School—school 1 or school 2
Math test score—continuous
Language test score—continuous
Bilingual status—4 bilingual categories

Gender	Ethnicity	School	Math Score	Language Score	Bilingual Status	Days Absent
2	4	1	56.98	42.45	2	4
2	4	1	37.09	46.82	2	4
1	4	1	32.37	43.57	2	2
1	4	1	29.06	43.57	2	3
1	4	1	6.75	27.25	3	3
1	4	1	61.65	48.41	0	13
1	4	1	56.99	40.74	2	11
2	4	1	10.39	15.36	2	7
2	4	1	50.52	51.12	2	10
2	6	1	49.47	42.45	0	9

Projects 6.3

1. Fit the following nonlinear model with the provided data:
Model: $y = ax^b$
Data:

t	7	14	21	28	35	42
y	8	41	133	250	280	297

2. Fit the following model, $y = ax^b$, with the provided data:

Year	0	1	2	3	4	5	6	7	8	9	10
Quantity	15	150	250	275	270	280	290	650	1200	1550	2750

6.4 Conclusion and Summary

We showed some of the common misconceptions by decision-makers concerning correlation and regression. Our purpose of this presentation is to help prepare more competent and confident problem-solvers for the twenty-first century. Data can be found using part of a sine curve in which the correlation is quite poor, close to zero but the decision-maker can describe the pattern. Decision-makers see the relationship in the data as periodic or oscillating. Examples such as these should dispel the idea that correlation of almost zero implies no relationship. Decision-makers need to see and believe concepts concerning correlation, linear relationships, and nonlinear (or no) relationship.

We recommended the following steps.

Step 1: Insure you under the problem and what answers are required.

Step 2: Get the data that are available. Identify the dependent and independent variables.

Step 3: Plot the dependent versus an independent variable and note trends.

Step 4: If the dependent variable is binary {0,1} then use binary logistics regression. If the dependent variables are counts that follow a Poisson distribution, then use Poisson regression. Otherwise, try linear, multiple, or nonlinear regression as needed.

Step 5: Insure your model produces results that are acceptable.

References and Suggested Reading

Affi, A. and S. Azen. 1979. *Statistical Analysis: A Computer Oriented Approach* (2nd ed.). New York: Academic Press, pp. 143–144.

Armed Forces of the Philippines. 2011. *Internal Peace and Security Plan*, p. 1.

Devore, J. 2012. *Probability and Statistics for Engineering and the Sciences* (8th ed.). Boston, MA: Cengage Publisher, pp. 211–217.

Fox, W. 2012a. *Mathematical Modeling with Maple*. Boston, MA: Cengage Publishers.

Fox, W. P. 2011. Using the EXCEL solver for nonlinear regression. *Computers in Education Journal*, 2(4), 77–86.

Fox, W. P. 2012b. Issues and importance of "good" starting points for nonlinear regression for mathematical modeling with maple: Basic model fitting to make predictions with oscillating data. *Journal of Computers in Mathematics and Science Teaching*, 31(1), 1–16.

Fox, W. P. and J. Durante. 2015. Modeling violence in the Philippines. *Journal of Mathematical Science*, 2(4) Serial 5, 127–140.

Fox, W. P. and C. Fowler. 1996. Understanding covariance and correlation. *PRIMUS*, VI(3), 235–244.

Frome, E. L. 1983. The analysis of rates using Poisson regression models. *Biometrics*, 39, 665–674.

Giordano, F., W. Fox, and S. Horton. 2014. *A First Course in Mathematical Modeling* (5th ed.). Boston, MA: Cengage Publishers.

Johnson, I. 2012. *An Introductory Handbook on Probability, Statistics, and Excel.* http://records.viu.ca/~johnstoi/maybe/maybe4.htm (accessed July 11, 2012).

Kreutz, J. 2010. How and when armed conflicts end: Introducing the UCDP conflict termination dataset. *Journal of Peace Research*, 47(2), 243–250. doi:10.1177/0022343309353108.

Melander, E., M. Oberg, and J. Hall. 2006. The "new wars" debate revisited: An empirical evaluation of the atrociousness of "new wars." Uppsala Peace Research Papers No. 9, Department of Peace and Conflict.

Neter, J., M. Kutner, C. Nachtsheim, and W. Wasserman. 1996. *Applied Linear Statistical Models* (4th ed.). Homewood, IL: Irwin Press, pp. 531–547.

Sivard, R. L. 1989. *World Military and Social Expenditures 1987–1988*. Washington, DC: World Priorities.

Sriram, C., J. Large, and S. Brown. 2009. *Governance in Conflict Prevention and Recovery: A Guidance Note*, United Nations Development Programme.

7

Discrete Dynamical System Models

OBJECTIVES

1. Define, model, solve, and interpret systems of discrete dynamical systems (DDS).
2. Analyze the long-term behavior of systems of DDS.
3. Understand the concepts of equilibrium and stability in systems.
4. Model both linear and nonlinear systems of DDS.
5. Model systems of DDS.

Consider an inventory system in which the management has restocking questions (Meerschaert, 1999). A store has a limited number of 20-gallon aquarium tanks. At the end of each week, the manager inventories and places orders. Store policy is to order 3 new 20-gallon tanks at the end of each week if all the inventory has been sold. If even 1 20-gallon aquarium tank remains in stock, no new units are ordered that week. This policy is based on the observation that the store sells on average 1 of the 20-gallon aquarium tanks each week. Is this policy adequate to guard against lost sales?

Historical data are used to compute the probabilities of demand given the number of aquarium tanks on hand. From this analysis, we obtain the following change diagram as shown in Figure 7.1.

Problem Statement: Determine the number of 20-gallon aquariums to order each week.

Assumptions and Variables: Let n represent the number of time period in weeks. We define

$A(n)$ as the probability that one aquarium is in demand in week n.

$B(n)$ as the probability that two aquariums are in demand in week n.

$C(n)$ as the probability that three aquariums are in demand in week n.

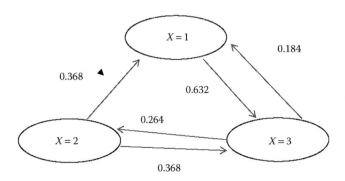

FIGURE 7.1
Flow diagram for inventory.

We assume that no other incentives are given to the demand for aquariums over this time frame. Can we build a model to analyze this inventory issues? We will solve this problem later in this chapter.

7.1 Introduction to Modeling with Dynamical Systems and Difference Equations

Consider financing a new car. Once you have looked at the makes and models to determine what type of car you like, it is time to consider the *costs* and financial packages that lure potential buyers into the car dealerships. This process can be modeled as a dynamical system. Payments are made typically at the end of each month. The amount owed is predetermined as we will see later in this chapter.

You are the chief executive officer (CEO) of a large company. You are considering replacing your fleet of cars. Should you buy or lease the fleet of new cars? What criteria should you use? How do you compare the alternatives?

We will see how to proceed in this chapter. We state up front that a nice way to examine all DDS problem is through iteration and graphs.

7.2 Modeling Discrete Change

We are interested in modeling discrete *change*. Modeling with DDS employs a method to explain certain discrete behaviors or make long-term predictions. A powerful paradigm that we use to model with DDS systems is

Future value = present value + change

The dynamical systems that we will study with this paradigm will differ in appearance and composition, but we will be able to solve a large class of these *seemingly* different dynamical systems with similar methods. In this chapter, we will use iteration and graphical methods to answer questions about the DDS.

We will use flow diagrams to help us to see how the dependent variable changes. These flow diagrams help to see the paradigm and put it into mathematical terms. Let us consider financing a new Ford Mustang. The cost is $25,000 and you can put down $2,000, so you need to finance $23,000. The dealership offers you 2% financing over 72 months. Consider the flow diagram for financing the car as shown in Figure 7.2 that depicts this situation.

We use this flow diagram to help build the discrete dynamical model. Let $a(n)$ = the amount owed after n months. Notice that the arrow pointing into the circle is the interest to the unpaid balance. This increases your debt. The arrow pointing out of the circle is your monthly payment that decreases your debt.

$$a(n+1) = \text{the amount owed in the future}$$

$$a(n) = \text{amount currently owed}$$

$$a(n+1) = a(n) + ia(n) - P$$

where:
 i is the monthly interest rate and
 P is the monthly payment

We will model dynamical systems that have only *constant coefficients*. A dynamical system with constant coefficients may be written in the form

$$a(n+3) = b_2 a(n+2) + b_1 a(n+1) + b_0 a(n)$$

where b_0, b_1, and b_2 are arbitrary constants.

A *DDS* is a *changing system*, where the change of the system at each discrete iteration depends on (is related to) its previous state (or states) of the system. For a prescription drug dosage problem that we will also model, the amount of drug in the bloodstream after n hours depends on the amount of drug in the bloodstream after $n-1$ hours. For financial matters, such as a mortgage balance or credit card balance, the amount you still owe after n months

FIGURE 7.2
Flow diagram for financing a new Ford Mustang.

depends on the amount you owed after $n-1$ months. You will also find this process concerning states that are useful when we discuss Markov chains as an example of DDS.

Example 7.1: Drug Dosage Problem

Suppose that a doctor prescribes that their patient takes a pill containing 100 mg of a certain drug every hour. Assume that the drug is immediately ingested into the bloodstream once taken. In addition, assume that every hour the patient's body eliminates 25% of the drug that is in his or her bloodstream. Suppose that the patient had 0 mg of the drug in his/her bloodstream before taking the first pill. How much of the drug will be in his or her bloodstream after 72 hours?

> *Problem Statement*: Determine the relationship between the amount of drug in the bloodstream and time.
>
> *Assumptions*: The system can be modeled by a DDS. The patient is of normal size and health. There are no other drugs being taken that will affect the prescribed drug. There are no internal or external factors that will affect the drug absorption rate. The patient always takes the prescribed dosage at the correct time.
>
> *Variables*: Define $a(n)$ to be the amount of drug in the bloodstream after a period n, $n = 0,1,2,\ldots$hours.
>
> *Flow Diagram*: We create the input–output flow diagram as displayed in Figure 7.3.
>
> *Model Construction*:

Let us define the following variables:

$a(n+1) = $ *amount of drug in the system in the future*
$a(n) = $ *amount currently in system*

We define change as follows: change = dose−loss in system

$$\text{change} = 100 - 0.25\, a(n)$$

so, Future = Present + Change is

$$a(n+1) = a(n) - 0.25\, a(n) + 100$$

or

$$a(n+1) = 0.75\, a(n) + 100 \text{ with } a(0) = 0.$$

Since the body loses 25% of the amount of drug in the bloodstream every hour, there would be 75% of the amount of drug in the bloodstream

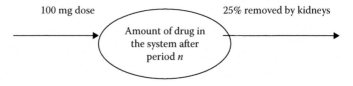

100 mg dose   25% removed by kidneys

Amount of drug in the system after period n

FIGURE 7.3
Flow diagram for drugs in system.

remaining every hour. After 1 hour, the body has 75% of the initial amount, 0 mg, to 100 mg that is added every hour. So the body has 100 mg of drug in the bloodstream after one hour. After 2 hours the body has 75% of the amount of drug that was in the bloodstream after 1 hour (100 mg), plus an additional 100 mg of drug added to the bloodstream. So, there would be 175 mg of drug in the bloodstream after 2 hours. After three hours the body has 75% of the amount of drug that was in the bloodstream after 2 hours (175 mg), plus an additional 100 mg of drug added to the bloodstream. So, there would be 231.25 mg of drug in the bloodstream after 3 hours.

An easy way to examine this is using MS-EXCEL. We will describe the process to iterate and graph the results shown in Figure 7.4.

Put the labels on cells, A3 and B3. In cell a4 put a 0 for time period 0 and a) for the initial value of a(0). In cell A5 type = a4 + 1, this increments the time period by 1. In cell B5, type = 0.75 * b4 + 100, which represents the DDS. Then copy down for 30 time period.

Next, highlight the two columns, on the Command bar; go to INSERT, for the graphs choose the Scatter plot.

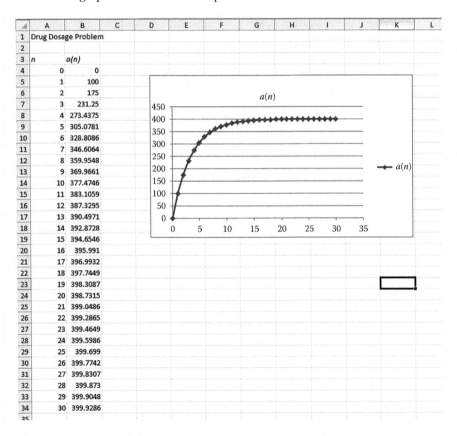

FIGURE 7.4
EXCEL output for DDS model.

Interpretation of results: The DDS shows that the drug reaches a value in which *change stops* and eventually the concentration in the bloodstream levels at 400 mg. If 400 mg is both a safe and effective dosage level then this dosage schedule is acceptable. We discuss this concept of change stopping (equilibrium) later in this chapter.

Example 7.2: Time Value of Money

You want to purchase a $1000 savings certificate, which pays 12% interest a year compounded monthly at 1% per month. In our research, we have not found a bank or a money facility that yields continuous interest, so a discrete model would be better and more accurate.

Problem Statement: Find a relationship between the amount of money invested and the time over which it is invested.

Assumptions: The interest rate is constant over the entire time period. No additional money is added or withdrawn other than by interest.

The following represents the worth of the certificate with interest accumulated each month:

Variables: Let $a(n)$ = the amount of money in the certificate after month n where $n = 1, 2, 3,...$

Flow Diagram: The flow diagram is displayed in Figure 7.5.

Model Construction:

$a(n + 1)$ is the future
$a(n)$ = present
$a(n + 1) = a(n) + 0.12/12\, a(n)$ or $a(n + 1) = a(n) + 0.01\, a(n)$

or

$$a(n + 1) = 1.01\, a(n)$$

We know then initially, we had $1000, so

$$a(0) = 1000$$

Remark: We always divide the annual interest rate for the compounding period in order to recompute the actual interest rate being used in the problem. Here, the annual rate is 12% or 0.12 and we pay monthly (12 months per year). So, we use $0.12/12 = 0.01$ as the interest rate.

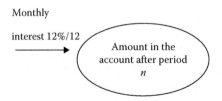

Monthly

interest 12%/12

Amount in the account after period n

FIGURE 7.5
Flow diagram for money in a certificate.

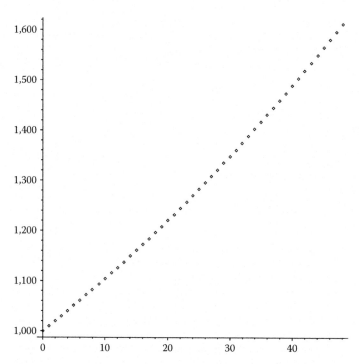

FIGURE 7.6
Plot of DDS for money in a certificate that grows over time.

Figure 7.6 shows that we will keep making money forever. Can we ever reach $10,000 in our account? We can iterate using technology to see that it will take about 19.33 years to have $10,000 in the account. The money continues to grow as long as the money is left in the account. You will have $1,000,000 in about 58 years.

Example 7.3: Simple Mortgage

Five years ago, your parents purchased a home by financing $80,000 for 20 years, paying monthly payments of $880.87 with a monthly interest of 1%. They have made 60 payments and wish to know what they actually owe on the house at this time. They can use this information to decide whether or not they should refinance their house at a lower interest rate for the next 15 or 20 years. The change in the amount owed each period increases by the amount of the interest and decreases by the amount of the payment.

> *Problem Statement*: Build a model that relates the time with the amount owed on a mortgage for a home.
>
> *Assumptions*: Initial interest was 12%. Payments are made on time each month. The current rate for refinancing is 6.25% for 15 years and 6.50% for 20 years.

FIGURE 7.7
Flow diagram for mortgage example.

Variables: Let $b(n)$ = amount owed on the home after n months
Flow Diagram: The flow diagram is displayed in Figure 7.7.
Model Construction:

$$b(n + 1) = b(n) + 0.12/12\, b(n) - 880.87,\ b(0) = 80,000$$
$$b(n + 1) = 1.01\, b(n) - 880.87,\ b(0) = 80,000$$

Model Solution: Graphical: First, we plot the DDS over the entire 20 years (240 months). Note that the graph in Figure 7.8 shows that it reaches 0 in 240 months.

After paying for 60 months, your parents still owe $73,395 out of the original $80,000. They have paid in $52,852.20 and only $6,605 went toward the principal payment of the home. The rest of the money went toward paying only the interest. If the family continues with this loan, then they will make 240 payments of $880.87 or $211,400.80 total in payments. This is 133,400.80 in interest. They have already paid $46,647.20 in interest.

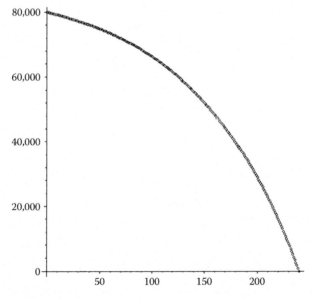

FIGURE 7.8
Mortgage payments over time.

They would pay an additional $86,753.60 in interest over the next 15 years. What should they do?
Alternative 1: Refinance the home at 6.50% for 15 years. Assume no closing costs for this refinancing. Only refinance what they owe, that is, $73,395.
Model Construction:

$$B(n + 1) = B(n) + 0.065/12 \ B(n) - \text{payment}, \ B(0) = 73,395$$

The monthly payment is now $639.34. We will show later how this was calculated.
The total payment over the lifetime of the refinance is $115,081.20. We have previously paid for 5 years a total of $52,852.20. The new total of payments is $167,933.40 as compared to $211,400.80. Refinancing is the smarter move because not only are the monthly payments $241.53 a month less ($639.34 versus $880.87) the total package saves the family a total of $43,467.40 over the lifetime of the mortgage.

Exercises 7.2

Iterate and graph the following DDS. Explain their long-term behavior. Try to find realistic scenarios that these DDS might explain.

1. $a(n + 1) = 0.5 \ a(n) + 0.1, \ a(0) = 0.1$
2. $a(n + 1) = 0.5 \ a(n) + 0.1, \ a(0) = 0.2$
3. $a(n + 1) = 0.5 \ a(n) + 0.1, \ a(0) = 0.3$
4. $a(n + 1) = 1.01 \ a(n) - 1000, \ a(0) = 90,000$
5. $a(n + 1) = 1.01 \ a(n) - 1000, \ a(0) = 100,000$
6. $a(n + 1) = 1.01 \ a(n) - 1000, \ a(0) = 110,000$
7. $a(n + 1) = -1.3 \ a(n) + 20, \ a(0) = 9$
8. $a(n + 1) = -4 \ a(n) + 50, \ a(0) = 9$
9. $a(n + 1) = 0.1 \ a(n) + 1000, \ a(0) = 9$
10. $a(n + 1) = 0.9987 \ a(n) + 0.335, \ a(0) = 72$

Projects 7.2

1. *Sewage Plant*: A sewage treatment plant processes raw sewage to produce useable fertilizer and clean water by removing all the contaminants. The process is such that each hour 12% of the remaining contaminants in the process tank are removed. What percentage is left after one day? How long would it take to lower the amount of sewage by about half? How long until the level of sewage is down to 10% of its original level?

Problem Statement:

Assumptions:

Variables:

Model:

2. *Cooling a Hot Soda:* A can of soda has been sitting out at about room temperature for days (about 72°). If that same soda is placed in a refrigerator that is set at 40°, how long will it take the soda to reach 50°, 45°, 40°? Assume a cooling rate of 0.0081 degrees per minute (obtained experimentally).

   Problem Statement:

   Assumptions:

   Variables:

   Model:

3. *Buying a New Car:*

   Part 1: You wish to buy a new car soon. You initially narrow your choices to a Saturn, Focus, and a Toyota Corolla. Each dealership offers you their prime deal:

Saturn	$11,900	$1,000 down	3.5% interest for up to 60 months
Focus	$11,500	$1,500 down	4.5% interest for up to 60 months
Corolla	$10,900	$500 down	6.5% interest for up to 48 months

   You have allocated at most $475 a month on a car payment.

   Use a dynamical system to compare the alternatives, choose a car, and establish your exact monthly payment.

   Part 2: Nissan is offering a first-time-buyers special. Nissan is offering 6.9% for 24 months or $500 *cash back*. If the regular interest rate is 9.00%, determine the amount financed at 6.9% so that these two options are equivalent for a new car buyer.

   Check out your local paper or dealership, what kind of car could you get from Nissan with the equivalent option?

4. You are also a bit worried about the comfort of a smaller car and its ride. You are in luck; *Consumer Report* did a study and compared the rides of four small cars in terms of the suspension system. Consider the following alternative models for suspension systems available for a small car. Analyze each alternative, describe the behavior of each suspension system over time, and choose an alternative suspension system. Each car sits at 0.2 initially. It is determined that under 0.6 it maintains a smooth ride.

Alternative 1: $S(n + 1) = 2 S(n) (1 - S(n))$

Alternative 2: $S(n + 1) = 3 S(n) (1 - S(n))$

Alternative 3: $S(n + 1) = 3.6 S(n) (1 - S(n))$

Alternative 4: $S(n + 1) = 3.7 S(n) (1 - S(n))$

5. D.B. Cooper: On November 24, 1971, a cold rainy Thanksgiving evening, a middle-aged man giving the name Dan B. Cooper purchased a plane ticket on Northwest Airlines Flight 305 from Portland, Oregon, to Seattle, Washington. He boarded the plane and flew into history.

After Cooper was seated, he demanded that the flight attendant bring him a drink and a $200,000 ransom. His note read: "Miss, I have a bomb in my suitcase and I want you to sit beside me." When she hesitated, Cooper pulled her into the seat next to him and opened his case, revealing several sticks of dynamite connected to a battery. He then ordered her to have the captain relay his ransom demand for the money and four parachutes: Two front-packs and two backpacks. The FBI delivered the parachutes and ransom to Cooper when the Boeing 727 landed in Seattle to be refueled. Cooper, in turn, allowed the 32 other passengers he had held captive to go free.

When the 727 was once again airborne, Cooper instructed the pilot to fly at an altitude of 10,000 feet on a course destined for Reno, Nevada. He then forced the flight attendant to enter the cockpit with the remaining three crewmembers and told her to stay there until landing. Alone in the rear cabin of the plane, he lowered the airstairs beneath the tail. Then, in the dark of the night, somewhere over Oregon, D. B. Cooper stepped with the money into history. When the plane landed in Reno, the only thing the FBI found in the cabin was one of the backpack chutes. Despite the massive efforts of manhunts conducted by the FBI and scores of local and state police groups, no evidence was ever found of D. B. Cooper, the first person to hijack a plane for ransom and parachute from it. One package of marked bills from the ransom was found along the Columbia River near Portland in 1980. Some surmise that Cooper might have been killed in his jump to fame and part of the ransom washed downstream.

Consider the following items dealing with Flight 305 and its famous passenger. Using Excel to assist you, complete each one as directed.

1. Suppose Cooper invested the ransom over a period of 10 years, one $20,000 unit per year, in a small local bank in rural Utah, at a fixed rate of 6%, compounded annually, starting on January 1, 1972. Further, if he left the money to accumulate, what would be the value of the account on December 31, 2004?

2. Suppose that instead of playing it safe, Cooper gambled half of the money away in Reno the night after the hijacking and then invested the other half with Night Wings Federal in Reno on January 1, 1972. If the account paid 6% interest compounded quarterly, what would this money be worth at the end of this year?

3. Another possibility for Cooper would have been to invest 75% of the total with a *loco* investor, Smiling Pete's Federal Credit Union, at a rate of 3% compounded monthly. What would this investment be worth at the end of this year?

4. Compare these three methods of investing. Contrast their different patterns of growth over time.

7.3 Equilibrium Values and Long-Term Behavior

7.3.1 Equilibrium Values

Let us go back to our original paradigm,

$$\text{Future} = \text{present} + \text{change}$$

When change stops, the change equals zero and future equals the present. The value for which this happens, if any, is the equilibrium value. This gives us a context for the concept of the equilibrium value.

We will define the equilibrium value (or fixed point) as the value in which change stops. The value $a(n)$ is an equilibrium value for the DDS, $a(n + 1) = f(a(n), a(n - 1),...)$ if for a value $a(0) = A_0$ all future values equal A_0.

Formally, we define the equilibrium value as follows:

The number ev is called an *equilibrium value* or *fixed point* for a DDS if $a(k) = ev$ for all values of k when the initial value $a(0)$ is set at ev. That is, $a(k) = ev$ is a constant solution to the recurrence relation for the dynamical system.

Another way of characterizing such values is to note that the number a is an equilibrium value for a dynamical system $a(n + 1) = f(a(n), a(n - 1),..., n)$ if and only if a satisfies the equation $ev = f(ev, ev,..., ev)$.

Using this definition, we can show that a linear homogeneous dynamical system of order 1 only has the value 0 as an equilibrium value.

In general, dynamical systems may have no equilibrium values, a single equilibrium value, or multiple equilibrium values. Linear systems have unique equilibrium values. The more nonlinear a dynamical system is, the more equilibrium values it may have.

Not all DDS's have equilibrium values, and many DDS's that have equilibrium values that the system will never achieve. However, we already know that for a first-order equation, if $a(0) = ev$, and every subsequent iteration value of the DDS is equal to ev, that is, $a(k) = ev$ for all value of k, then ev is an equilibrium value. For example, the DDS $a(n + 1) = 2\ a(n) + 1$ (Tower of Hanoi) has an equilibrium value of $ev = -1$. If we begin with $a(0) = -1$ and iterate, we get

$a(1) = 2\ a(0) + 1 = 2\ (-1) + 1 = -2 + 1 = -1$, so $a(1) = -1$

$a(2) = 2\ a(1) + 1 = 2\ (-1) + 1 = -2 + 1 = -1$, so $a(2) = -1$

$a(3) = 2\ a(2) + 1 = 2\ (-1) + 1 = -2 + 1 = -1$, so $a(3) = -1$

etc.

So, we say that when $a(0) = -1$, $a(k) = -1$ for all values of k, or in general, when $a(0) = ev$, then $a(k) = ev$ for all values of k.

We can use this observation to find equilibrium values and to find out whether or not a DDS has an equilibrium value. Let us look at the DDS $a(n + 1) = 2\ a(n) + 1$, when $a(0) = ev$, $a(k) = ev$ for all value of k so that $a(1) = ev$, $a(2) = ev$, $a(3) = ev,...,$ $a(n) = ev$, $a(n + 1) = ev$. Substituting $a(n) = ev$ and $a(n + 1) = ev$ into our DDS yields

$a(n + 1) = 2\ a(n) + 1$

$ev = 2\ ev + 1$

$-ev = 1$

$ev = -1$

So, the equilibrium value for the DDS is -1.

Now, let us consider the DDS $a(n + 1) = a(n) + 1$. Using our definition of equilibrium values, we write

$a(n + 1) = a(n) + 1$

$ev = ev + 1$

$ev - ev = 1$

$0 = 1$

The statement $0 = 1$ is not true, so the DDS *does not have an equilibrium value.*

Example 7.4: DDS and equilibriums

Consider the following DDS and find their equilibrium value, if they exist.

  a. $a(n + 1) = 0.3\,a(n) - 10$
  b. $a(n + 1) = 1.3\,a(n) + 20$
  c. $a(n + 1) = 0.5\,a(n)$
  d. $a(n + 1) = -0.1\,a(n) + 11$

Solutions:

  a. $ev = 0.3\,ev - 10$
     $0.7\,ev = -10$
     $ev = -10/0.7 = -14.29$
  b. $ev = 1.3\,ev + 20$
     $-0.3\,ev = 20$
     $ev = -20/0.3 = -66.66667$
  c. $ev = 0.5ev,\ ev = 0$
  d. $ev = -0.1\,ev + 11$
     $ev = 11$
     $ev = 10$

As mentioned earlier, we said that DDSs that have equilibrium values *may not ever attain* their equilibrium value, given some initial condition $a(0)$. For the *Tower of Hanoi*, the equilibrium value was −1. Suppose that we begin iterating the DDS with an initial value $a(0) = 0$:

$a(1) = 2(0) + 1 = 3$
$a(2) = 2(3) + 1 = 7$
$a(3) = 2(7) + 1 = 15$
$a(4) = 2(15) + 1 = 31$
etc.

The values continue to get larger and will never reach the value of −1. Since $a(n)$ represents the number of moves of the disks, the value of −1 makes no real sense in the contest of the number of disk to move.

We will study equilibrium values in many of the applications of DDS. In general, a linear, nonhomogeneous DDS, where the nonhomogeneous part is a constant, will have an equilibrium value. (Can you find any exceptions?) Linear, homogeneous DDS will have an equilibrium value of zero. (Why?)

7.3.2 A Graphical Approach to Equilibrium Values

We can examine the plot of the iterations using technology. If the values reach a specific value and remain constant then that value is an equilibrium value (change has stopped).

As seen, dynamical systems often represent real-world behavior that we are trying to understand. At times, we want to predict future behavior and gain deeper insights into how to influence or alter the behavior. Thus, we have great interest in the predictions of the model. How does it change in the future?

Models of the form: $a(n + 1) = r\,a(n)$, r is a constant

Let us revisit our savings account problem where we invest $1000 at 12% a year compounded monthly.

$$a(n+1) = 1.01\,a(n), a(0) = 1000$$

This sequence, with $r > 1$, grows without bound. The graph in Figure 7.9 suggests that there is no equilibrium value (where the graph levels out to a constant value). Analytically, $ev = 1.1ev + 1000$, $ev = -909.09$. There is an equilibrium value of -909.09 but that value will never be reached in our savings account problem.

Now, what happens if r is less than 0? We replace $r = 1.01$ with $r = -1.01$ in the previous example. First, we can analytically solve for the equilibrium value.

$ev = -1.01\ ev + 1000$

$2.01\ ev = 1000$

$ev = 497.5124378$

The definition implies that if we start at $a(0) = 497.5124378$ we stay there forever. If we plot the solution, we note the oscillations between positive and negative numbers, each growing without bound as the oscillations fan out as shown in Figure 7.10. Although there is an equilibrium value, the solution to our example does not tend toward this equilibrium value.

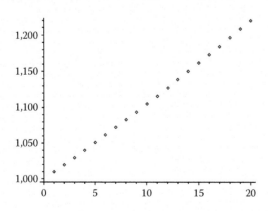

FIGURE 7.9
Saving account revisit.

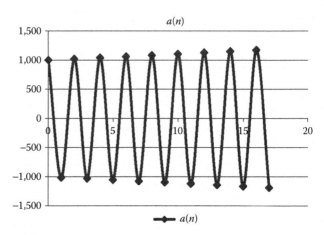

FIGURE 7.10
DDS graph using $r = 1.01$ and $r = -1.01$.

Let us examine values of r, $0 < r < 1$. Let us take a look at a drug dosage model, $a(n + 1) = 0.5\ a(n)$ with $a(0) = 20$, where half of what is in the system is discarded at each time period.

The equilibrium value is zero.

Models of the form: $a(n + 1) = r\ a(n) + b$, where r and b are constants

Let us return to our drug dosage problem and consider adding the constant dosage at each time period (time periods might be 4 hours). Our model is $a(n + 1) = 0.5\ a(n) + 16$ mg. We will also assume that there is an initial dosage applied before beginning the regime. We will let these initial values be as follows as we plot the results in Figure 7.11a–c.

Regardless of the starting value, the future terms of $a(n)$ approach 32. Thus, 32 is the equilibrium value. We could have solved for this algebraically as well.

$$a(n+1) = 0.5\ a(n) + 16$$
$$ev = 0.5\ ev + 16$$
$$0.5\ ev = 16$$
$$ev = 32$$

Another method of finding the equilibrium values involves solving the equation

$a = ra + b$ and solving for a (where a is ev) we find

$$a = \frac{b}{1-r}, \text{if } r \neq 1$$

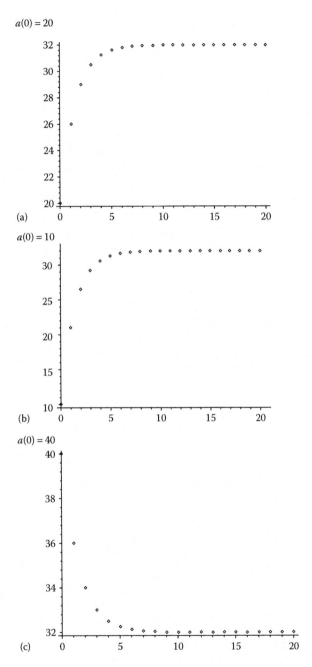

FIGURE 7.11

(a) $a(0) = 20$, (b) $a(0) = 10$, and (c) $a(0) = 40$. Plot of drugs in our systems with different starting conditions.

Using this formula in our previous example, the equilibrium value is

$$a = \frac{16}{1 - 0.5} = 32$$

7.3.3 Stability and Long-Term Behavior

For a dynamical system, $a(n + 1)$ with a specific initial condition, $a(0) = a_0$, we have shown that we can compute $a(1)$, $a(2)$, and so forth. Often these particular values are not as important as the long-term behavior. By long-term behavior, we refer to what will eventually happen to $a(n)$ for larger values of n. There are many types of long-term behavior that can occur with DDS, we will only discuss a few here.

If the $a(n)$ values for a DDS eventually get close to the equilibrium value, ev, no matter the initial condition, then the equilibrium value is called a *stable equilibrium value* or *an attracting fixed point*.

Example 7.5: Consider the following DDS

$A(n + 1) = 0.5 \, A(n) + 64$, with initial conditions $A(0) = 0$ or $A(0) = 150$, the ev is 128. The ev is stable as shown in Figure 7.12.

Notice that both sequences are converging on 128 as the attracting fixed point or equilibrium value.

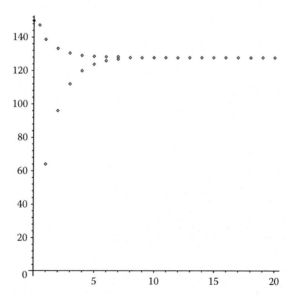

FIGURE 7.12
Stable equilibrium value.

Example 7.6: **Consider the DDS, $A(n + 1) = -0.5\,A(n) + 10$**
with $ev = 6.66667$

With $A(0) = 100$, this is the plot of behavior, Figure 7.13.

Example 7.7: **A financial model, $A(n + 1) = 1.1\,A(n) + 100$, $A(0) = 100$.**

Consider the DDS for the financial model, $A(n + 1) = 1.1\,A(n) + 100$, $A(0) = 100$.

The ev value is -1000. If the DDS ever achieves an input of -1000 then the systems stay at -1000 forever. However, when we start at typical values such as \$100 to begin the process we find the values tend to move away from the ev. When this occurs, we say that the ev is unstable or at a repelling fixed point.

The values tend to increase over time and never move toward -1000. Therefore, the ev is unstable.

Often we characterize the long-term behavior of the system in terms of its stability. If a DDS has an equilibrium value and if the DDS tends to the equilibrium value from starting values near the equilibrium value, then the DDS is said to be stable.

Thus, for the dynamical system $a(n + 1) = r\,a(n) + b$, where $b \neq 0$ we provide Table 7.1 to show the stability results.

If $r \neq 1$, an equilibrium exists at $a = b/(1 - r)$.
If $r = 1$, no equilibrium value exists.

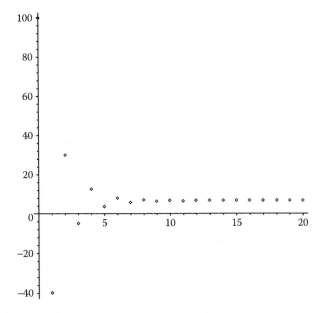

FIGURE 7.13
Plot of $A(n + 1) = -0.5\,A(n) + 10$ with different starting values showing stability of the equilibrium value.

TABLE 7.1

Stability of $a(n + 1) = r\, a(n) + b$, where $b \neq 0$

Value of r	DDS Form	Equilibrium	Stability of Solution	Long-Term Behavior
$r = 0$	$a(n + 1) = b$	b	Stable	Stable equilibrium
$r = 1$	$a(n + 1) = a(n) + b$	*None*	Unstable	
$r < 0$	$a(n + 1) = r{*}a(n) + b$	$b/(1 - r)$	Depends on $|r|$	Oscillations
$|r| < 1$	$a(n + 1) = r{*}a(n) + b$	$b/(1 - r)$	Stable	Approaches $b/(1 - r)$
$|r| > 1$	$a(n + 1) = r{*}a(n) + b$	$b/(1 - r)$	Unstable	Unbounded

RELATIONSHIP TO ANALYTICAL SOLUTIONS

If a DDS has an *ev* value, we can use the *ev* value to find the analytical solution.

Example 7.8: Mortgage Revisited

Recall the mortgage example from Section 7.2.

$B(n + 1) = 1.00541667\, B(n) - 639.34$,
$B(0) = 73{,}395$

The *ev* value is found as 118031.9274.
The analytical solution may be found using the following form:

$B(k) = (1.00541667^k)C + D$ where D is the *ev*.
$B(k) = (1.00541667^k)C + 118031.9274$, $B(0) = 73.395$
Since $B(0) = 73395 = 1.00541667^0\,(C) + 118031.9274$
$C = -44636.92736$

Thus, the closed form model is

$$B(k) = -44636.92737\,(1.00541667^k) + 118031.9274$$

Let us assume we did not know the payment was \$639.34 month. We could use the analytical solutions to help find the payment.

$$B(k) = (1.00541667^k)C + D$$

We build a system of two equations and two unknowns.

$$B(k) = (1.00541667^k)C + D$$

$$B(0) = 73395 = C + D$$

$$B(180) = 0 = 1.00541667^{180}C + D$$

$$C = -44638.70, D = 118033.7$$

$$B(k) = -44638.70(1.00541667)^k + 118033.7$$

D represents the equilibrium value and we accepted some round-off error. From our model form:

$B(n + 1) = 1.00541667\ B(n) - P$, we can find P.
Solving analytically for the equilibrium value,

$$X - 1.00541667X = -P$$

$$X = \frac{P}{0.00541667}$$

X is 118033.70

So,

$$118033.70 = \frac{P}{0.00541667}$$

$$P = 639.34$$

Exercises 7.3

1. For the following DDS, find the equilibrium value if one exists. Classify the DDS as stable or unstable.
   a. $a(n + 1) = 1.23\ a(n)$
   b. $a(n + 1) = 0.99\ a(n)$
   c. $a(n + 1) = -0.8\ a(n)$
   d. $a(n + 1) = a(n) + 1$
   e. $a(n + 1) = 0.75\ a(n) + 21$
   f. $a(n + 1) = 0.80\ a(n) + 100$
   g. $a(n + 1) = 0.80\ a(n) - 100$
   h. $a(n + 1) = -0.80\ a(n) + 100$

2. Build a numerical table for the following initial value DDS problems. Observe the patterns and provide information on equilibrium values and stability.
   a. $a(n + 1) = 1.1\ a(n) + 50, a(0) = 1010$
   b. $a(n + 1) = 0.85\ a(n) + 100, a(0) = 10$
   c. $a(n + 1) = 0.75\ a(n) - 100, a(0) = -25$
   d. $a(n + 1) = a(n) + 100, a(0) = 500$

Projects 7.3

Determine if there is an equilibrium value in each project from Section 7.2 Project. If there is an equilibrium value, determine if it is stable or unstable.

7.4 Modeling Nonlinear Discrete Dynamical Systems

7.4.1 Introduction and Nonlinear Models

In this section, we build nonlinear DDS to describe the change in behavior of the quantities we study. We will also study systems of DDS to describe the changes in various systems that act together in some way or ways. We define a nonlinear DDS. If the function of $a(n)$ involves powers of $a(n)$ (like $a^2(n)$), or a functional relationship (like $a(n)/a(n-1)$), we will say that the DDS is *nonlinear*. A *sequence* is a function whose domain is the set of nonnegative integers ($n = 0, 1, 2,...$). We will restrict our model solution to the numerical and graphical solutions. Analytical solutions may be studied in more advanced mathematics courses.

Example 7.9: Population Growth: Growth of a Yeast Culture (Giordano et al., 2014)

We often model population growth by assuming that the change in population is directly proportional to the current size of the given population. This produces a simple, first-order DDS similar to those seen earlier. It might appear reasonable at first examination, but the long-term behavior of growth without bound is disturbing. Why would growth without bound of a yeast culture in a jar (or controlled space) be alarming?

There are certain factors that affect the population growth. Things include resources (food, oxygen, space, etc.). These resources can support some maximum population. As this number is approached, the change (or growth rate) should decrease and the population should never exceed its resource supported amount.

> *Problem Statement*: Predict the growth of yeast in a controlled environment as a function of the resources available and the current population.
>
> *Assumptions and Variables.* We assume that the population size is best described by the weight of the biomass of the culture. We define $y(n)$ as the population size of the yeast culture after a period n. There exists a maximum carrying capacity, M, that is sustainable by the resources available. The yeast culture is growing under the conditions established.
> *Model*:

$$y(n+1) = y(n) + k\,y(n)\,(M - y(n))$$

where:
$y(n)$ is the population size after period n
n is the time period measured in hours
k is the constant of proportionality
M is the carrying capacity of our system

Table 7.2 shows the data collected on the culture.
We plot the time versus the Biomass, $P(n)$, and see that as time increases
it appears to stabilize at 1000as shown in Figure 7.14.

TABLE 7.2

Growth of Yeast in a Culture

Time	Yeast Biomass, $P(n)$	$P(n+1) - P(n)$	$P(n)*(1000 - P(n))$
0	14.496	13.137	14285.86598
1	27.633	16.157	26869.41731
2	43.79	27.482	41872.4359
3	71.272	36.089	66192.30202
4	107.361	72.48	95834.61568
5	179.841	83.805	147498.2147
6	263.646	124.877	194136.7867
7	388.523	141.034	237572.8785
8	529.557	136.353	249126.3838
9	665.91	109.173	222473.8719
10	775.083	70.064	174329.3431
11	845.147	53.001	130873.5484
12	898.148	52.246	91478.1701
13	950.394	17.214	47145.24476
14	967.608	15.553	31342.75834
15	983.161	7.248	16555.44808
16	990.409	5.587	9499.012719
17	995.996	3.314	3987.967984
18	999.31		

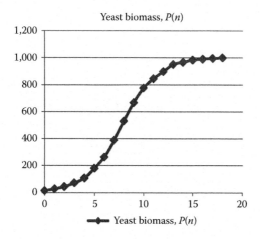

FIGURE 7.14

Biomass as a function of time.

Next, we plot $y(n + 1) - y(n)$ versus $y(n) (1000 - y(n))$ to find the slope, k, is approximately 0.000544 as shown in Figure 7.15. With $k = 0.000544$ and the carrying capacity in biomass is 1000. This model is

$$y(n+1) = y(n) + 0.00082\, y(n)(1000 - y(n))$$

Again, this is nonlinear because of the $y^2(n)$ term. The solution iterated (there is no closed form analytical solution for this equation) from an initial condition, biomass, of 14.496:

The model and plot in Figure 7.16 show stability in that the population (biomass) of the yeast culture approaches 1000 as n gets large. Thus, the population is eventually stable at approximately 1000 units.

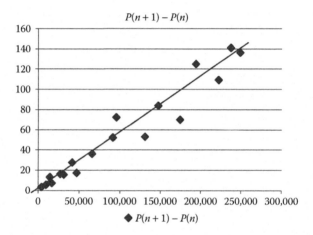

FIGURE 7.15
Plot of $y(n + 1) - y(n)$ versus $y(n) (1000 - y(n))$.

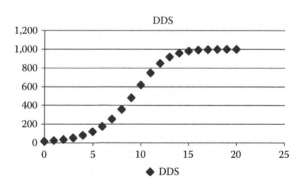

FIGURE 7.16
Plot of DDS from growth of a yeast culture.

Example 7.10: Spread of a Contagious Disease

Suppose that there are 1000 students in a college dormitory, and some students have been diagnosed with meningitis, a highly contagious disease. The health center wants to build a model to determine how fast the disease will spread.

> *Problem Statement*: Predict the number of students affected with meningitis as a function of time.
>
> *Assumptions and Variables*: Let $m(n)$ be the number of students affected with meningitis after n days. We assume all students are susceptible to the disease. The possible interactions of infected and susceptible students are proportional to their product (as an interaction term).
>
> *The model is*

$$m(n+1) - m(n) = k\,m(n)(1000 - m(n))\,\text{or}$$

$$m(n+1) = m(n) + k\,m(n)(1000 - m(n))$$

Two students returned from spring break with meningitis. The rate of spreading per day is characterized by $k = 0.0090$. It is assumed that a vaccine can be in place and the students are vaccinated within 1–2 weeks.

$$m(n+1) = m(n) + 0.00090\,m(n)(1000 - m(n))$$

> *Interpretation*: The results show that most students will be affected within 2 weeks as depicted from Figure 7.17. Since only about 10% will be affected within one week, every effort must be made to get the vaccination at the school and get the students vaccinated within 1 week.

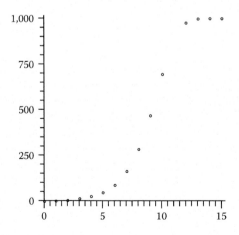

FIGURE 7.17
Plot of DDS for the spread of a disease.

Exercises 7.4

Consider the model $a(n + 1) = r\, a(n)\, (1 - a(n))$. Let $a(0) = 0.2$. Determine the numerical and graphical solution for the following values of r. Find the pattern in the solution.

    1. $r = 2$

    2. $r = 3$

    3. $r = 3.6$

    4. $r = 3.7$

For Exercises 5 through 8 find the equilibrium value by iteration and determine if it is stable or unstable.

    5. $a(n + 1) = 1.7\, a(n) - 0.14\, a(n)^2$

    6. $a(n + 1) = 0.8\, a(n) + 0.1\, a(n)^2$

    7. $a(n + 1) = 0.2\, a(n) - 0.2\, a(n)^3$

    8. $a(n + 1) = 0.1\, a(n)^2 + 0.9\, a(n) - 0.2$

    9. Consider spreading a rumor through a company of 1000 employees all working in the same building. We assume that the spread of a rumor is similar to the spread of a contagious disease in that the number of people hearing the rumor each day is proportional to the product of the number hearing the rumor and the number who have not heard the rumor. This is given by

$$r(n+1)=r(n)+1000\,k\,r(n)-k\,r(n)^2$$

where k is the parameter that depends on how fast the rumor spreads. Assume $k = 0.001$ and further assume that four people initially know the rumor. How soon will everyone know the rumor?

Projects 7.4

    1. Consider the contagious disease as the Ebola virus. Look up on the Internet and find out some information about how deadly this virus actually is. Now consider an animal research laboratory in Restin, VA., a suburb of Washington, DC, with population of 856,900 people. A monkey with the Ebola virus had escaped its captivity and infected one employee (unknown at the time) during its escape. This employee reports to the University hospital later with Ebola symptoms. The Infectious Disease Center (IDC) in Atlanta gets a call and begins to model the spread of this disease. Build a model for the IDC with the following growth rates to determine the number infected after 2 weeks:

      a. $k = 0.00025$

      b. $k = 0.000025$

   c. $k = 0.00005$

   d. $k = 0.000009$

   e. List some ways of controlling the spread of the virus.

2. Consider the spread of a rumor concerning termination among 1000 employees of a major company. Assume that the spreading of a rumor is similar to the spread of contagious disease in that the number hearing the rumor each day is proportional to the product of those who have heard the rumor and those who have not heard the rumor. Build a model for the company with the following rumor growth rates to determine the number having heard the rumor after 1 week:

   a. $k = 0.25$

   b. $k = 0.025$

   c. $k = 0.0025$

   d. $k = 0.00025$

   e. List some ways of controlling the spread of the rumor.

7.5 Modeling Systems of Discrete Dynamical Systems

In this section, we examine models of systems of difference equations (DDS). For selected initial conditions, we build numerical solutions to get a sense of long-term behavior for the system. For the systems that we will study, we will find their equilibrium values. We then explore starting values near the equilibrium values to see if by starting close to an equilibrium value, the system will

a. Remain close

b. Approach the equilibrium value

c. Not remain close

What happens near these values gives great insight concerning the long-term behavior of the system. We study the resulting patterns of the numerical solutions.

Example 7.11: Inventory System Analysis

A store has a limited number of 20-gallon aquarium tanks. At the end of each week, the manager inventories and places orders. Store policy is to order 3 new 20-gallon tanks at the end of each week if all the inventory has been sold. If even 1 20-gallon aquarium tank remains in stock, no new units are ordered that week. This policy is based on the observation that the store sells on average 1 of the 20-gallon aquarium tanks each week. Is this policy adequate to guard against lost sales?

Historical data are used to compute the probabilities of demand given the number of aquarium tanks on hand. From this analysis, we obtain the change diagram as shown in Figure 7.1.

Problem Statement: Determine the number of 20-gallon aquariums to order each week.

Assumptions and Variables: Let n represent the number of time period in weeks. We define

$A(n)$ is the the probability that one aquarium is in demand in week n.

$B(n)$ is the the probability that two aquariums are in demand in week n.

$C(n)$ is the the probability that three aquariums are in demand in week n.

We assume that no other incentives are given to the demand for aquariums over this time frame.

The Model: The number of aquariums demanded in each time period is found using the paradigm, Future = present + change. Mathematically, this is written as

$A(n + 1) = A(n) - 0.632A(n) + 0.368\ B(n) + 0.184\ C(n)$
$B(n + 1) = B(n) - 0.632\ B(n) + 0.368\ C(n)$
$C(n + 1) = C(n) - 0.552\ C(n) + 0.632\ A(n) + 0.264\ B(n)$

We seek to find the long-term behavior of this system. We iterate the DDS

A	B	C
1	0	0
0.368	0	0.632
0.251712	0.232576	0.515712
0.273109	0.27537	0.451521
0.28492	0.267496	0.447584
0.285645	0.263149	0.451206
0.284978	0.262883	0.452139
0.284806	0.263128	0.452066
0.28482	0.263191	0.451989
0.284834	0.263186	0.45198
0.284836	0.263181	0.451983
0.284835	0.26318	0.451984
0.284835	0.263181	0.451984
0.284835	0.263181	0.451984
0.284835	0.263181	0.451984
0.284835	0.263181	0.451984
0.284835	0.263181	0.451984
0.284835	0.263181	0.451984
0.284835	0.263181	0.451984
0.284835	0.263181	0.451984
0.284835	0.263181	0.451984
0.284835	0.263181	0.451984

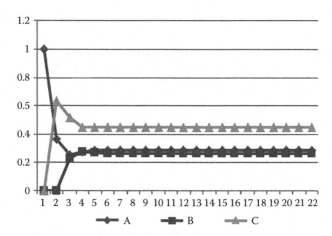

FIGURE 7.18
Plot of DDS for the aquarium inventory example.

The long-term probabilities for aquarium demand are $P(D = 1) = 0.284835$, $P(D = 2) = 0.263181$, and $P(D = 3) = 0.451984$. These are shown in Figure 7.18.

We iterate from near those equilibrium values and we find the sequences tend toward those values. We conclude that the system has *stable* equilibrium values.

You should go back and change the initial conditions and see what behavior follows.

Interpretation: The long-term behavior show that eventually (without other influences) the probabilities are $P(D = 1) = 0.284835$, $P(D = 2) = 0.263181$, and $P(D = 3) = 0.451984$.

We might want to try to attract new sales for aquariums by adding incentives for purchasing the aquarium.

Source: Adapted from Meerschaert, M., *Mathematical Modeling* (2nd ed.), Academic Press, San Diego, CA, 1999.

Example 7.12: Competitive Hunter Models

Competitive hunter models involve species vying for the same resources (such as food or living space) in the habitat. The effect of the presence of a second species diminishes the growth rate of the first species. We now consider a specific example concerning trout and bass in a small pond. Hugh Ketum owns a small pond that he uses to stock fish and eventually allows fishing. He has decided to stock both bass and trout. The fish and game warden tells Hugh that after inspecting his pond for environmental conditions he has a solid pond for growth of his fish. In isolation, bass grow at a rate of 20% and trout at a rate of 30%. The warden tells Hugh that the interactions for the food affect trout more than bass. They estimate the interaction affecting the bass is 0.0010 bass*trout and for trout is 0.0020 bass*trout. Assume that no changes in the habitant occur.

Model: Let us define the following variables:

$B(n)$ is the the number of bass in the pond after a period n.
$T(n)$ is the the number of trout in the pond after a period n.
$B(n) * T(n)$ is the interaction of the two species.

$$B(n+1) = 1.20\,B(n) - 0.0010\,B(n) * T(n)$$

$$T(n+1) = 1.30\,T(n) - 0.0020\,B(n) * T(n)$$

The equilibrium values can be found by allowing $X = B(n)$ and $Y = T(n)$ and solving for X and Y.

$$X = 1.2X - 0.001X * Y \tag{7.1}$$

$$Y = 1.3Y - 0.00201X * Y \tag{7.2}$$

We rewrite Equations 7.1 and 7.2 as

$$0.2\,X - 0.001\,X * Y = 0 \tag{7.3}$$

$$0.3Y - 0.002X * Y = 0 \tag{7.4}$$

We factor X out of Equation 7.3 and Y out of Equations 7.4 to obtain

$$X(0.2 - 0.001Y) = 0$$

$$Y(0.3 - 0.002X) = 0$$

Solving, we find $X = 0$ or $Y = 2000$ and $Y = 0$ or $X = 1500$.

We want to know the long-term behavior of the system and the stability of the equilibrium points.

Hugh initially considers 151 bass and 199 trout for his pond. From Hugh's initial conditions, bass will grow without bound and trout will eventually die out. This is certainly not what Hugh had in mind.

Example 7.13: Lanchester Combat Models

History is filled with examples of the unparalleled heroism and barbarism of war. Specific battles such as Bunker Hill, the Alamo, Gettysburg, Little Big Horn, Iwo Jima, and the Battle of the Bulge are a part of our culture and heritage. Campaigns such as the Cuban Revolution, Vietnam, and now the conflicts in Afghanistan and Iraq are a part of our personal history. Although combat is continuous, the models of combat usually employ discrete time simulation. For years, Lanchester equations were the norm for computer simulations of combat. The diagram of simple combat as modeled by Lanchester is illustrated in Figure 7.1. We investigate the use of a discrete version of these equations. We will use models of DDS through difference equations to model these conflicts and gain insight into the different methods of *directed fire* conflicts such as Nelson's

Battle at Trafalgar and the Alamo, and Iwo Jima. We employ *difference equations,* which allow for a complete numerical and graphical solution to be analyzed and do not require the mathematical rigor of differential equations. We further investigate the analytical form of the *direct fire* solutions to provide a solution template to be used in modeling efforts.

Lanchester model stated that *under conditions of modern warfare* that combat between two homogeneous forces could be modeled from the state condition of this diagram. We will call this diagram (Figure 7.19) the change diagram.

We will use the paradigm,

Future = present + change

to build our mathematical models. This will be paramount as eventually models will be built that cannot be solved analytically but can be solved by numerical (iteration) methods

We begin by defining the following variables:

$x(n)$ = the number of combatants in the X-force after a period n.
$y(n)$ = the number of combatants in the Y-force after a period n.

Future is then $x(n + 1)$ and $y(n + 1)$, respectively.
So, we have

$$x(n+1) = x(n) + \text{change}$$

$$y(n+1) = y(n) + \text{change}$$

Figure 7.19 provides the information of the change diagram that reflects change. Our dynamical system of equations is

$$x(n+1) = x(n) - k_1 y(n)$$

$$y(n+1) = y(n) - k_2 x(n)$$

(7.5)

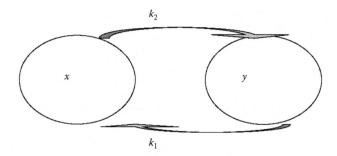

FIGURE 7.19
Change diagram of combat modeled by F.W. Lanchester.

We define our starting conditions as the size of the combatant forces at time period zero: $x(0) = x_0$ and $y(0) = y_0$.

Dynamical systems can always be solved by iteration, which make them quite attractive for use in computer modeling and simulations of combat. However, we can gain some powerful insights into those equations that have analytical solutions. This particular dynamical system of equations for Lanchester's direct fire model does have an analytical solution.

DISCRETE FORM OF LANCHESTER'S DIRECT FIRE EQUATIONS

We return to the typical system of equations of Lanchester's direct fire equations in difference equation form from Equation 7.5:

$$x(n+1) = x(n) - k_1 y(n)$$

$$y(n+1) = -k_2 x(n) + y(n)$$

Now, we write these in matrix form, Equation 7.6:

$$X_{n+1} = \begin{bmatrix} 1 & -k_1 \\ -k_2 & 1 \end{bmatrix} X_n, X_0 = \begin{bmatrix} x_0 \\ y_0 \end{bmatrix} \tag{7.6}$$

We need to take advantage of both eigenvalues and eigenvectors.

Let A be a $n \times n$ matrix. The real number λ is called an eigenvalue of A if there exists a nonzero vector x in R^n such that

$$Ax = \lambda x \tag{7.7}$$

The nonzero vector x is called an eigenvector of A associated with the eigenvalue λ.

Equation 7.7 is written as $Ax - \lambda x = 0$, or $(A - \lambda I)x = 0$
where I is a 2×2 identity matrix. The solution to finding λ comes from taking the determinant of the $(A-\lambda I)$ matrix, setting it equal to zero, and solving for λ.

We know that matrix A is $\begin{bmatrix} 1 & k_1 \\ -k_2 & 1 \end{bmatrix}$.

We set up the form for the use of eigenvalues:

$\det \begin{bmatrix} 1-\lambda & -k_1 \\ -k_2 & 1-\lambda \end{bmatrix} = 0$ that yields the characteristic equation

$$(1-\lambda) \cdot (1-\lambda) - k_1 k_2 = \lambda^2 - 2\lambda + 1 - k_1 k_2 = 0$$

We solve for λ. Although not intuitively obvious to the casual observer, the two eigenvalues are

$$\lambda_1 = 1 + \sqrt{k_1 k_2}$$

$$\lambda_2 = 1 - \sqrt{k_1 k_2} \tag{7.8}$$

Therefore, we have the eigenvalues from the initial form of the equation. We note that the eigenvalues are a function of the kill rates, k_1 and k_2. If you know the kill rates then you can easily obtain the two eigenvalues. Equation 7.8 yields the two eigenvalues.

We also note two other characteristics of the eigenvalues: (1) $\lambda_1 + \lambda_2 = 2$ and (2) $\lambda_1 \geq \lambda_2$. For most of these combat models one eigenvalue will be >1 and the other eigenvalue will be <1. The equation shows how attrition is being affected by the larger value of k_1 or k_2 has the eigenvalue, $\lambda > 1$.

Most literature on dynamical systems suggests that the dominant eigenvalue (that eigenvalue is the largest eigenvalue and greater than 1 in this case) will control the system. However, these combat models are observed to be controlled by the smaller eigenvalue.

The general form of the solution, Equation 7.9 is as follows:

$$X(k) = c_1 V_1 (\lambda_1)^k + c_2 V_2 (\lambda_2)^k \tag{7.9}$$

where the vector V_1 and V_2 are the corresponding eigenvectors.

These eigenvectors, interestingly enough, are in a ratio of the attrition coefficients, k_1 and k_2. The vector for the dominant eigenvalue always has both a positive and a negative component as its eigenvector, whereas the vector for the other smaller of the two eigenvalues always has two positive entries in this same ratio. This is because the equation for finding the eigenvector comes from

$$\sqrt{k_1 k_2}\, c_1 - k_1 c_2 = 0 \text{ and } -\sqrt{k_1 k_2}\, d_1 - k_1 d_2 = 0$$

So, $\tag{7.10}$

$$c_1 = k_1, \ c_2 = \sqrt{k_1 k_2} \text{ and } d_1 = -k_1, \ d_2 = \sqrt{k_1 k_2}$$

As shown in Equations 7.6. Having simplified formulas for obtaining eigenvalues and eigenvectors allows us to quickly obtain the general form of the analytical solution. We can then use the initial conditions to obtain the particular solution.

Example 7.14: Red and Blue Force

For example, consider a battle between a red force, $R(n)$, and a blue force, $B(n)$, as follows:

$$B(n+1) = B(n) - 0.1 \cdot R(n), B(0) = 100$$

$$R(n+1) = R(n) - 0.05 \cdot B(n), R(0) = 50$$

The ratio of $B(0)/R(0) = 100/50 = 2$.

We are given the attrition coefficients, $k_1 = -0.1$ and $k_2 = -0.050$.

Using the formulas that we just presented, we can quickly obtain the analytical solution.

$$\sqrt{k_1 k_2} = \sqrt{-0.1 - 0.05} = 0.0707$$

The eigenvalues are 1.0707 and 0.9293. We could build the closed form solution with the ratio of the vectors as ± 1 and $\left[\left(\sqrt{k_1 k_2}\right)/k_1\right]$. We find $\left[\left(\sqrt{k_1 k_2}\right)/k_1\right] = 0.7070$. Our general solution would be

$$X(k) = c_1 \begin{pmatrix} -1 \\ 0.707 \end{pmatrix}(1.0707)^k + c_2 \begin{pmatrix} 1 \\ 0.707 \end{pmatrix}(0.9293)^k$$

With our initial conditions of (100,50) at period 0, we have the particular solution

$$X(k) = -14.64 \begin{pmatrix} -1 \\ 0.707 \end{pmatrix}(1.0707)^k + 85.36 \begin{pmatrix} 1 \\ 0.707 \end{pmatrix}(0.9293)^k$$

We can graph these separately as shown in Figure 7.20, and observe the behavior:

a. Blue force over time
b. Red force over time

These two graphs (Figures 7.20a and b) of the analytical solution show as the y force (initially at size 50) approaches 0 as the x force (initially at 100) is slightly below 70. Thus, we know the x force or the blue force wins.

We can also develop a relationship for this *win* and quickly see that when $\sqrt{k_1 k_2} \cdot x_0 > k_1 \cdot y_0$ then the X force wins.

For our example, we find $\sqrt{k_1 k_2} \cdot x_0$ and $k_1 \cdot y_0$.

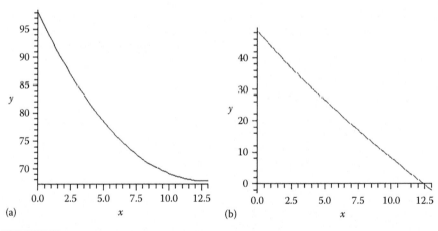

(a)

(b)

FIGURE 7.20
Solution graphs for (a) Blue force over time and (b) Red force over time.

$$x_0 \cdot \sqrt{k_1 k_2} = 100 \cdot 0.0707 = 7.07$$

$$k_2 \cdot y_0 = 0.1 \cdot 50 = 5$$

$$7.07 > 5.0$$

Since 7.07 is greater than 5.0 then the X force wins.
In general, the relationship can be $<$, $=$, or $>$. So, we state that

$$\sqrt{k_1 k_2}\, x_0 \begin{Bmatrix} > \\ = \\ < \end{Bmatrix} k_1 y_0 \qquad (7.11)$$

When the relationship in Equation 7.11 is $>$ then X wins, when the relationship is $<$ then Y wins, and when the relationship is an equality then we have a draw.

DEFINING A FAIR FIGHT: PARITY

The concept of parity in combat modeling is important. We define parity as a fight to finish that ends in a draw—neither side wins. We can find *parity* by either manipulating one of the initial conditions, x_0 or y_0, or one of the attrition coefficients k_1 or k_2.

Again the knowledge of the solution is critical to finding or obtaining these *parity* values. It turns out under *parity* that the eigenvectors are in a ratio of the square of the initial conditions.

One eigenvector is $\left(\dfrac{k_1}{\sqrt{k_1 k_2}} \right)$. So $\dfrac{k_1}{\sqrt{k_1 k_2}} = \dfrac{X_0}{Y_0}$ or $\sqrt{k_1 k_2}\, X_0 = k_1 Y_0$

Example 7.15: Parity between Red and Blue

Let us return to our example. Let us assume that blue force starts with 100 combatants and the red force with 50 combatants. Further let us fix k_1 at 0.1. What value is required for k_2 so that the red force fights a draw?

We find $\sqrt{(0.1)k_2}\,(100) = (0.1) \cdot (50)$

Thus, $k_2 = 0.025$.

If we fix k_2 at 0.05 and hold the initial number of combatants as fixed constants then k_1 would equal $k_1 = 0.2$.

If x starts with 100 soldiers and the kill rates are fixed, how many soldiers would y need? The y force needs 71 combatants to win.

We are able to quickly determine not only who wins the engagement but we can find values that allow both sides to fight to a draw. This is important because any deviation away from the parity values allows for one side to win the engagement. This helps a force that could be facing defeat to either increase their force enough to win or obtain better weaponry to improve their kill rates enough to win.

QUALITATIVE AND QUANTITATIVE APPROACH

We develop a few qualitative insights with the direct fire approach. First, we return to the forms:

$$\Delta X = -k_1 Y$$

$$\Delta Y = -k_2 X$$

We set both equal to zero and solve for the equations that make both equal to 0. This yields two lines $X = Y = 0$ that intersect at (0,0) the equilibrium point. Vectors point toward (0,0) but (0,0) is not stable. Our assumptions imply that trajectories terminate when it reaches either coordinate axis indicating one variable has gone to zero. Figures 7.21 and 7.22 illustrate the vectors and then the regions where the curves result in wins for X, wins for Y, or a draw (the solid line).

Recall our parity form: $\sqrt{k_1 k_2} \cdot x_0 = k_1 \cdot y_0$. This yields a nice line through the origin of the form: $y = \left(\sqrt{k_1 k_2}/k_1\right)x$ along which we have a draw. Above this line, we have the region where y wins and below we have the region where x wins. We plot our solution for y versus x and it is shown in Figure 7.22 that we are in the region where y wins.

ILLUSTRATIVE *DIRECT FIRE* EXAMPLES
AND HISTORICAL PERSPECTIVE

Let us use the theory and relationships developed to investigate some historical examples.

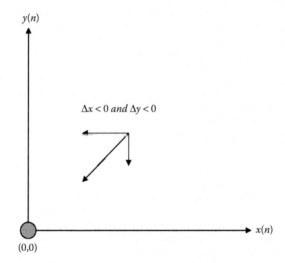

FIGURE 7.21
Rest point (0,0) of Y versus X for the direct fire model.

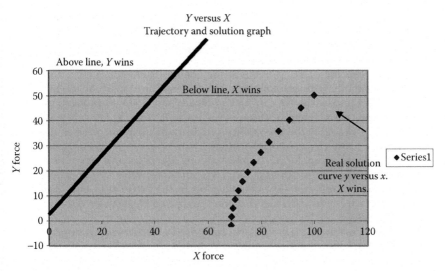

FIGURE 7.22
Trajectories for the basic direct fire model.

Example 7.16: The Battle of the Alamo

First, consider the situation at the Alamo. According to some historical records, there were approximately 189 Texans barricaded in the Alamo being attacked by 2000 Mexicans in the open fields surrounding the Alamo. We are interested in describing the loss of the combatants in each force over the course of the engagement. We will do this by measuring or defining change. We define $T(n)$ to be the number of Texans after period n and $M(n)$ to be the number of Mexican soldiers after time period n. That is, we want to devise a way to express $\Delta T = T(n+1) - T(n)$ (the loss of Texican combatants over time) and $\Delta M = M(n+1) - M(n)$ (the loss of Mexican combatants over time). The Battle of the Alamo is an example of a *directed fire* battle. The combatants on each side can see their opponents and can direct their fire at them. The Texans hiding behind the barricades were the more difficult target, and we need to have our models reflect this fact.

First, consider ΔM. On what does this depend? It depends on the number of bullets being fired by the Texican defenders and how accurately they are being fired at the Mexican army. We can use a proportionality model, $\Delta M \propto$ (number of bullets)(probability of hit). The number of bullets capable of being fired depends on how many men are firing and how rapidly each can fire. Given the weaponry at the time, it might be more effective to have only a portion of the combatants firing with the rest loading for them. This might increase the intensity of fire. There is also an issue of what portion of the force is in a position to fire on the enemy. If the force is in a rectangular formation, with several lines of combatants one behind the other, only the first one or two rows may be capable of firing freely at the enemy. Thus, ΔM = (Texicans) (% firing)(bullets/Texican/min)(probhit/bullet)(Mexicans disabled/hit)

All of these variables can be combined into a single proportionality constant k. Some of these variables will vary over distance or time. For example, the probability of a hit will likely increase as the Mexican army closes in on the Alamo. However, our model assumes each of these except for the number of combatants is constant over the course of the battle. Consequently, we can write, $\Delta M = -kT(n)$, where $T(n)$ is the number of Texans remaining in the battle after period n. The negative sign indicates that the number of Mexican combatants is decreasing.

Now, consider ΔT. It is similarly composed of terms such as number of Mexicans, percent firing, number of bullets per Mexican combatant per minute, the probability of a hit, and the number of Texans disabled per hit. We would expect that the rate of fire for the Mexican army to be smaller than the Texans, since they will be reloading while marching instead of reloading while standing still. Similarly, the probability of a hit will also be higher for the Texans shooting from a stance behind a wall than for the Mexicans shooting while marching in the open fields. So,

$$\Delta M = -k_1 T(n) \text{ and } \Delta T = -k_2 M(n)$$

but the values of k_1 and k_2 will be very different for the two forces. The constants k and c are known as the coefficients of combat effectiveness.

The Battle of the Alamo is actually two battles. The first battle was waged while the Mexicans were in the open field and the effectiveness constant k_2 was very much smaller than k_1 was to the advantage of the Texans. Once the Alamo walls were breached, the values of k_1 and k_2 were vastly altered, and the battle ended in a very short time. We model only the first battle as if it were a fight to the finish.

The model as described is

$$\begin{pmatrix} T(n+1) \\ M(n+1) \end{pmatrix} = \begin{pmatrix} 1 & -0.06 \\ -0.5 & 1 \end{pmatrix} \begin{pmatrix} T(n) \\ M(n) \end{pmatrix}, \begin{pmatrix} T(0) = 200 \\ M(0) = 1200 \end{pmatrix}$$

From our equation $\sqrt{k_1 k_2} \cdot T_0 < k_1 \cdot M_0$ we have
0.1732 (200) < 0.5(1200) and we know that the Mexican army wins decisively. In Table 7.1, we obtained the values to achieve parity in each case. We can easily see that many of these values are unrealistic for the event. The Texans were going to lose this battle without outside help.

Parity	k_1	k_2	$T(0)$	$M(0)$
k_1 variable	2.16 (very unrealistic value)	0.06	200	1200
k_2 variable	0.5	0.01388	200	1200
$T(0)$ variable	0.5	0.06	200	578
$M(0)$ variable	0.5	0.06	3464 (unrealistic)	1200

Example 7.17: The Battle of Trafalgar

Another classic example of the directed fire model of combat is the Battle of Trafalgar. In classical naval warfare, two fleets would sail parallel to each other and fire broadside at one another until one fleet was annihilated or gave up (see Figure 7.23). The white fleet represents the British and the black fleet represents the French–Spanish fleet.

In such an engagement, the fleet with superior firepower will inevitably win. To model this battle, we begin with the system of difference equations that models the interaction of two fleets in combat. Suppose we have two opposing forces with A_0 and B_0 ships initially, and $A(t)$ and $B(t)$ ships t units of time after the battle is engaged. Given the style of combat at the time of Trafalgar, the losses for each fleet will be proportional to the effective firepower of the opposing fleet. That is,

$$\Delta A = -bB \text{ and } \Delta B = -aA$$

where a and b are positive constants that measure the effectiveness of the ship's cannonry and personnel and A and B are both functions of time. In preparing for the Battle at Trafalgar, Admiral Nelson assumed that the coefficients of effectiveness of the two fleets were approximately equal. To keep things simple initially, we let $a = b = 0.05$. The figure and numerical listing mentioned in the following allow us to look at many different initial settings and try to ascertain a pattern in the results of the battles.

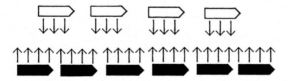

FIGURE 7.23
The white fleet takes a beating.

We could iterate these numbers to find who wins, as those with the larger numbers win.

Stage	British force	French force
1	27	33
2	25.35	31.65
3	23.7675	30.3825
4	22.248375	29.194125
5	20.7886688	28.08170625
6	19.3845834	27.04227281
7	18.0324698	26.07304364
8	16.7288176	25.17142015
9	15.4702466	24.33497927
10	14.2534976	23.56146694
11	13.0754243	22.84879206
12	11.9329847	22.19502084
13	10.8232337	21.59837161
14	9.74331507	21.05720993
15	8.69045458	20.57004417
16	7.66195237	20.13552144
17	6.65517629	19.75242382
18	5.6675551	19.41966501
19	4.69657185	19.13628725
20	3.73975749	18.90145866
21	2.79468456	18.71447079
22	1.85896102	18.57473656
23	0.93022419	18.48178851
24	0.00613476	18.4352773

In this example, Admiral Nelson has 27 ships, whereas the allied French and Spanish fleet had 33 ships. As we can see in Figure 7.24, Admiral Nelson is expected to lose all 27 of his ships, whereas the allied fleet will lose only about 14 ships.

Now, let us return to our equations that we developed earlier

$$(0.05)(33) > (0.05)(27)$$

$$1.65 > 1.35$$

Since $\sqrt{k_1 k_2} \cdot FS_0 > k_1 \cdot B_0$ then the French–Spanish fleet wins. The analytical solution can be easily developed as

$$X(k) = -3\begin{pmatrix} -1 \\ 1 \end{pmatrix}(1.05)^k + 30\begin{pmatrix} 1 \\ 1 \end{pmatrix}(0.95)^k$$

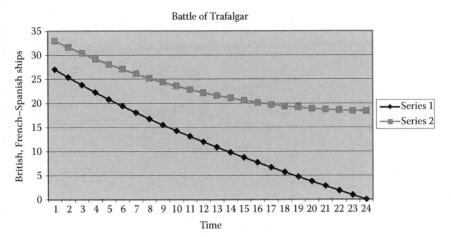

FIGURE 7.24
Battle of Trafalgar under normal battle strategies.

In order for the British to win, we first find the values that provide them with a draw. We find the British would require 33 ships to have obtained parity. In addition, we find that the British would have to increase their kill effectiveness to 0.07469 to obtain a draw. Increases just beyond these values, give the British the theoretical edge. However, there were no more ships and the armaments were in place on the ships already. The only option would be a change in strategy.

We can also test this new strategy was used by Admiral Nelson at the Battle of Trafalgar. Admiral Nelson decided to move away from the course of linear battle of the day and use a *divide and conquer* strategy. Nelson decided to break his fleet into two groups of size 13 and size 14. He also divided the enemy fleet into three groups: (1) a force of 17 ships (called B), (2) a force of 3 ships (called A), and (3) a force of 13 ships (called C). We can assume these as the head, middle, and tail of the enemy fleet. His plan was to take the 13 ships and attack the middle 3 ships. Then he had his reserve 14 ships to rejoin the attack and then to attack the larger force B, and then turn to attack the smaller force C. How did Nelson's strategy prevail?

Assuming that all other variables remain constant other than the order of the attacks against the differing size forces, we find the Admiral Nelson and the British fleet now win the battle sinking all the French–Spanish ships with 13 to 14 ships remaining, as shown in Figure 7.25.

How did we obtain these results? The easiest method was iteration and used three battle formulas. We stop each battle when one of the values gets close to zero (before going negative).

Source: Modified from Giordano et al., 2014, pp. 40–43.

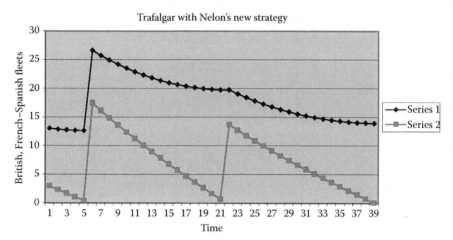

FIGURE 7.25
British prevail with new strategy.

Example 7.18: Determining the Length of the Battle of Trafalgar

To determine the length of the battle, we need to see something in the solution system of equations $\Delta A = A(n+1) - A(n) = -k_1 B(n)$ and $\Delta B = B(n+1) - B(n) = -k_2 A(n)$ that should be obvious. Recall the solution to our first example:

$$X(k) = -14.64 \begin{pmatrix} -1 \\ 0.707 \end{pmatrix}(1.0707)^k + 83.36 \begin{pmatrix} 1 \\ 0.707 \end{pmatrix}(0.9293)^k. \text{ This simplifies to}$$

$$X(k) = -\begin{pmatrix} 14.64 \\ -10.35 \end{pmatrix}(1.0707)^k + \begin{pmatrix} 83.36 \\ 58.9355 \end{pmatrix}(0.9293)^k$$

The graph shows that x wins (as our other analysis), so the time parameter we are interested in is *when does y go o zero*? If you try to use the x equation, we end up with trying to take the ln of a negative number, which is not possible.

We use $y(k) = -10.35 \cdot 1.0707^k + 58.9355 \cdot 0.9293^k$ and set $y(k) = 0$.

The solution for k (which is our time parameter) is

$$k = \frac{\ln\left(\dfrac{58.9335}{10.35}\right)}{\ln\left(\dfrac{1.0707}{0.9293}\right)} = 12.28 \text{ time periods.}$$

In general, the time parameter is either of the following two equations:

$$\frac{\ln\left(\frac{c_1 v_{11}}{c_2 v_{12}}\right)}{\ln\left(\frac{\lambda_1}{\lambda_2}\right)} \quad \text{or} \quad \frac{\ln\left(\frac{c_1 v_{21}}{c_2 v_{22}}\right)}{\ln\left(\frac{\lambda_1}{\lambda_2}\right)} \tag{7.12}$$

depending on which form yields the ln(positive number) in the numerator.

If our Red–Blue combat data were in kills/hour, then the battle will last for 12.28 hours. Often we are interested in the approximate time or length of the battle. These formulas in Equation 7.12 allow for a quick computation.

Example 7.19: Battle of Iwo Jima

At Iwo Jima in WWII, the Japanese had 21,500 soldiers and the United States had 73,000 soldiers. We assume that all forces were initially in place. The combatants were engaged in conventional direct warfare, but the Japanese were fighting from reinforced entrenchments. The coefficient of effectiveness for the Japanese was assumed to be 0.0544, whereas that of the United States side was assumed to be 0.0106 (based on data after the battle). If these values are approximately correct, which side should win? How many should remain on the winning side when the other side has only 1500 remaining?

We move directly to both the winning conditions and an analytical solution to answer these questions.

$$\sqrt{k_1 k_2}\, x_0 \begin{Bmatrix} > \\ = \\ < \end{Bmatrix} k_1 y_0$$

$$0.02401(73{,}000) \begin{Bmatrix} > \\ = \\ < \end{Bmatrix} (0.0106)*(21{,}500)$$

1752.97 > 227.90, so we know that the United States wins decisively. The analytical solution is

$$X(k) = -12145.67 \begin{pmatrix} -1 \\ 0.4414 \end{pmatrix}(1.024)^k + 60854.33 \begin{pmatrix} 1 \\ 0.4414 \end{pmatrix}(0.976)^k$$

or

$$X(k) = \begin{pmatrix} 12145.67 \\ -5361.1 \end{pmatrix}(1.024)^k + \begin{pmatrix} 60854.33 \\ 26861.1 \end{pmatrix}(0.976)^k$$

We solve the equation for time it takes the Japanese to reach 1500 soldiers. We find that it takes 30.922 time periods for the Japanese to reach 1500 soldiers. Thus, the model shows that U.S. had approximately 53,999 soldiers remaining.

The battle actually ended with 1500 Japanese survivors and 44314 U.S. survivors and took approximately 33–34 days. Our model's approximations are not too bad. We are off by about 6% in the time and by 21.8% in the number of surviving U.S. soldiers. The error in surviving soldiers should cause us to revisit the model's assumptions for an explanation. The United States actually used a phased landing over 15 days of actual combat to reach their final force of 73,000 soldiers. We could have treated this like the Trafalgar battle with at least 15 different battles to be more accurate.

Example 7.20: Insurgency and Counterinsurgency Operations

Today's warfare is different. The dynamics of today's battlefield is quite different. Consider the later stages of the war in Iraq that has become a multiring conflict. The issues of regular warfare, civil war, and insurgency warfare are all taking place.

Insurgency and counterinsurgency operations can be modeled in a simplified sense using the following discrete Lanchester model using a modified version of Brackney's Mixed law (also called the Parabolic Law was developed in 1959). This can be used to represent Guerilla warfare and can now be used to represent insurgency and counterinsurgency operations.

We define $Y(n)$ to be the insurgent strength after a period n and we define $X(n)$ to be the government troop strength after a period n.

Then,

$$X(n+1) = X(n) - k_1 * X(n) * Y(n)$$

$$Y(n+1) = Y(n) - k_2 * X(n)$$

where k_1 and k_2 are kill rates.

Further, If we model the total conflict with both growth and attrition we could use the following models:

$$X(n+1) = X(n) + a * (K_1 - X(n)) * X(n) - k_1 * X(n) * Y(n)$$

$$Y(n+1) = Y(n) + b * (K_2 - Y(n)) * Y(n) - k_2 * X(n)$$

where:
k_1 and k_2 are kill rates
a and b are positive constants
K_1 and K_2 are carrying capacities

This is a combination of the growth model and the combat model and represents when the conflict is ongoing and growth is still part of the insurgency operation.

These types of equations can only be solved and analyzed using numerical iteration. Having laptops with Excel enables soldiers/ decision-makers to characterize the situations and get quick *results*.

COMPARISON TO STANDARD LANCHESTER EQUATIONS THROUGH DIFFERENTIAL EQUATIONS

Let us revisit the red and blue force illustrative example now as a differential equation.

$$\frac{dx(t)}{dt} = -0.1 \cdot y(t)$$

$$\frac{dy(t)}{dt} = -0.05 \cdot x(t)$$

$$x(0) = 100, y(0) = 50$$

This system of differential equations yields the solution to three decimals places:

$$x(t) = 14.644 \cdot e^{0.0707 t} + 85.355 \cdot e^{-0.707 t}$$

$$y(t) = -10.355 \cdot e^{0.0707 t} + 60.355 \cdot e^{-0.707 t}$$

We provide a plot of the solution through differential equations and the solution through difference equations in Figure 7.26. Note how close they align.

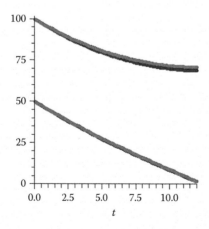

FIGURE 7.26
Solution plots of blue versus red forces through DE and DDS methods.

CONCLUSIONS ABOUT LANCHESTER EQUATIONS

The use of difference equations in combat modeling has a practical value. Not only do analytical solutions allow analysts to provide decision-makers with quantitative information to quickly analyze potential results but every difference equation has a numerical solution that can be achieved easily. For the decision-maker in the field a differential equation is an abstract concept and the tools for analysis are not available. However, an EXCEL spreadsheet is a powerful tool for decision-makers who are available in the field. The systems of difference equations based on *Future = Present + Change* are an intuitive, nonevasive approach for which every combat model has a numerical solution and some combat models such as the direct fire models have analytical solutions that directly lend themselves to analysis and results.

Example 7.21: A Predator–Prey Model: Foxes and rabbits

In the study of the dynamics of a single population, we typically take into consideration such factors as the *natural* growth rate and the *carrying capacity* of the environment. Mathematical ecology requires the study of populations that interact, thereby affecting each other's growth rates. In this module, we study a very special case of such an interaction, in which there are exactly two species, one of which the predators eat the prey. Such pairs exist throughout the nature:

- Lions and gazelles
- Birds and insects
- Pandas and eucalyptus trees
- Venus fly traps and flies

To keep our model simple, we will make some assumptions that would be unrealistic in most of these predator–prey situations. Specifically, we will assume that

- The predator species is totally dependent on a single prey species as its only food supply
- The prey species has an unlimited food supply
- There are no other threats to the prey other than the specific predator.

THE DISCRETE LOTKA–VOLTERRA MODEL FOR PREDATOR–PREY MODELS

Vito Volterra (1860–1940) was a famous Italian mathematician who retired from a distinguished career in pure mathematics in the early 1920s. His son-in-law, Humberto D'Ancona, was a biologist who studied the populations of various species of fish in the Adriatic Sea. In 1926, D'Ancona completed a statistical study of the numbers of each species sold on the fish markets of three ports: (1) Fiume, (2) Trieste, and (3) Venice. The percentages of predator species (sharks, skates, rays, etc.) in the Fiume catch are shown in Table 7.3.

TABLE 7.3

Percentages of Fiume Fish Catch

Percentages of Predators in the Fiume Fish Catch									
1914	1915	1916	1917	1918	1919	1920	1921	1922	1923
12	21	22	21	36	27	16	16	15	11

We may assume that proportions within the *harvested* population reflect those in the total population. D'Ancona observed that the highest percentages of predators occurred during and just after World War I, when fishing was drastically curtailed. He concluded that the predator–prey balance was at its natural state during the war and that intense fishing before and after the war disturbed this natural balance—to the detriment of predators. Having no biological or ecological explanation for this phenomenon, D'Ancona asked Volterra if he could come up with a mathematical model that might explain what was going on. In a matter of months, Volterra developed a series of models for interactions of two or more species. The first and simplest of these models is the model that we will use for this scenario.

Alfred J. Lotka (1880–1949) was an American mathematical biologist (and later actuary) who formulated many of the same models as Volterra independently and at about the same time. His primary example of a predator–prey system comprised a plant population and an herbivorous animal dependent on that plant for food.

We repeat our two key assumptions:

- The predator species is totally dependent on the prey species as its only food supply.
- The prey species has an unlimited food supply and no threat to its growth other than the specific predator.

If there were no predators, the second assumption would imply that the prey species grows exponentially without bound, that is, if $x = x(n)$ is the size of the prey population after a discrete time period n, then we would have $x(n + 1) = a\, x(n)$.

However, there *are* predators, which must account for a negative component in the prey growth rate. Suppose we write $y = y(n)$ for the size of the predator population at time t. Following are the crucial assumptions for completing the model:

- The rate at which predators encounter prey is jointly proportional to the sizes of the two populations.
- A fixed proportion of encounters lead to the death of the prey.

These assumptions lead to the conclusion that the negative component of the prey growth rate is proportional to the product xy of the population sizes, that is,

$$x(n+1) = x(n) + ax(n) - bx(n)y(n)$$

Now, we consider the predator population. If there were no food supply, the population would die out at a rate proportional to its size, that is,
we would find $y(n+1) = -cy(n)$.

We assume that is the simple case that the *natural growth rate* is a
composite of birth and death rates, both presumably proportional to
the population size. In the absence of food, there is no energy supply
to support the birth rate. However, there is a food supply: the prey.
And what is bad for hares is good for lynx. That is, the energy to support growth of the predator population is proportional to deaths of
prey, so

$$y(n+1) = y(n) - cy(n) + px(n)y(n)$$

This discussion leads to the discrete version of the Lotka–Volterra
predator–prey Model shown as Equation 7.13:

$$x(n+1) = (1+a)x(n) - bx(n)y(n)$$

$$y(n+1) = (1-c)y(n) + px(n)y(n) \tag{7.13}$$

$$n = 0,1,2,\ldots$$

where a, b, c, and p are positive constants.

The Lotka–Volterra model as shown in Equation 7.13 consists of a
system of linked difference equations that cannot be separated from
each other and that cannot be solved in closed form. Nevertheless, they
can be solved numerically through iteration and graphed in order to
obtain insights into the scenario being studied. In our foxes and hares
scenario, let us assume this discrete model explained earlier. Further,
data investigation yields the following estimates for the parameters
$\{a, b, c, p\} = \{0.039, 0.0003, 0.12, 0.0001\}$. We plot the results in Figures 7.27
through 7.29.

If we ran this model for more iterations then we would find the plot
of foxes versus rabbits continue spiral in a similar fashion as mentioned
earlier. We conclude that the model appears reasonable. We could find
the equilibrium values for the system. There are two sets of equilibrium
points for rabbits and foxes: (0,0) and (1200,130). The orbits of the spiral
indicate that the system is moving away from (1200,130), so we conclude
that the system is not stable. In the exercise set, you will be asked to do
more explorations.

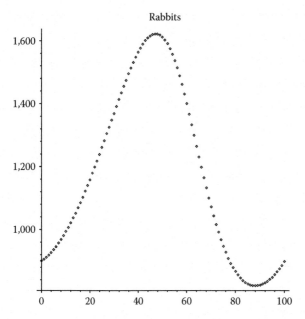

FIGURE 7.27
Rabbits over time.

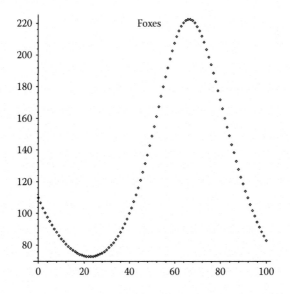

FIGURE 7.28
Foxes over time.

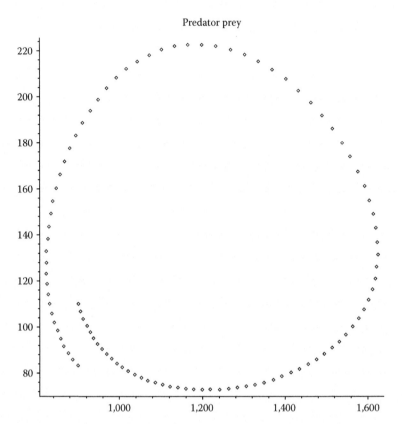

FIGURE 7.29
Foxes versus Rabbits over time.

Example 7.22: Discrete Susceptible (S), Infected (I), and Resistant (R) Model (SIR Model) of Epidemics

Consider a disease that is spreading throughout the United States such as the new flu or Ebola. The Center for Disease Control (CDC) is interested in knowing and experimenting with a model for this new disease before it actually becomes a *real* epidemic. Let us consider the population being divided into three categories: (1) susceptible, (2) infected, and (3) removed. We make the following assumptions for our model:

- No one enters or leaves the community and there is no contact outside the community.
- Each person is either susceptible, *S* (able to catch this new flu); infected, *I* (currently has the flu and can spread the flu); or removed, *R* (already had the flu and will not get it again that includes death).
- Initially every person is either *S* or *I*.
- Once someone gets the flu this year they cannot get it again.

- The average length of the disease is 2 weeks over which the person is deemed infected and can spread the disease.
- Our time period for the model will be per week.

The model we will consider is the SIR model (Allman and Rhodes, 2004, pp. 276–314).
Let us assume the following definition for our variables.

$S(n)$ = number in the population susceptible after a period n.
$I(n)$ = number infected after a period n.
$R(n)$ = number removed after a period n.

Let us start our modeling process with $R(n)$. Our assumption for the length of time someone has the flu is 2 weeks. Thus, half the infected people will be removed each week:

$$R(n+1) = R(n) + 0.5 I(n)$$

The value, 0.5, is called the removal rate per week. It represents the proportion of infected persons who are removed from infection each week. If real data are available, then we could do *data analysis* in order to obtain the removal rate.

$I(n)$ will have terms that both increase and decrease its amount over time. It is decreased by the number that is removed each week, $0.5*I(n)$. It is increased by the numbers of susceptible that come into contact with an infected person and catch the disease, $aS(n)I(n)$. We define the rate, a, as the rate at which the disease is spread or the transmission coefficient. We realize this is a probabilistic coefficient. We will assume, initially, that this rate is a constant value that can be found from initial conditions.

Let us illustrate as follows. Assume we have a population of 1000 students in the dorms. Our nurse found three students reporting to the infirmary initially. The next week, five students came in to the infirmary with flu-like symptoms. $I(0) = 3$, $S(0) = 997$. In week 1, the number of newly infected is 30.

$$5 = a I(n)S(n) = a(3)*(995)$$
$$a = 0.00167$$

Let us consider $S(n)$. This number is decreased only by the number that becomes infected. We may use the same rate, a, as before to obtain the model:

$$S(n+1) = S(n) - aS(n)I(n)$$

Our coupled SIR model is

$$R(n+1) = R(n) + 0.5I(n)$$
$$I(n+1) = I(n) - 0.5I(n) + 0.00167I(n)S(n)$$
$$S(n+1) = S(n) - 0.00167S(n)I(n)$$
$$I(0) = 3, S(0) = 997, R(0) = 0$$

$$(7.14)$$

The SIR model in Equation 7.14 can be solved iteratively and viewed graphically. Let us iterate the solution and obtain the graph to observe the behavior to obtain some insights, see Figures 7.30 through 7.33.

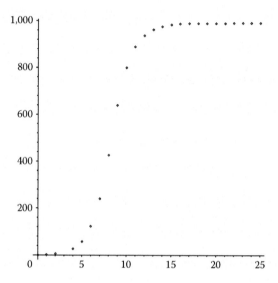

FIGURE 7.30
Plot of $R(n)$ versus n.

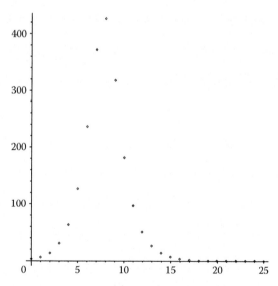

FIGURE 7.31
Plot of $I(n)$ versus n.

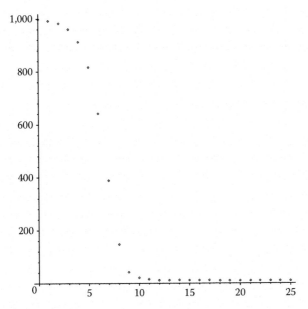

FIGURE 7.32
Plot of $S(n)$ *versus n.*

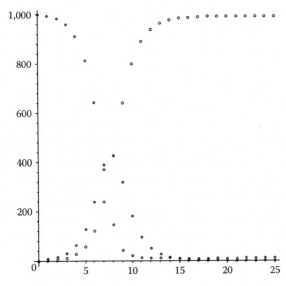

FIGURE 7.33
Plot of $R(n)$, $I(n)$, $S(n)$ versus n together.

The worse of the flu epidemic occurs around week 8, at the maximum of the infected graph. The maximum number is slightly larger than 400, from the table it is 427. After 25 weeks, slightly more than 9 people never get the flu.

Example 7.23: Modeling Military Insurgencies

Insurgent forces have a strong foothold in the city of Urbania, a major metropolis in the center of the country of Ibestan. Intelligence estimates that they currently have a force of about 1000 fighters. The local police force has approximately 1300 officers, many of which have had no formal training in law enforcement methods or modern tactics for addressing insurgent activity. Based on data collected over the past year, approximately 8% of insurgents switch sides and join the police each week, whereas about 11% of police switch sides and join the insurgents. Intelligence also estimates that around 120 new insurgents arrive from the neighboring country of Moronka each week. Recruiting efforts in Ibestan yield about 85 new police recruits each week as well. In armed conflict with insurgent forces, the local police are able to capture or kill approximately 10% of the insurgent force each week on average while losing about 3% of their force.

> *Problem Statement*: Build a mathematical model of this insurgency. Determine the equilibrium state (if it exists) for this DDS.

We define the variables

$P(n)$ = the number of police in the system after a time period n.
$I(n)$ = the number of insurgents in the system after a time period n.
$n = 0,1,2,3,\dots$ weeks

Model:

$$P(n+1) = P(n) - 0.03\,P(n) - 0.11\,P(n) + 0.08\,I(n) + 85, P(0) = 1300$$

$$I(n+1) = I(n) - 0.11\,P(n) - 0.08\,I(n) - 0.1\,I(n) + 120\,I(0) = 1000$$

In this insurgency model, we see that under the current conditions, the insurgency overtakes the government within 5 to 10 weeks through an examination of Figures 7.34 through 7.36. If this is unacceptable, then we must modify conditions that affect the parameters in such a way to obtain a governmental victory. You can use the model to explore strategies and their effect on the results in order to find a course of action.

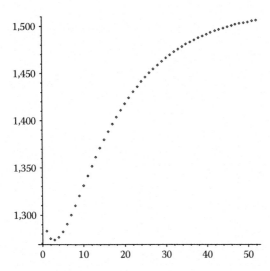

FIGURE 7.34
Plot of $P(n)$ versus n.

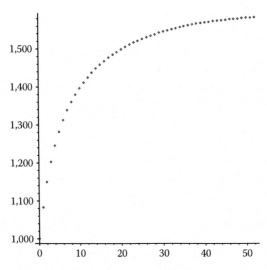

FIGURE 7.35
Plot of $I(n)$ versus n.

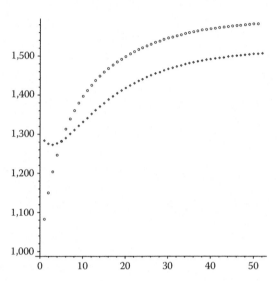

FIGURE 7.36
Solution plots of both $P(n)$ and $I(n)$ versus time period n for Example 7.23.

Exercises 7.5

1. What happens to the merchant problem if 200 were initially in the mall and 50 were in the downtown portion?

2. Determine the equilibrium values of the bass and trout. Can these levels ever be achieved and maintained? Explain.

3. Test the fast food models with different starting conditions summing to 14,000 students. What happens? Obtain a graphical output and analyze the graph in terms of long-term behavior.

4. The Battle of Iwo Jima

   The validity of Lanchester's equation can be demonstrated in an actual situation where U.S. forces captured the island of Iwo Jima. Information required for the verification and *what if* analysis is the number of friendly troops put ashore each day and the number of friendly causalities for each days' engagement, knowledge that the enemy troops were not reinforced or withdrawn, the number of enemy troops at the start of the battle, and the number at the end of the engagement, and the length of the engagement. The enemy was well entrenched into the rocks on the island. The U.S. forces were attacking into the enemy's prepared defenses. In an idealized situation, the U.S. forces would be considered to follow a modified Lanchester's square law with replacement troops landing each day, whereas the enemy could be considered to follow the standard square law.

Part 1: Since the enemy is entrenched and looking down onto the U.S. troops attacking it is easier to hit and kill U.S. forces. The P(hit U.S. troops with an enemy weapon) = 0.54 and the P(kill U.S. troops|a hit) = 0.1. We assume these events are independent and that their product represents the kill coefficient of the aggregated Japanese forces against U.S. forces. The P(hit Japanese troops with a U.S. weapon) = 0.12 and we will assume P(kill Japanese troops | a hit) is also 0.1. We assume that these events are independent and that their product represents the kill coefficient of the aggregated U.S. forces against the Japanese forces.

1. Determine the kill rates for the U.S. and the Japanese forces. Explain from your knowledge of probability and statistics why they might be reasonable.

2. Determine who wins a fight to the finish.

3. *Parity*: Is parity possible in this problem? Can we find it easy? Is it possible or easier to find parity after all the U.S. troops have landed and now assumed that the battle is new? Under this scenario of what kill ratio could the enemy have reached parity? Is that value feasible? Explain.

4. The real battle ended with 1,500 Japanese survivors and 44,314 U.S. survivors and took approximately 33–34 days. Relate your result with these real results. If different, why do you think these results are different?

5. Reflect on your use of Lanchester equations to adequately explain the results of the Battle of Iwo Jima.

    Enemy initial strength was 21,000 troops in fortified positions on the island.

    Friendly troop strength was modified by landing as follows:

Day 1	30,000
Day 2	1,200
Day 3	6,735
Day 4	6,626
Day 5	5,158
Day 6	13,227
Day 7	3,054
Day 8	3,359
Day 9	3,180
Day 10	1,456
Day 11	250
Thereafter	0
Total troops	71,245

5. Find the equilibrium values for the predator–prey model presented.
6. Find the equilibrium values for the SIR model presented.
7. Find the equilibrium values for the combat model presented.
8. In the predator–prey model determine the outcomes with the following sets of parameters:

   a. Initial foxes are 200 and initial rabbits are 400.

   b. Initial foxes are 2,000 and initial rabbits are 10,000.

   c. Birth rate of rabbits increases to 0.1.
9. In the SIR model determine the outcome with the following parameters changed.

   a. Initially 5 are sick and 10 the next week.

   b. The flu lasts 1 week.

   c. The flu lasts 4 weeks.

   d. There are 4,000 students in the dorm and 5 are initially infected and 30 more the next week.

Projects 7.5

1. *Owls and Hawks*

   Problem Statement. We need to predict the number of owls and hawks in the same environment at a function of time.

   Assumptions:

   The variables:

   $$O(n) = \text{number of owls at the end of period } n$$
   $$H(n) = \text{number of hawks at the end of period } n$$

   Model:

   $$O(n+1) = 1.2O(n) - 0.001O(n)H(n)$$

   $$H(n+1) = 1.3\,H(n) - 0.002\,H(n)O(n)$$

   1. Find the equilibrium values of the system.
   2. Iterate the system from the following initial conditions and determine what happens to the hawks and owls in the long term:

Owls	Hawks
150	200
151	199
149	210
200	300
300	200

2. Ricky and Franky play racquetball very often and are very competitive. Their racquetball match consists of two games. When Ricky wins the first game, he wins the second game 65% of the time. When Franky wins the first game, she wins the second only 45% of the time. Model this as a DDS and determine the long-term percentages of their racquetball games. What assumptions are necessary?

3. It is getting close to the election day. The influence of the new Independent Party is of concern to the Republicans and Democrats. Assume that in the next election 75% of those who vote Republican vote Republican again, 5% vote Democratic, and 20% vote Independent. Of those that voted Democratic before, 20% vote Republican, 60% vote Democratic, and 20% vote Independent. Of those that voted for Independent, 40% vote Republican, 20% vote Democratic, and 40% vote Independent.

   a. Formulate and write the system of DDS that models this situation.

   b. Assume that there are 399,998 voters initially in the system, how many will vote Republican, Democratic, and Independent in the long run? (*Hint*: You can break down the 399,998 voters in any manner that you desire as initial conditions.)

   c. (New Scenario) In addition to the earlier, the community is growing. (18-year-olds + new people − deaths − losses to the community, etc.) Republicans prjedict a gain of 2,000 voters between elections. Democrats estimate a gain of 2,000 voters between elections. The independents estimate a gain of 1000 voters between elections. If this rate of growth continues, what will be the long-term distribution of the voters?

7.6 Summary

In this chapter, we have explored the use of DDS using the paradigm, Future = Present + Change. We showed the importance of the change diagram to assist in the model equation formulation. We illustrated a solution process through iteration as well as visual approach. We introduced the concept of equilibrium and stable equilibrium to assist in analyzing the behavior of the DDS. We provided linear, nonlinear, and systems of DDS examples.

References and Suggested Further Readings

Allman, E. and J. Rhodes. 2004. *Mathematical Models in Biology: An Introduction.* New York: Cambridge University Press, pp. 276–314.

Giordano, F., W. Fox, and S. Horton. 2014. *A First Course in Mathematical Modeling* (5th ed.). Boston, MA: Cengage Publishing.

Sandefur, J. 2003. *Elementary Mathematical Modeling.* Pacific Grove, CA: Thompson Publishing.

Bonder, S. 1981. Mathematical modeling of military conflict situations. *Proceedings of Symposia in Applied Mathematics, Volume 25, Operations Research, Mathematics and Models,* American Mathematical Society.

Braun, M. 1981. *Differential Equations and Their Applications* (3rd ed.). Berlin, Germany: Springer-Verlag.

Coleman, C. S. 1983. Combat models. In M. Braun, C. Coleman, D. Drew and W. Lucas (Eds.), *Differential Equation Models,* Vol. 1 of *Models in Applied Mathematics,* New York: Springer-Verlag.

Engel, J. H. 1954. A verification of Lanchester's Law. *Journal of Operations Research Society of America,* 2(2), 163–171.

Lanchester, F.W. 1956. Mathematics in warfare. In J. Newman (Ed.), *The World of Mathematics,* (Vol. 4). New York: Simon and Shuster.

Taylor, J. G. 1980. *Force-on-Force Attrition Modeling.* Arlington, VA: Military Applications Sections (ORSA).

Taylor, J. G. 1983. *Lanchester Models of Warfare* (Vol. 2). Arlington, VA: Military Applications Sections (ORSA).

Teague, D. 2005. Combat models. *Teaching Contemporary Mathematics Conference,* Durham, NC.

Meerschaert, M. 1999. *Mathematical Modeling* (2nd ed.). San Diego, CA: Academic Press.

8

Simulation Modeling

OBJECTIVES

1. Understand the power and limitation to simulations.
2. Understand random numbers.
3. Understand the concept of simulation algorithms or flow charts.
4. Build simple deterministic and stochastic simulations using technology.
5. Understand the law of large numbers in simulations.

8.1 Introduction

Consider an engineering company that conducts vehicle inspections for a specific state. We have data for times of vehicle arrivals and departures, service times for inspectors under various conditions, numbers of inspection stations, and penalties levied for failure to meet state inspection standards in terms of waiting time for customers. The company wants to know how it can improve its inspection process throughout the state in order to both maximize its profit and minimize the penalties it receives. This type of analysis of a complex system has many variables, and we could use a computer simulation to model this operation.

A modeler may encounter situations in which the construction of an analytic model is infeasible because of the complexity of the situation. In instances where the behavior cannot be modeled analytically or where data are collected directly, the modeler might simulate the behavior indirectly and then test various alternatives to estimate how each affects the behavior. Data can then be collected to determine which alternative is best. Monte Carlo simulation is a common simulation method that a modeler can use, usually with the aid of a computer. The proliferation of today's computers in the academic and business worlds makes Monte Carlo simulation very attractive. It is imperative that students have at least a basic understanding of how to use and interpret Monte Carlo simulations as a modeling tool.

There are many forms of simulation ranging from building scale models such as those used by scientists or designers in experimentation to various types of computer simulations. One preferred type of simulation is the Monte Carlo simulation. The Monte Carlo simulation deals with the use of random numbers. There are many serious mathematical concerns associated with the construction and interpretation of Monte Carlo simulations. Here, we are concerned only with reinforcing the techniques of simulations with these random variates.

A principal advantage of Monte Carlo simulation is the ease with which it can be used to approximate the behavior of very complex systems. Often, simplifying assumptions must be made to reduce this complex system into a manageable model. In the environment forced on the system, the modeler attempts to represent the real system as closely as possible. This system is probably a stochastic system; however, simulation can allow either a deterministic or stochastic approach. We will concentrate on the stochastic modeling approach to deterministic behavior.

Many undergraduate mathematical science, engineering, and operations research programs currently require or offer a course involving simulation. Typically, such a course will use a high-level simulation language such as C++, Java, FORTRAN, SLAM, Prolog, STELLA, Siman, or GPSS as the tool to teach simulation. Here, we will use Excel to simulate some simple modeling scenarios.

Our emphasis is twofold. First, we want you (the student) to think in terms of an algorithm, not a specific language. Second, we want you to understand that *more* is better in Monte Carlo simulations. The *More Is Better* rule is based on the law of large numbers where probabilities are assigned to events in accordance with their limiting relative frequencies.

In this chapter our focus is on Monte Carlo simulation. The concept of Monte Carlo simulation stems from the study of games of chance. These types of simulations can be accomplished using three distinct steps: (1) generate a random number, (2) define how the random numbers relate to an event, and (3) execute the event as shown in Figure 8.1. These three steps are repeated lots of times as we will illustrate in our examples.

One advantage of dealing with simulations is their ease to examine *what if* analysis to the systems with actually altering the real system. For example, if we want to design a sensor to detect an illness it is easier to test on a computer simulation than to affect many people and actually the experiment.

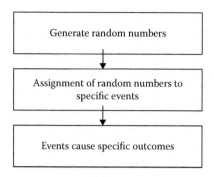

FIGURE 8.1
Three main steps in Monte Carlo Simulations.

8.2 Random Number and Monte Carlo Simulation

A Monte Carlo simulation model is a model that uses random numbers to simulate the behavior of a situation. Using a known probability distribution (such as uniform, exponential, or normal) or an empirical probability distribution, a modeler assigns a behavior to a specific range of random numbers. The behavior returned from the random number generated is then used in analyzing the problem. For example, if a modeler is simulating the tossing of a fair coin using a uniform random number generator that gives numbers in the range $0 \leq x < 1$, then he or she may assign all numbers less than 0.5 to be a head, whereas numbers from 0.5 to 1 are tails.

A Monte Carlo simulation can be used to model either stochastic or deterministic behavior. It is possible to use a Monte Carlo simulation to determine the area under a curve (a deterministic problem) or stochastic behavior such as the probability of winning in craps (a stochastic problem). In this chapter, we will introduce both a deterministic problem and a stochastic problem. We discuss how to create algorithms to solve both. We will start with the deterministic simulation modeling.

First, Monte Carlo simulation deals with the use of generated random numbers to cause specific events to occur within the simulation according to a specific scheme. Basically, the flow from

$$\text{Random number} \rightarrow \text{Assignment} \rightarrow \text{Event}$$

is observed within the simulation. The most important aspect of the simulation process is the algorithm. The algorithm is the step-by-step process to go from INPUTS to OUTPUTS. We will illustrate with a few examples in class as well as the use of EXCEL.

Steps of a Monte Carlo simulation include the following:

1. Establish a probability distribution for each variable that is subject to chance. Obtain the cumulative distribution function (CDF) of the distribution.
2. Generate a random number from this distribution for each variable in Step 1.
3. Make assignments from random numbers to the appropriate events.
4. Repeat the process for a series of replications (trials).

8.2.1 Random-Number Generators in Excel

Using random numbers is of paramount importance in running Monte Carlo simulations, so a good random-number generator is critical. In particular, a modeler must have a method of generating uniform, $U(0,1)$, random numbers—that is, numbers that are uniformly distributed between 0 and 1. All other distributions, known and empirical, can be derived from the $U(0,1)$ distribution. At the graduate level, a lot of class time is spent on the theory behind good and bad random-number generators, and the tests that can be made on them. More and more is being learned about what does and does not make up a true random-number generator. At the undergraduate level, this is not necessary, provided the students have access to either random numbers or a good algorithm for generating pseudo-random numbers.

In addition, most computer languages now use good pseudo-random number generators (although this has not always been the case—the old RANDU generator distributed by IBM was statistically unsound). These good generators use the recursive sequence $X_i = (aX_{i-1} + c)$ where a, c, and m determine the statistical quality of the generator. As we do not discuss the testing of random-number generators in our course, we trust the generators provided by our software packages. Serious study of simulation must, of course, include a study of random-number generators because a bad generator will provide output from which a modeler may make poor conclusions.

In EXCEL, here are commands to obtain random numbers

To Simulate	EXCEL Formula to Use
Random number, uniform [0,1]	=rand()
Random number between [a, b]	=a + (b − a)*rand()
Discrete integer random number between [a, b]	=randbetween(a, b)
Normal random number	=NORMINV(rand(),μ,σ)
Exponential random number with mean rate μ	=(−1/μ)* ln(rand())
Discrete general distribution with only two outcomes (such as a flip of a coin): A and B Probability of outcome is p.	=if(rand()<p, A, B)
Discrete general distribution for more than two outcomes. Range1 = cell range for lower limits of the random number intervals Range2 = cell range containing the variable values	=lookup(RAND(), Range1,Range2)

Note: in the command RAND() there is not space between the two parentheses.

8.2.2 Examples in Excel

We need uniform random number between [0,1]. To get these we type =rand() in cell D6. We obtained a random number, 0.317638748. We can copy this down for as many random numbers as we need. In cell E6 we create a random number between [1,10] using $1 + (10 - 1)*$rand(). We obtained 1.157956 and we copy down for as many random numbers as we need in our simulation. Figure 8.2 provides a screen shot of the Excel formulas and then the values for obtaining 10 random numbers.

The following algorithms might be helpful to obtain other types of random numbers.

1. Uniform [a, b]

   a. Generate a random uniform number U from [0,1]

   b. Return $X = a + (b - a)*U$

   c. $X = a + (b - a)*$rand()

Random Numbers in

				Uniform [0,1]	Uniform [a,b]	a	1
				rand()	a+(b-a)*rand()	b	10
				=RAND()	=$H$3+($H$4-$H$3)*RAND()		
				=RAND()	=$H$3+($H$4-$H$3)*RAND()		
				=RAND()	=$H$3+($H$4-$H$3)*RAND()		
				=RAND()	=$H$3+($H$4-$H$3)*RAND()		
				=RAND()	=$H$3+($H$4-$H$3)*RAND()		
				=RAND()	=$H$3+($H$4-$H$3)*RAND()		
				=RAND()	=$H$3+($H$4-$H$3)*RAND()		
				=RAND()	=$H$3+($H$4-$H$3)*RAND()		
				=RAND()	=$H$3+($H$4-$H$3)*RAND()		
				=RAND()	=$H$3+($H$4-$H$3)*RAND()		

A	B	C	D	E	F	G	H
Random Numbers in Excel							
						a	1
			Uniform [0,1]	*Uniform [a,b]*		b	10
			rand()	a+(b-a)*rand()			
			0.224316498	7.536124847			
			0.231291676	6.141957603			
			0.812110885	9.964171051			
			0.588337604	2.622892726			
			0.452714581	2.716438128			
			0.760120157	9.143616067			
			0.83754106	8.076745654			
			0.33676385	6.830805207			
			0.930387388	1.970512054			
			0.43170986	8.869520106			

FIGURE 8.2
Screenshots to obtain random numbers in Excel.

2. Exponential with mean β
   a. Generate a random uniform number U from [0,1]
   b. Return $X = -\beta \ln(U)$
   c. $X = -\beta \ln(\text{rand}())$
3. Normal(0, 1)
   a. Generate U_1 and U_2 from uniform [0,1].
   b. Let $V_i = 2U_i - 1$ for $i = 1,2$.
   c. Let $W = V_1^2 + V_2^2$
   d. If $W > 1$, go back to Step a. Otherwise, let $Y = \sqrt{(-2\ln(W))/W}$, $X_1 = V_1 Y, X_2 = V_2 Y$
   e. X_1 and X_2 are normal (0,1).

Exercises 8.2

For each, generate 20 random numbers.

1. Uniform (0,1)
2. Uniform (−10,10)
3. Exponential (λ = 0.5)
4. Normal (0,1)
5. Normal (5,0.5)

8.3 Probability and Monte Carlo Simulation Using Deterministic Behavior

One key to good Monte Carlo simulation is an understanding of the axioms of probability discussed briefly in Chapter 4. Probability is a long-term average. For example, if the probability of an event occurring is 1/5, this means that "in the long term, the chance of the event happening is 1/5 = 0.2" and not that it will occur exactly once out of every five trials.

8.3.1 Deterministic Simulation Examples

Let us consider the following deterministic examples:
   Compute the area under a nonnegative curve.

1. The curve $y = x^3$ from $0 \le x \le 2$.
2. The curve (which does not have a closed-form solution to $\int_{x=0}^{1.4} \cos(x^2) \cdot \sqrt{x} \cdot e^{x^2} dx$ from [0,1.4]).
3. Compute the volume in the first octant of $x^2 + y^2 + z^2 \ge 1$.

We will present algorithms for their models as well as produce output of the Monte Carlo simulation to analyze. These algorithms are important to the understanding of simulation as a mathematical modeling tool.

Here is a generic framework for an algorithm. This framework includes inputs, outputs, and the steps required to achieve the desired output.

Example 8.1: Monte Carlo Algorithm Area under the Nonnegative Curve (for EXCEL)

Input: Total number of points
Output: AREA = approximate area under a specified curve $y = f(x)$ over the given interval $a \leq x \leq b$, where $0 \leq f(x) \leq M$.

Step 1: In Column 1, list $n = 1,2,...,N$ from cell $a1$ to aN. Create columns 2–5.
Step 2: In Column 2, generate a random x_i between a and b using, $a+(b-a)$*rand(). These are listed in cells $b1$ to bN.
Step 3: In Column 3, generate a random y_i between 0 and M using, $0 + (M - 0)$*rand(). These are listed in cells $c1$ to cN.
Step 4: In Column 4, compute $f(x_i)$. These are listed in cells $d1$ to dN.
Step 5: In Column 5, check to see if each random coordinate (x_i, y_i) point is below curve. Compute $f(x_i)$ and see if $y_i \leq f(x_i)$. Use a logical IF statement, If $y_i \leq f(x_i)$ then let the cell value = 1, otherwise let the cell value equal 0. In cells $d1$ to dn, put, IF(cell $c1 <= d1$, 1,0). These are listed in cells $e1$ to eN.
Step 6: Count the cell values that equal 1, use Sum($e1$:eN).
Step 7: Calculate area in g4. Area = $M(b - a)$ Sum/N.

Repeat the process and increase N to get better approximations. You can plot the (x_i, y_i) coordinate and $f(x_i)$ for a visual representation.

In Maple, we developed a procedure called Area for doing this procedure. You enter the function and the domain and range.

```
> Area:=proc(expr::algebraic, an::numeric, bn::numeric, ymin::numeric, ymax::numeric,

N::posint)option remember;
```

```
> local A, B,y1,y2,areaR, count, i,area, xpt, ypt, xrpt, yrpt, gen_x, gen_y, c1,c2,c3,c4,c5;
```

```
> y1:=ymin;
```

```
> y2:=ymax;
```

```
> A:=an;
```

```
> B:=bn;
```

```
> areaR:=(B–A)*(y2–y1);

> count:=0;

> i:=0;

> gen_x:=evalf(an+(bn-an)*rand(0..10^5)/10^5):

> gen_y:=evalf(ymin+(ymax-ymin)*rand(0..10^5)/10^5):

> xrpt:=[]:

> yrpt:=[]:

> while i<N do
```

Step 1

```
> xpt[i]:=A+(B–A)*(rand()/10^12);

> ypt[i]:=y1+(y2–y1)*(rand()/10^12);
```

Step 2

```
> xrpt:=[op(xrpt),gen_x()]:

> yrpt:=[op(yrpt),gen_y()]:

> if ypt[i]<evalf(f1(xpt[i])) then

> count:=count+1;

> i:=i+1;

> else

> i:=i+1;

> fi;
```

```
> od;

> area:=areaR*count/N;

> print(areaR, count, N,area);

> with(plots):

> c1:=pointplot(zip((x, y)->[x, y], xrpt, yrpt)):

> c2:=plot(expr, x=A..B, thickness=4, colour=black):

> c3:=area:

> c4:=plot(y2,x=A..B, thickness=4, colour=black):

> c5:=plot(y1,x=A..B, thickness=4,colour=black):

> display(c1,c2,c4,c5);

> end:
```

Example 8.2: $y = x^3$ **from** $0 \le x \le 2$ **Using 100 Random Numbers in Excel and 2000 in Maple**

We applied our area under the curve algorithm to $y = x^3$ from [0, 2]. We see a visual representation of this in Figure 8.3. In this example, with only 100 random points, we find our simulated area is about 4.64 in Figure 8.4. Using Maple with 2000 random points its approximation is 3.872. The real area is found by integration, $\int_0^2 x^3 dx = 4$, when our function can be integrated (Figure 8.5).

We present a method for repeating the process in Excel in order to obtain more iteration for the simulation.

For example, go to cell M1 and enter 1 and iterate to cell M1000 for 1000 trials. They are number from 1 to 1000. In cell N1, reference your cell g4 (see Figure 8.4). Highlight cells M1 to N1000. Go to Data → What if Analysis → Data Table and enter. In the dialog box that come up, put nothing in rows and put an used cell reference in the column (such as P1). Press OK. The table fills in running the area simulation previously written 1000 times. For this example, we now have 100 runs 1,000 times or 10,000 results. Copy M1 to N1000 and paste as *values* into another location such as AA1. You do this, so the values do not keep changing. Then highlight the column of simulated area values and obtain their description statistics.

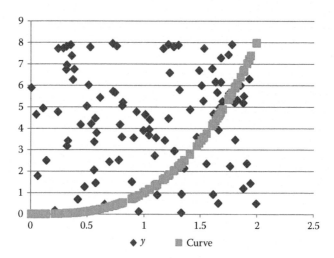

FIGURE 8.3
Area under curve graphical representation for $y = x^3$ from [0, 2].

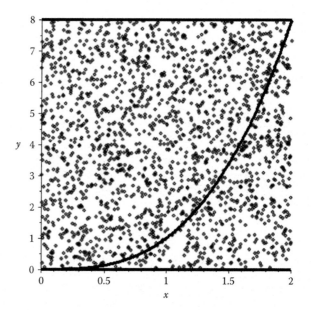

FIGURE 8.4
Area under curve graphical representation for $y = x^3$ from [0, 2] in Maple.

Function y=x^3 from [0,2]						
		Rand_X		Rand_Y		Area
n	rand()	x	y	y	count	
1	0.245527	0.491055	0.118410534	4.487261	0	4.64
2	0.980981	1.961963	7.552178557	7.117628	1	
3	0.830105	1.66021	4.576028846	4.358678	1	
4	0.921794	1.843587	6.266011943	6.465696	0	
5	0.164428	0.328856	0.035564438	5.399423	0	
6	0.065181	0.130363	0.002215439	5.388175	0	
7	0.292613	0.585226	0.200433862	5.11966	0	
8	0.453498	0.906995	0.746131043	5.937811	0	
9	0.224417	0.448834	0.090418732	4.260345	0	
10	0.497616	0.995232	0.985763098	5.59896	0	
11	0.081892	0.163784	0.004393533	2.951267	0	
12	0.708622	1.417245	2.846652288	2.954609	0	
13	0.824472	1.648944	4.483507308	2.02574	1	
14	0.459856	0.919711	0.777954851	6.481868	0	
15	0.981421	1.962841	7.562327224	1.76143	1	

FIGURE 8.5
Screenshot of simulation of area showing only 1–15 random trials.

The descriptive statistics table in Figure 8.6 would look like this:

Now we are ready to *approximate* the area by using Monte Carlo simulation. The simulation only approximates the solution. We increase the number of trials attempting to get closer to the value. We present the results in Table 8.1. Recall that we introduced randomness into the procedure with the Monte Carlo simulation area algorithm. In our output, we provide graphical output as well so that the algorithm may be seen as a process. In our graphical output, each generated coordinate (x_i, y_i) is a point on the graph. Points are randomly generated in our intervals $[a, b]$ for x and $[0, M]$ for y. The curve for the function $f(x)$ is overlaid with the points. The output also includes the approximate area.

We need to stress that in modeling deterministic behavior with stochastic features, we (not nature) have introduced the randomness into the problem. Although more runs are generally better, it is not true that the deterministic solution becomes closer to reality as we increase the number of trials, $N, \rightarrow \infty$. It is generally true that more runs are better than a small number of runs (16% was the worst by almost an order of magnitude, and that occurred at $N = 100$). In general, more trials are better.

	A	B	C
	Column1		
	Mean	3.98896	
	Standard Error	0.022034274	
	Median	4	
	Mode	3.68	
	Standard Deviation	0.696784922	
	Sample Variance	0.485509228	
	Kurtosis	-0.228529708	
	Skewness	0.197260113	
	Range	4.32	
	Minimum	1.76	
	Maximum	6.08	
	Sum	3988.96	
	Count	1000	

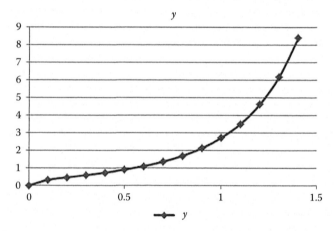

FIGURE 8.6
Screenshot of descriptive statistics for our simulated areas.

TABLE 8.1

Summary of Output for the Area under x^3 from 0 to 2

Number of Trials	Approximate Area	Absolute Percent Error (%)
100	3.36	16
500	3.872	3.2
1.000	4.32	8
5.000	4.1056	2.64
10,000	3.98896	0.275

Example 8.3: The Curve (Which Does Not Have a Closed-Form Solution) $\int_{x=0}^{1.4} \cos(x^2) \cdot \sqrt{x} \cdot e^{x^2} dx$ **from [0,1.4]**

We can tell from the integral that x varies from 0 to 1.4. But what about y? Take the function for y and obtain the plot as x varies from 0 to 1.4. In Figure 8.7, we can estimate the maximum value as about 9. Thus, we generate random values for y from 0 to 9.

We ran 1000 iterations and obtained a numerical approximation for the area of 2.9736. Since we cannot find the integral solution directly in this case, we use the trapezoidal method to approximate the solution to see how well our simulation faired. The numerical method provides an approximate solution of 3.0414. Our simulation's error compared to the trapezoidal method was within 2.29%.

Monte Carlo Volume Algorithm

INPUT   The total number of random points, N. The nonnegative function, $f(x)$, the interval for $x[a,b]$, interval for $y[c,d]$ and an interval for $z[0,M]$ where $M > \text{Max } f(x,y)$, $a < x < b$, $c < y < d$

OUTPUT   The approximate volume enclosed for the function $f(x,y)$ in the first octant, $x > 0$, $y > 0$, and $z > 0$.

Step 1.   Set all counters at 0

Step 2.   For i from 1 to N do step 3–5

Step 3.   Calculation random coordinates in the rectangular region:

$$a < x_i < b, \ c < y_i < d, \ 0 < z_i < M$$

Step 4.   Calculate $f(x_i,y_i)$

Step 5.   Compare $f(x_i,y_i)$ and z_i. If $z_i < f(x_i,y_i)$ then increment counter by 1.

Otherwise, do not increment counter.

Step 6.   Estimate the Volume by $V = (M - 0) \cdot (c - d) \cdot (b - a) \cdot \dfrac{counter}{N}$
Stop

Our algorithm to produce random numbers for a simulation is:

INPUT: Number of random numbers

OUTPUT: Probability of getting a {0,1,2,3,4,5,6}

Step 1   Initialize all counters (counter 1 through counter 6) to 0.

Step 2   For $i = 1, 2, ..., n$, do steps 3 and 4.

Step 3   Obtain a random number x from [0,1].

Step 4   Apply logic sequence of

FIGURE 8.7

Plot of $\cos(x^2)\sqrt{x} \, e^{x^2}$ from 0 to 1.4.

TABLE 8.2

Percent Errors in Finding the Volume in First Octant

Number of Points	Approximate Volume	Percent Error ($|\%|$)
100	0.47	10.24
200	0.595	13.64
300	0.5030	3.93
500	0.514	1.833
1,000	0.518	1.069
2,000	0.512	2.21
5,000	0.518	1.069
10,000	0.5234	0.13368
20,000	0.5242	0.11459

Example 8.4: Finding the Volume in the First Octant

We can also extend this concept to multiple dimensions. We develop an algorithm for the volume under a surface in the first octant.

Table 8.2 provides the numerical output. The actual volume in the first octant is $\pi/6$ (with radius as 1). We take $\pi/6$ to four decimals as 0.5236 cubic units. Figure 8.8 graphically displays the algorithm.

Generally, though not uniformly, the percentage errors become smaller as the number of points, N, is increased.

Exercises 8.3

1. Use Monte Carlo simulation to approximate the area under the curve $f(x) = 1 + \sin x$ over the interval $(-\pi/2) \le x \le (\pi/2)$.

2. Use Monte Carlo simulation to approximate the area under the curve $f(x) = x^{0.5}$ over the interval $(1/2) \le x \le (3/2)$.

3. Use Monte Carlo simulation to approximate the area under the curve $f(x) = \sqrt{1-x^2}$ over the interval $0 \le x \le (\pi/2)$.

4. How would you modify question 3 to obtain an approximation to π?

5. Use Monte Carlo simulation to approximate the volume under the surface $f(z) = x^2 + y^2$, the first octant.

6. Determine the area under the following nonnegative curves:

   a. $y = \sqrt{1-x^2}, 0 \le x \le 1$

   b. $y = \sqrt{4-x^2}, 0 \le x \le 2$

   c. $y = \sin(x), 0 \le x \le \text{pi}/2$

   d. $y = x^3, 0 \le x \le 4$

7. Find the area between the following two curves and the two axes by simulation:
$$y = 2x + 1 \text{ and } y = -2x^2 + 4x + 8$$

	A	B	C
1	Trial Number	rand()	result
2	1	=RAND()	=IF(B2<=0.33,0,IF(B2<=0.58,1,IF(B2<=0.77,2,IF(B2<=0.86,3,IF(B2<=0.91,4,IF(B2<=0.96,5,6))))))
3	=A2+1	=RAND()	=IF(B3<=0.33,0,IF(B3<=0.58,1,IF(B3<=0.77,2,IF(B3<=0.86,3,IF(B3<=0.91,4,IF(B3<=0.96,5,6))))))
4	=A3+1	=RAND()	=IF(B4<=0.33,0,IF(B4<=0.58,1,IF(B4<=0.77,2,IF(B4<=0.86,3,IF(B4<=0.91,4,IF(B4<=0.96,5,6))))))
5	=A4+1	=RAND()	=IF(B5<=0.33,0,IF(B5<=0.58,1,IF(B5<=0.77,2,IF(B5<=0.86,3,IF(B5<=0.91,4,IF(B5<=0.96,5,6))))))
6	=A5+1	=RAND()	=IF(B6<=0.33,0,IF(B6<=0.58,1,IF(B6<=0.77,2,IF(B6<=0.86,3,IF(B6<=0.91,4,IF(B6<=0.96,5,6))))))
7	=A6+1	=RAND()	=IF(B7<=0.33,0,IF(B7<=0.58,1,IF(B7<=0.77,2,IF(B7<=0.86,3,IF(B7<=0.91,4,IF(B7<=0.96,5,6))))))
8	=A7+1	=RAND()	=IF(B8<=0.33,0,IF(B8<=0.58,1,IF(B8<=0.77,2,IF(B8<=0.86,3,IF(B8<=0.91,4,IF(B8<=0.96,5,6))))))
9	=A8+1	=RAND()	=IF(B9<=0.33,0,IF(B9<=0.58,1,IF(B9<=0.77,2,IF(B9<=0.86,3,IF(B9<=0.91,4,IF(B9<=0.96,5,6))))))
10	=A9+1	=RAND()	=IF(B10<=0.33,0,IF(B10<=0.58,1,IF(B10<=0.77,2,IF(B10<=0.86,3,IF(B10<=0.91,4,IF(B10<=0.96,5,6))))))
11	=A10+1	=RAND()	=IF(B11<=0.33,0,IF(B11<=0.58,1,IF(B11<=0.77,2,IF(B11<=0.86,3,IF(B11<=0.91,4,IF(B11<=0.96,5,6))))))
12	=A11+1	=RAND()	=IF(B12<=0.33,0,IF(B12<=0.58,1,IF(B12<=0.77,2,IF(B12<=0.86,3,IF(B12<=0.91,4,IF(B12<=0.96,5,6))))))
13	=A12+1	=RAND()	=IF(B13<=0.33,0,IF(B13<=0.58,1,IF(B13<=0.77,2,IF(B13<=0.86,3,IF(B13<=0.91,4,IF(B13<=0.96,5,6))))))
14	=A13+1	=RAND()	=IF(B14<=0.33,0,IF(B14<=0.58,1,IF(B14<=0.77,2,IF(B14<=0.86,3,IF(B14<=0.91,4,IF(B14<=0.96,5,6))))))
15	=A14+1	=RAND()	=IF(B15<=0.33,0,IF(B15<=0.58,1,IF(B15<=0.77,2,IF(B15<=0.86,3,IF(B15<=0.91,4,IF(B15<=0.96,5,6))))))
16	=A15+1	=RAND()	=IF(B16<=0.33,0,IF(B16<=0.58,1,IF(B16<=0.77,2,IF(B16<=0.86,3,IF(B16<=0.91,4,IF(B16<=0.96,5,6))))))
17			

A	B	C
Trial Number	rand()	result
1	0.207393982	0
2	0.127847961	0
3	0.560264474	1
4	0.567940969	1
5	0.16113914	0
6	0.417507081	1
7	0.538948905	1
8	0.387619174	1
9	0.747548053	2
10	0.476099542	1
11	0.932428906	5
12	0.577012606	1
13	0.598187466	2
14	0.986144446	6
15	0.408557927	1

FIGURE 8.8
Algorithm for volume in first octant.

8.4 Probability and Monte Carlo Simulation Using Probabilistic Behavior

Let us consider the following simple probabilistic examples.

1. Compute the probability of getting a head or a tail if you flip a fair coin.
2. Compute the probability of rolling a number from 1 to 6 using a fair die.

Example 8.5: Flip a Fair Coin

Algorithm

Input: The number of trials, N
Output: The probability of a head or a tail

Step 1: Initialize counters to 0.
Step 2: For $i = 1, 2, ..., N$ do.
Step 3: Generate a random number, x, $U(0,1)$.
Step 4: If $0 \leq x < 0.5$ increment heads, $H = H + 1$, otherwise $T = T + 1$.
Step 5: Output H/N and T/N, the probabilities for heads and tails.

Example 8.6: Rolling a Fair Die

Rolling a fair die adds the additional process of multiple assignments (six for a six-sided die). The probability will be the number of occurrences of each number divided by the total number of trials.

    INPUT: Number of rolls
    Output: Probability of getting a {1,2,3,4,5,6}
Step 1: Initialize all counters (counter1 through counter 6) to 0.
Step 2: For $i = 1, 2, ..., n$, do Steps 3 and 4.
Step 3: Obtain a random number j from integers $(1,6)$.
Step 4: Increment the counter for the value of j so that

Counter j = counter $j + 1$.

Step 5: Calculate the probability of each roll {1,2,3,4,5,6} by

Counter j/n

Step 6: Output probabilities.
Step 7: Stop.

ROLL-A-FAIR-DIE PROGRAM

The expected probability is 1/6 or 0.1667. We note that as the number of trials increases, the closer our probabilities are to the expected long-run values. We offer the following concluding remarks. When you have to run simulations, run them for a very large number of trials.

Example 8.7: Discrete Probability Distribution

Assume we have a distribution as follows:

X	0	1	2	3	4	5	6
P(X = x)	0.33	0.25	0.19	.09	0.05	0.05	0.04

Our algorithm to produce random numbers for a simulation is

    INPUT: Number of random numbers
    Output: Probability of getting a {0, 1, 2, 3, 4, 5, 6}
Step 1: Initialize all counters (counter1 through counter 6) to 0.
Step 2: For $i = 1, 2, ..., n$, do Steps 3 and 4.
Step 3: Obtain a random number x from $[0, 1]$.
Step 4: Apply logic sequence of
    if $x < 0.33$ then $y = 0$
    if $0.33 < x \leq 0.58$ then $y = 1$

					p	0.5	T	0.9	P*T	0
18		Initial S		Bombers	N	q	0.3			
19										
20	S	0			15	Quess			S > 99	good
21								S_Final	0.99313666	
22	i	B	P	P*B	New S					
23	0	0.004747562	0.9999	0.004747	0.004747					
24	1	0.030520038	0.9998	0.030513	0.03526					
25	2	0.091560115	0.9996	0.091522	0.126781					
26	3	0.170040213	0.9992	0.16991	0.296691					
27	4	0.218623131	0.9986	0.218319	0.51501					
28	5	0.206130381	0.9975	0.205608	0.720618					
29	6	0.147235986	0.9954	0.146558	0.867176					
30	7	0.081130033	0.9916	0.080451	0.947627					
31	8	0.034770014	0.9848	0.034241	0.981867					
32	9	0.011590005	0.9723	0.011269	0.993137					
33	10	0.002980287	0.9497	0.00283	0.995967					
34	11	0.000580575	0.9085	0.000527	0.996494					
35	12	8.29393E-05	0.8336	6.91E-05	0.996564					
36	13	8.20279E-06	0.6975	5.72E-06	0.996569					
37	14	5.02212E-07	0.45	2.26E-07	0.996569					
38	15	1.43489E-08	0	0	0.996569					

FIGURE 8.9
Screenshot for Example 8.7.

if $0.58 < x \leq 0.77$ then $y = 2$
if $0.77 < x \leq 0.86$ then $y = 3$
if $0.86 < x \leq 0.91$ then $y = 4$
if $0.91 < x \leq 0.96$ then $y = 5$
if $0.96 < x \leq 1.0$ then $y = 6$
Step 5: Use the probability found in the simulation as needed.

Figure 8.9 displays the formulas followed by the values.

We note that as the number of trials increases, the closer our probabilities are to the expected long-run values in the probability table. We offer the following concluding remarks. When you have to run simulations, run them for a very large number of trials.

Exercises 8.4

1. Develop an algorithm for an unfair coin that yields a head 55% of the time.
2. Develop an algorithm for an 8-sided die with sides {1, 2, 3, 4, 5, 6, 7, 8}.

Projects 8.4

1. *The Price Is Right*. On the popular TV game show *The Price Is Right*, at the end of each half hour, the three winning contestants face off in what is called the *Showcase Showdown*. The game consists of spinning a large wheel with 20 spaces on which the pointer can land; the spaces are numbered from $0.05 to $1.00 in 5-cent increments. The contestant who has won the least amount of money at this point in the show spins first, followed by the one who has won the next most, and then by the biggest winner for that half hour.

The objective of the game is to obtain as close to $1.00 as possible with-out going over that amount with an allowed maximum of two spins. Naturally, if the first player does not go over, the other two will use one or both spins in their attempt to overtake the leader.

But what of the person spinning first? If he or she is an expected-value decision-maker, how high a value on the first spin does he or she need to not want to take a second spin? Remember, the person can lose if

a. Either of the other two players surpasses the player's total

b. The player spins again and goes over

2. *Let us Make a Deal*. You are *dressed to kill* in your favorite costume, and the host Monte Hall picks you out of the audience. You are offered the choice of three wallets. Two wallets contain a single $50 bill, and the third contains a $1000 bill. You choose one of the three wal-lets. Monte knows which wallet contains the $1000, so he shows you one of the other two wallets—one with one of the two $50 bills inside. Monte does this on purpose because he must have at least one wallet with $50 inside. If he holds the $1000 wallet, he shows you the other wallet, the one with $50. Otherwise, he just shows you one of his two $50 wallets. Monte then asks you if you want to trade your choice for the one he is still holding. Should you trade?

Develop an algorithm and construct a computer simulation to support your answer.

8.5 Applied Simulations and Queuing Models

In this section, we present algorithms and Excel output for the following simulations:

1. An aircraft missile attack
2. The amount of gas that a series of gas stations will need
3. A simple single barber in a barbershop queue

Example 8.8: Missile Attack

An analyst plans a missile strike using F-15 aircraft. The F-15 must fly through air-defense sites that hold a maximum of eight missiles. It is vital to ensure success early in the attack. Each aircraft has a probability of 0.5 of destroying the target, assuming it can get to the target through the air-defense systems and then acquire and attack its target. The probability that a single F-15 will acquire a target is approximately 0.9. The target is protected by air-defense equipment with a 0.30 probability of stopping the F-15 from either arriving at or acquiring the target. How many F-15 are needed to have a successful mission assuming we need a 99% success rate?

Algorithm: Missiles
Inputs: N = number of F-15s
 M = number of missiles fired
 P = probability that one F-15 can destroy the target
 Q = probability that air defense can disable an F-15
Output: S = probability of mission success
Step 1: Initialize $S = 0$
Step 2: For $I = 0$ to M do
Step 3: $P(i) = [1 - (1 - P)^{N-I}]$
Step 4: $B(i)$ = binomial distribution for (m, i, q)
Step 5: Compute $S = S + P(i) * B(i)$
Step 6: Output S.
Step 7: Stop.

We run the simulation letting the number of F-15s vary and calculate the probability of success. We guess $N = 15$ and find that we have probably

	D	E	F	G	H
	Customer number	Time between arrivals	Arrival time	Start time	Service time
1		=-(1/$B$1)*LN(1-RAND())	=E2	=F2	=-(1/$B$2)*LN(1-RAND())
	=D2+1	=-(1/$B$1)*LN(1-RAND())	=F2+E3	=MAX(I2,F3)	=-(1/$B$2)*LN(1-RAND())
	=D3+1	=-(1/$B$1)*LN(1-RAND())	=F3+E4	=MAX(I3,F4)	=-(1/$B$2)*LN(1-RAND())
	=D4+1	=-(1/$B$1)*LN(1-RAND())	=F4+E5	=MAX(I4,F5)	=-(1/$B$2)*LN(1-RAND())
	=D5+1	=-(1/$B$1)*LN(1-RAND())	=F5+E6	=MAX(I5,F6)	=-(1/$B$2)*LN(1-RAND())
	=D6+1	=-(1/$B$1)*LN(1-RAND())	=F6+E7	=MAX(I6,F7)	=-(1/$B$2)*LN(1-RAND())
	=D7+1	=-(1/$B$1)*LN(1-RAND())	=F7+E8	=MAX(I7,F8)	=-(1/$B$2)*LN(1-RAND())
	=D8+1	=-(1/$B$1)*LN(1-RAND())	=F8+E9	=MAX(I8,F9)	=-(1/$B$2)*LN(1-RAND())
	=D9+1	=-(1/$B$1)*LN(1-RAND())	=F9+E10	=MAX(I9,F10)	=-(1/$B$2)*LN(1-RAND())

	I	J	K	
	Completion time	wait time	Cumulative wait time	
	=G2+H2	=G2-F2	=J2	=K2/D2
	=G3+H3	=G3-F3	=K2+J3	=K3/D3
	=G4+H4	=G4-F4	=K3+J4	=K4/D4
	=G5+H5	=G5-F5	=K4+J5	=K5/D5
	=G6+H6	=G6-F6	=K5+J6	=K6/D6
	=G7+H7	=G7-F7	=K6+J7	=K7/D7

B	C	D	E	F	G	H	I	J	K	L
2		Customer number	Time between arrivals	Arrival time	Start time	Service time	Completion time	wait time	Cumulative wait time	average wait
3		1	1.934408754	1.934408754	1.9344088	0.071524068	2.005933422	0	0	0
		2	0.116601281	2.051010035	2.05101	0.714047959	2.765957994	0	0	0
		3	0.055768834	2.106778869	2.765958	0.36811946	3.134077454	0.659179	0.659179125	0.219726375
		4	0.879801355	2.986580224	3.3140775	0.206478930	3.340556303	0.147467	0.806676355	0.201669089
		5	0.095844504	3.082424728	3.3405564	1.055590089	4.396146462	0.258132	1.06480802	0.212961604
		6	1.043432803	4.125857531	4.3961465	0.63308224	5.029228702	0.270289	1.335096951	0.222516159
		7	0.223185659	4.34904319	5.0292287	0.818579146	5.847807847	0.580186	2.015282463	0.287897495
		8	2.251848324	6.600891514	6.6008915	0.393204228	6.994095741	0	2.015282463	0.251910308
		9	0.384299775	6.985191288	6.9940957	0.320344496	7.314440237	0.008904	2.024186916	0.224909657
		10	0.163595249	7.148786537	7.3144402	0.066657268	7.381097506	0.165654	2.189840616	0.218984062
		11	0.000502847	7.149289384	7.3810975	0.792646337	8.173743842	0.231808	2.421648738	0.220149885
		12	0.102456472	7.251745856	8.1737438	1.062486091	9.236230734	0.921998	3.343646724	0.278637227
		13	0.384817067	7.636562923	9.2362307	0.2167925	9.453023234	1.599668	4.943314534	0.380254964
		14	0.525581112	8.262144036	9.4530232	0.342525761	9.795548994	1.190879	6.134193732	0.438156695
		15	0.548688836	8.811132636	9.795549	0.387117518	10.18266651	0.384416	7.118610041	0.474574006
		16	0.540099845	9.351232481	10.187667	0.500628779	10.68829529	0.836434	7.955044122	0.497190258
		17	0.025796165	9.377028547	10.688295	0.051349505	10.7396448	1.311267	9.266310766	0.545077104
		18	0.199860228	9.576888375	10.739645	0.487924558	11.23756935	1.162756	10.42906669	0.579392594
		19	0.422003799	9.998892574	11.237569	0.610221593	11.84779095	1.238677	11.66774337	0.614091756
		20	1.086043979	11.08553465	11.847791	0.139853034	11.98764398	0.762250	12.42999960	0.621499983
		21	0.085067941	11.17060259	11.987644	0.304856573	12.29250065	0.817041	13.24704105	0.630811478
		22	0.888452558	11.85905515	12.292501	0.13728232	12.42978297	0.433446	13.68048655	0.621840298

FIGURE 8.10
Screenshot for missile attack example.

a success greater than 0.99 when we send nine planes. Thus, any number greater than nine works.

We find that nine F-15s gives us $P(s) = 0.99313$.

Actually, any number of F-15 greater than nine provides a result with the probability of success we desire. Fifteen F-15s yielding a $P(s) = 0.996569$. Any more would overkill.

Example 8.9: Gasoline-Inventory Simulation

You are a consultant to an owner of a chain of gasoline stations along a freeway. The owner wants to maximize profits and meet consumer demand for gasoline. You decide to look at the following problem.

Problem Identification Statement: Minimize the average daily cost of delivering and storing sufficient gasoline at each station to meet the consumer demand.

Assumptions: For an initial model, consider that, in the short run, the average daily cost is a function of demand rate, storage costs, and delivery costs. You also assume that you need a model for the demand rate. You decide that historical date will assist you. Data used from Giordano et al. (2014). This is displayed in Tables 8.3 through 8.5.

Model Formulation: We convert the number of days into probabilities by dividing by the total and we use the midpoint of the interval of demand for simplification.

As cumulative probabilities will be more useful we convert to a CDF.

We might use cubic splines to model the function for demand (see additional readings for a discussion of cubic splines).

Inventory Algorithm

TABLE 8.3

Demand Table Intervals

Demand: Number of Gallons	Number of Occurrences (Days)
1000–1099	10
1100–1199	20
1200–1299	50
1300–1399	120
1400–1499	200
1500–1599	270
1600–1699	180
1700–1799	80
1800–1899	40
1900–1999	30
Total number of days =	1000

TABLE 8.4

Demand Table Probabilities

Demand: Number of Gallons	Probabilities
1000	0.010
1150	0.020
1250	0.050
1350	0.120
1450	0.200
1550	0.270
1650	0.180
1750	0.080
1850	0.040
2000	0.030
Total number of days =	1.000

TABLE 8.5

CDF of Demand

Demand: Number of Gallons	Probabilities
1000	0.010
1150	0.030
1250	0.080
1350	0.20
1450	0.4
1550	0.670
1650	0.850
1750	0.93
1850	0.97
2000	1.0

Inputs: Q = delivery quantity in gallons
T = time between deliveries in days
D = delivery cost in dollars per delivery
S = storage costs in dollars per gallons
N = number of days in the simulation
Output: C = average daily cost
Step 1: Initialize: Inventory $\rightarrow I = 0$ and $C = 0$.
Step 2: Begin the next cycle with a delivery:

$$I = I + Q$$

$$C = C + D$$

Step 3: Simulate each day of the cycle.

For $i = 1, 2, ..., T$, do Steps 4–6.

Step 4: Generate a demand, q_i. Use cubic splines to generate a demand based on a random CDF value, x_i.

Step 5: Update the inventory: $I = I - q^i$.

Step 6: Calculate the updated cost: $C = C + s * I$ if the inventory is positive.

If the inventory is ≤ 0, then set $I = 0$ and go to Step 7.

Step 7: Return to Step 2 until the simulation cycle is completed.

Step 8: Compute the average daily cost: $C = C/n$.

Step 9: Output C.

Stop.

We ran the simulation and find that the average cost is about \$5,753.04, and the inventory on hand is about 199,862.4518 gallons.

Example 8.10: Queuing Model

A queue is a waiting line. An example would be people in line to purchase a movie ticket or in a drive through line to order fast food. There are two important entities in a queue: (1) customers and (2) servers. There are some important parameters to describe a queue:

1. The number of servers available.
2. Customer arrival rate: average number of customers arriving to be serviced in a time unit.
3. Server rate: average number of customers processed in a time unit.
4. Time.

In many simple queuing simulations, as well as theoretical approaches, assume that arrivals and service times are exponentially distributed with a mean arrival rate of λ_1 and a mean service time of λ_2.

Theorem 8.1 If the arrival rate is exponential and the service rate is given by any distribution, then the expected number of customers waiting in line, L_q, and the expected waiting time, W_q, are given by

$$L_q = \frac{\lambda^2 \sigma^2 + \rho^2}{2(1-\rho)} \quad \text{and} \quad W_q = \frac{L_q}{\lambda}$$

where:

λ is the mean number of arrival per time period

μ is the mean number of customers serviced per time unit, $\rho = \lambda/\mu$

σ is the standard deviation of the service time

Here, we have a barber shop where we have two customers arriving every 30 minutes. The service rate of the barber is 3 customers every 60 minutes. This implies that the time between arrivals is 15 minutes and the mean service time is one customer every 20 minutes. How many customers will be in the queue and what is their average waiting time?

POSSIBLE SOLUTION WITH SIMULATION

We provide an algorithm for possible use.
Algorithm:

For each customer 1...N
Step 1: Generate an interarrival time, an arrival time, start time
based on finish time of the previous customer, service time,
completion time, amount of time waiting in a line, cumulative
wait time, average wait time, number in queue, average queue
length.
Step 2: Repeat N times.
Step 3: Output average wait time and queue length
Stop

You will be asked to calculate the theoretical solution in the exercise set.
We illustrate the simulation.
We will use the following to generate exponential random numbers:

$$x = \frac{-1}{\lambda \ln(1 - \text{rand}())}$$

We generate a sample of 5000 runs and plot customers versus average
weight time.

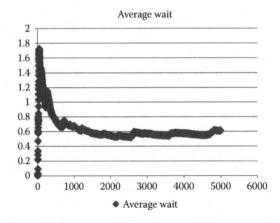

We note that the plot appears to be converging at values slightly higher
than 0.66. Thus, we will run 100 more trials of 5000 and recomputed the
average.

We obtain the descriptive statistics from Excel. We note that the
mean is 0.6601 that is very close to our theoretical mean. The theoretical expected queue length and expected waiting times are 4/3 and 2/3,
respectively.

Column1	
Mean	0.660147135
Standard error	0.006315375
Median	0.658168429
Mode	#N/A
Standard deviation	0.063153753
Sample variance	0.003988397
Kurtosis	−0.319393469
Skewness	0.155656707
Range	0.318586462
Minimum	0.500642393
Maximum	0.819228855
Sum	66.01471348
Count	100

Exercises 8.5

1. Solve for the theoretical L_q and W_q for the barber problem.
2. Modify the missile strike problem if the probability of S were only 0.95 and the probability of an F-15 being deterred by air defense were 0.3. Determine the number of F-15s needed to complete the mission.
3. What if in the missile attack problem the air-defense units were modified to carry 10 missiles each? What effect does that have on the number of F-15s needed?
4. Perform sensitivity analysis on the gasoline-inventory problem by modifying the delivery to 11,450 gallons per week. What effect does this have on the average daily cost?

Projects 8.5

1. *Tollbooths* Heavily traveled toll roads such as the Garden State Parkway, Interstate 95, and so forth, are multilane divided highways that are interrupted at intervals by toll plazas. As collecting tolls is usually unpopular, it is desirable to minimize the motorist annoyance by limiting the amount of traffic disruption caused by the toll plazas. Commonly, a much larger number of tollbooths are provided than the number of travel lanes entering the toll plaza. On entering the toll plaza, the flow of vehicles fans out to the larger number of tollbooths; when leaving the toll plaza, the flow of vehicles is forced to squeeze down to a number of travel lanes equal to the number of travel lanes before the toll plaza. Consequently, when traffic is heavy, congestion increases when vehicles leave the toll plaza. When traffic is very heavy, congestion also builds at the entry to the toll plaza because of the time required for each vehicle to pay the toll.

Construct a mathematical model to help you determine the optimal number of tollbooths to deploy in a barrier-toll plaza. Explicitly, first consider the scenario in which there is exactly one tollbooth per incoming travel lane. Then consider multiple tollbooths per incoming lane. Under what conditions is one tollbooth per lane more or less effective than the current practice? Note that the definition of optimal is up to you to determine.

2. *Major League Baseball.* Build a simulation to model a baseball game. Use your two favorite teams or favorite all-star players to play a regulation game.

3. *NBA Basketball.* Build a simulation to model the NBA basketball playoffs.

4. *Hospital Facilities.* Build a simulation to model surgical and recovery rooms for the hospital.

5. *Class Schedules.* Build a simulation to model the registrar's scheduling changes for students or final exam schedules.

6. *Automobile Emissions.* Consider a large engineering company that performs emissions control inspections on automobiles for the state. During the peak period, cars arrive at a single location that has four lanes for inspections following exponential arrivals with a mean of 15 minutes. Service times during the same period are uniform: between [15,30] minutes. Build a simulation for the length of the queue. If cars wait for more than 1 hour, the company pays a penalty of $200 per car. How much money, if any, does the company pay in penalties? Would more inspection lanes help? What costs associated with the inspection lanes need to be considered?

7. *Recruiting Simulation Model*

Monthly demand for recruits is provided in the following table.

Demand	Probability	CDF
300	0.05	0.05
320	0.10	0.15
340	0.20	0.35
360	0.30	0.65
380	0.25	0.90
400	0.10	1.0

In addition, depending on conditions the average cost per recruit is between $60 and $80 in integer values. Returns from higher HQ are between 20% and 30% of costs. There is a fixed cost of $2000 per month for the office, phones, and so on. Build a simulation model to determine the average monthly costs.

Assume Cost = demand* cost per recruit + fixed cost-return amount, where return = %*cost

8. *Inventory Model*

Demand of ammunition palette for resupply on a weekly basis is provided in the following table.

Demand	Frequency	Probability	CDF
0	15	0.05	0.05
1	30	0.10	0.15
2	60	0.20	0.35
3	120	0.40	0.75
4	45	0.15	0.90
5	30	0.10	1.00

Assumptions: Lead time if resupply requires between 1 and 3 days. Currently, we have seven palettes in stock and no orders due. Needs order quantity and order point to reduce COSTS. Fixed cost for placing an order is $20. The cost for holding the unused stock is $0.02 per palette per day. Each time we cannot satisfy a demand the unit goes elsewhere and assumes a loss of $8 to the company. We operate 24/7.

9. *Simple Queuing Problem*

The bank manager is trying to improve customer satisfaction by offering better service. They want the average customer to wait for less than 2 minutes and the average length of the queue (line) if 2 or fewer. The bank estimates about 150 customers per day. The existing service and arrival times are given in the following table.

Service Time	Probability	Time between Arrival	Probability
1	0.25	0	0.10
2	0.20	1	0.15
3	0.40	2	0.10
4	0.15	3	0.35
		4	0.25
		5	0.05

Determine if the current servers are satisfying the goals. If not, how much improvement is needed in service to accomplish the stated goals?

10. Intelligence gathering (Information Operations)

Currently, Intelligence reports come according to the following historical information:

Time between Reports	Probability
1	0.11
2	0.21
3	0.22
4	0.20
5	0.16
6	0.10

The time it takes to process these reports is given as follows:

Process Time	Probability
1	0.20
2	0.19
3	0.18
4	0.17
5	0.13
6	0.10
7	0.03

Further, if we employ sensors the reports come more often as follows:

Time between Reports	Probability
1	0.22
2	0.25
3	0.19
4	0.15
5	0.12
6	0.07

Advise the manager on the current system. Determine utilization and sensor satisfaction. How many report processors are needed to insure that reports are processed in a timely manner?

Further Reading and References

Giordano, F.R., W. P. Fox, and S. Horton 2014. *A First Course in Mathematical Modeling* (5th ed.). Boston, MA: Cengage Publishers.

Law, A. and D. Kelton. 2007. *Simulation Modeling and Analysis* (4th ed.). New York: McGraw Hill.

Meerschaert, M. M. 1993. *Mathematical Modeling*. San Diego, CA: Academic Press.

Winston, W. 1994. *Operations Research: Applications and Algorithms* (3rd ed.). Boston, MA: Duxbury Press.

9

Mathematics of Finance with Discrete Dynamical System

OBJECTIVES

1. Define, model, solve, and interpret financial systems of discrete dynamical systems (DDS).
2. Analyze the long-term behavior of systems of DDS.
3. Understand the concepts of equilibrium and stability in systems.
4. Model both linear and nonlinear systems of DDS.
5. Model systems of DDS.

This chapter discusses mathematical methods and formulas that are used in many businesses and financial organizations. The discrete models in this chapter on DDS are used to derive many of these formulas. So why use discrete models? Yet we have to find a financial institution that gives continuous interest; they all use some discrete *time* method.

9.1 Developing a Mathematical Financial Model Formula

Now, consider $1000 being deposited in a money market that earns 2.75% per year. Using the DDS modeling paradigm, *Future = Present + Change*, we develop the following model.

Let $A(n)$ = Amount of money in the money market account after time period n, n is defined in years.

$$A(n+1) = A(n) + 0.0275 A(n), \ A(0) = 1000$$

or

$$A(n+1) = (1.0275) \ A(n), \ A(0) = 1000$$

By the closed form method, we can easily derive the particular solution as

$$A(k) = (1.0275^k) A(0) = 1000 \ (1.0275^k)$$

After 3 years, we have 1000 (1.0275 3) = \$1084.78 in the account.

So what exactly have we accomplished? We have derived a formula for *compound interest*. Let us remind ourselves that we can always iterate a solution. However, at times we only need one value and it might be easier to know and to use a formula if one is available.

9.1.1 Simple Interest and Compound Interest

Simple interest can be expressed in terms of three variables:

P is the principal (amount borrowed)
r is the interest rate (in decimals per time period)
t is the time (period over which to repay the principal)

By definition, simple interest is computed by the formula:

$$Prt = \text{principal} \times \text{interest rate} \times \text{time}$$

Thus if money is borrowed at a simple interest, the amount paid back A that must be repaid after t years is shown in Equation 9.1:

$$A = P = Prt \text{ or } P(1+rt) \tag{9.1}$$

Let us see how we use the formula in Equation 9.1.

Example 9.1: Borrowing Money

A principal of \$1500 is borrowed at 4.5% per year simple interest. Find the amount owed (or the future value) after:

a. 2 years
b. 4 months
c. 180 days

SOLUTION

a. $P = 1500$, $r = .045$ and $t = 2$, so

$$A = 1500\left[1 + 0.045(2)\right] = 1500(1.09) = 1635$$

b. Since 4 months is equivalent to 4/12 or 1/3 of a year $t = 1/3$, $r = 0.045$, and $P = 1500$ so that we obtain the following from the simple interest formula:

$$A = 1500\left[1 + 0.045\left(\frac{1}{3}\right)\right] = 1500(1.015) = \$1522.50$$

c. Since 180 days is 0.4931 years (180/365), we have $t = 0.4931$. From our simple interest formula,

$$A = 1500[1 + 0.045(0.4931)] = 1500(1.022191781) = \$1533.287$$

If you are paying this, round down to \$1533.28 (if requesting the amount to be paid, then round up to \$1533.29).

COMPOUND INTEREST

A more interesting and more widely accepted method uses compound interest. This is a method whereby the interest previously earned also earns interest.

Compound Interest Formula: If P dollars are invested at an annual interest rate r, compounded k times a year, then after n conversion periods (where a conversion period is the number of periods × time), the investment has grown to an amount A given by Equation 9.2:

$$A = P\left(1 + \frac{r}{k}\right)^n \tag{9.2}$$

Again this is a familiar form. It is the general solution to DDS of the form: $A(n + 1) = A(n)(1 + \Delta)$ where Δ represents change. Previously in Chapter 7, we learned that DDS are always solvable using iteration and graphical methods. Now, we are finding analytical solutions to certain forms of DDS that can be used in financial mathematics.

Example 9.2: Future Value

Find the future value after 2 years of \$1500 invested at 4.5% annual interest compounded

a. Quarterly
b. Semi-annually
c. Annually
d. Monthly
e. Daily

In each case, our general formula is $A = P(1 + rt)^k$. What we notice to be changing in each part of our example is the compounding period. Let us see the effect of the compounding period on the money earned using the analytical formula.

a. $k = 4$, $r/k = 0.045/4$, $P = \$1500$, $n = (4)(2) = 8$ conversion periods in 2 years
   $r/k = (0.045/4) = 0.01125$ or 1.125% each compounding period.

$$A = P\left(1 + \frac{r}{k}\right)^n = 1500(1 + 0.01125)^8 = 1500(1.01125)^8 = \$1640.43$$

b. $k = 2$, $r/k = 0.045/2 = 0.0225$, $P = \$1500$, $n = (2)(2) = 4$ conversion periods in 2 years

$$A = P\left(1 + \frac{r}{k}\right)^n = 1500(1 + .0225)^4 = 1500(1.0225)^4 = \$1639.62$$

c. $k = 1$, $r/k = 0.045/1 = 0.045$, $P = \$1500$, $n = (2)(1) = 2$ conversion periods in 2 years

$$A = P\left(1 + \frac{r}{k}\right)^n = 1500(1 + 0.045)^2 = 1500(1.045)^2 = \$1638.03$$

d. $k = 12$, $r/k = 0.045/12 = 0.00375$, $P = \$1500$, $n = (2)(12) = 24$ conversion periods in 2 years

$$A = P\left(1 + \frac{r}{k}\right)^n = 1500(1 + 0.00375)^{24} = 1500(1.00375)^{24} = \$1640.98$$

e. $k = 365$, $r/k = 0.045/365 = 0.0001232$, $P = \$1500$, $n = (2)(365) = 730$ conversion periods in 2 years

$$A = P\left(1 + \frac{r}{k}\right)^n = 1500(1 + 0.0001232)^{730} = 1500(1.0001232)^{730} = \$1641.14$$

Compounding Period	Future Money (based on $1500 deposited)
Daily	$1641.14
Monthly	$1640.98
Quarterly	$1640.43
Semi-Annually	$1639.62
Annually	$1638.03

Do you see a trend in the amount earned based on the compounding period?

CONTINUOUS COMPOUNDING

If we imagine a situation where the investment grows whereas the number of conversions increases indefinitely and the investment grows in proportion to its current value, then we have continuous compounding.

Of the 4,935 institutions, Bank rate surveys nationally that sell one-year certificate of deposits (CDs), 3,639 of them offer daily compounding, 697 offer monthly compounding, and 222 offer quarterly compounding. For consumers, more-frequent compounding produces a higher total yield. None offers continuous compounding. Finding continuous compounding might not be worth the effort in finding but we will introduce the concept nevertheless.

Continuous compounding interest formula is shown in Equation 9.3:

$$A = Pe^{rt} \tag{9.3}$$

Example 9.3: Continuous Compounding

Suppose $10,000 was invested at 5% per year compounded continuously. What is the value after 2 years? Compare the result to daily compounding after 2 years.

SOLUTION

By substitution into $A = Pe^{rt}$, where $P = 10,000$, $r = 0.05$, and $t = 2$ years, we have

$$A = 10,000e^{(0.05)(2)} = \$11051.70$$

If we compounded daily (assuming 365 days in year), $A = 10,000(1 + 0.05/365)^{720} = \$11036.50$.

However, we have not seen any banks or financial institutions who currently give continuous interest.

Exercises 9.1

1. A principal of $3,500 is borrowed at 6% per year simple interest. Find the value after
   a. 1 year
   b. 3 years
   c. 3 months
   d. 3 days

2. If $500 were deposited in a bank account earning 5% per year, find the value after
   a. 2 years
   b. 3 years
   c. 5 years

3. You are saving for a new set of golf clubs that cost $2500. You find a money market account that pays 6.5% per year simple interest. How much should be invested so that this purchase can be made within 2 years?

4. Find the interest on $5,500 for 1 year at
   a. 6.5% per year simple interest
   b. 6.5% per year compounded monthly
   c. 6.5% per year compounded quarterly
   d. 6.5% per year compounded semi-annually

5. Find the interest on $5000 for
   a. 2 years at 4.7% compounded monthly
   b. 3 years at 4.7% compounded quarterly
   c. 4 years at 4.7% compounded semi-annually
   d. 6 months at 4.7% per year compounded daily

6. Find the future value of $12,000 invested at a rate of 9% compounded
   a. Monthly
   b. Quarterly
   c. Semi-annually
   d. Daily

7. Suppose $5,500 is invested and compounded continuously at 5% per year. Find the value after
   a. 1 year
   b. 2 years
   c. 6 months
   d. 5 hours

8. Assume that $5,000 is deposited in an account that pays simple interest. If the account grows to $5,500 in 2 years, find the interest rate.

Projects

1. In 1945, Noah Sentz died in a car accident and his estate was handled by the local courts. The state law stated that 1/3 of all assets and property go to the wife and 2/3 of all assets go to the children. There were four children. Over the next 4 years, three of the four children sold their shares of the assets back to the mother for a sum of $1,300 each. The original total assets were mainly 75.43 acres of land. This week, the fourth child has sued the estate for his rightful inheritance from the original probate ruling. The judge has ruled in

favor of the fourth son and has determined that he is rightfully due monetary compensation. The judge has picked your group as the jury to determine the amount of compensation. Use the principles of financial mathematical modeling to build a model that enables you to determine the compensation. In addition, prepare a short one-page summary letter to the court that explains your results. Assume the date is November 10, 2003.

2. Buying a new car: Part 1—You wish to buy a new car soon. You initially narrow your choices to a Saturn, Cavalier, and a Toyota Tercel. Each dealership offers you their prime deal:

Saturn	$11,900	$1,000 down	3.5% interest for up to 60 months
Cavalier	$11,500	$1,500 down	4.5% interest for up to 60 months
Tercel	$10,900	$500 down	6.5% interest for up to 48 months

You have allocated at most $475 a month on a car payment.

Use a dynamical system to compare the alternatives, choose a car, and establish your exact monthly payment.

Part 2—Nissan is offering a first time buyers special. Nissan is offering 6.9% for 24 months or $500 *cash back*. If the regular interest rate is 9.00%, determine the amount financed at 6.9% so that these two options are equivalent for a new car buyer.

What kind of car could you get from Nissan with the equivalent option?

3. On November 24, 1971, a cold rainy Thanksgiving evening, a middle-aged man giving the name Dan B. Cooper purchased a plane ticket on Northwest Airlines flight 305 from Portland, Oregon, to Seattle, Washington. He boarded the plane and flew into history.

After Cooper was seated, he demanded the flight attendant bring him a drink and a $200,000 ransom. His note read: "Miss, I have a bomb in my suitcase and I want you to sit beside me." When she hesitated, Cooper pulled her into the seat next to him and opened his case, revealing several sticks of dynamite connected to a battery. He then ordered her to have the captain relay his ransom demand for the money and four parachutes: two front-packs and two backpacks. The Federal Bureau of Investigation (FBI) delivered the parachutes and ransom to Cooper when the 727 landed in Seattle to be refueled. Cooper, in turn, allowed the thirty-two other passengers he had held captive to go free.

When the 727 was once again airborne, Cooper instructed the pilot to fly at an altitude of 10,000 feet on a course destined for Reno, Nevada. He then forced the flight attendant to enter the cockpit with the remaining three crew members and told her to stay there until landing. Alone in the rear cabin of the plane, he lowered the aft stairs beneath the tail. Then, in the dark of the night, somewhere over Oregon, D. B. Cooper stepped with the money into history. When the plane landed in Reno, the only thing the FBI found in the cabin was one of the backpack chutes. Despite the massive efforts of manhunts conducted by the FBI and scores of local and state police groups, no evidence was ever found of D. B. Cooper, the first person to hijack a plane for ransom and parachute from it. One package of marked bills from the ransom was found along the Columbia River near Portland in 1980. Some surmise that Cooper might have been killed in his jump to fame and part of the ransom washed downstream.

Consider the following items dealing with Flight 305 and its famous passenger. Using Excel to assist you, complete each one as directed.

1. Suppose Cooper invested the ransom over a period of 10 years, one $20,000 unit per year, in a small local bank in rural Utah, at a fixed rate of 6%, compounded annually, starting on January 1, 1972. Further, if he left the money to accumulate, what would be the value of the account December 31, 1993?

2. Suppose that instead of playing it safe, Cooper gambled half of the money away in Reno the night after the hijacking and then invested the other half with Night wings Federal in Reno on January 1, 1972. If the account paid 6% interest compounded quarterly, what would this money be worth at the end of this year?

3. Another possibility for Cooper would have been to invest 75% of the total with a *loco* investor, Smiling Pete's Federal Credit Union, at a rate of 3% compounded monthly. What would this investment be worth at the end of this year?

4. Compare these three methods of investing. Contrast their different patterns of growth over time.

9.2 Rates of Interest, Discounting, and Depreciation

As we saw in the previous section, compound interest is affected by both the annual interest rate and the frequency of the compounding. Often it is difficult to compare options when the rates and compounding periods are different. For example, which is better to invest at 4% compounded monthly or 4.5% compounded semi-annually? To make these comparisons, it is common to use effective interest rates.

9.2.1 Annual Percentage Rate

Just watch any TV advertisement about buying a car and you will see in fine print the term *annual percentage rate (APR)*. Let us examine this more closely. The APR is a standard measure to compare interest rates when the compounding periods are different.

If interest is compounded once a year at APR rate, the investment would yield exactly the same return P_t at the end of t years as it would if interest were compounded m times a year at a nominal rate.

$P_t = P(1 + (r/m))^{mt}$ return after t years on a nominal rate r compounded m times per year.

$P_t = P(1 + APR)^t$ return after t years on APR rate, compounded once a year.

$$P\left(1+\left(\frac{r}{m}\right)\right)^{mt} = P(1+APR)^t$$

$$\ln(P) + mt \ln\left(1+\frac{r}{m}\right) = \ln(P) + t\ln(1+APR)$$

$$mt \ln\left(1+\frac{r}{m}\right) = t\ln(1+APR)$$

$$m \ln\left(1+\frac{r}{m}\right) = \ln(1+APR)$$

$$\ln\left(1+\frac{r}{m}\right)^m = \ln(1+APR)$$

$$e^{\ln\left(1+\frac{r}{m}\right)m} = e^{\ln(1+APR)}$$

$$\left(1+\frac{r}{m}\right)^m = (1+APR)$$

$$APR = \left(1+\frac{r}{m}\right)^m - 1$$

The *APR* formula is shown in Equation 9.4,

$$APR = \left(1+\frac{r}{m}\right)^m - 1 \qquad (9.4)$$

where:

APR is the effective interest rate

r is the annual interest rate

m is the number of conversion periods per year.

Example 9.4: Compounding

Calculate the *APR* for a 6.5% nominal interest rate, which is compounded

    a. Quarterly
    b. Monthly

SOLUTION

    a. $APR = \left(1 + \dfrac{0.065}{4}\right)^4 - 1 = 6.66\%$

    b. $APR = \left(1 + \dfrac{0.065}{12}\right)^{12} - 1 = 6.697\%$

Example 9.5: Investment Options

Which is a better option for investment, 2.5% per year compounded monthly or 2.6% simple interest?

SOLUTION

We will compute the APR for the first investment. We have $r = .025, m = 12$

$$APR = \left(1 + \left(\frac{0.025}{12}\right)\right)^{12} - 1 = 2.52\%$$

Since this value is less than 2.6% then the simple interest rate is better.

Example 9.6: Credit Plans

You are offered two credit plans. One has an interest rate of 9.8% annual interest compounded monthly. The other offers 10% compounded quarterly.

SOLUTION

$$APR_1 = \left(1 + \frac{0.098}{12}\right)^{12} - 1 = 10.25\%$$

$$APR_2 = \left(1 + \frac{0.1}{4}\right)^4 - 1 = 10.38\%$$

Since we are paying back the money on this credit plan, we choose the smaller APR option 1, the 9.8% annual interest compounded monthly.

9.2.2 APR for Continuous Compounding

Although more of a mathematical method, we present it as Equation 9.5:

$$APR = e^r - 1 \qquad (9.5)$$

Example 9.7: Continuous APR

Find the *APR* if the money is invested at 7.5% per year with continuous compounding.

SOLUTION

$$APR = e^{0.075} - 1 = 0.0778 \text{ or } 7.78\%$$

9.2.3 Discounts

It is a common business procedure for a lender to deduct the interest due in advance. This procedure is called discounting and the money deducted in advance is called the discount. The money received by the borrower is called the proceeds.

Let us define

P is the proceeds (amount received by the borrower)

d is the discount rate

t is the time in years

S is the amount paid back by the borrower

The simple discount is $S \times d \times t$ so that the proceeds are given by $P = S - Sdt$ (Note that this is also a *Future = Present + Change* form)

Simple discount formula, Equation 9.6, is:

$$P = S(1 - dt) \qquad (9.6)$$

Example 9.8: Discount Loan

Paul borrows $600 for 2 years at 6% simple discount.

$$P = S(1 - dt) = 600(1 - 0.06(2)) = \$528$$

Therefore, Paul receives $528 and pays back $600 in 2 years.

Example 9.9: Transaction APR

What is the APR on this transaction?
600–528 = $72 paid in interest over 2 years.

$$72 = 528(r)(2) = 1056\,r$$

$$r = \frac{72}{1056} = 0.06818 \text{ or } 6.18\% \text{ APR}$$

9.2.4 Depreciation

Reducing balance depreciation is the converse of compound interest with larger amounts being subtracted from the original asset value each year. The formula, Equation 9.7, is:

$$A_t = A_0(1 - r)^t \tag{9.7}$$

where:
A_t is the value of the asset after t years taking into account depreciation
A_0 is the original value of the asset
r is the depreciation rate
t is the number of years

Example 9.10: Machinery Depreciation

We purchase machinery for retooling for $100,000. This type machinery historically depreciates at 10% per year. How much will the machinery be worth in 10 years?

SOLUTION

$$A_{10} = 100,000(1 - 0.10)^{10} = \$34,867.84$$

Example 9.11: Car Depreciation

We purchase a new car for $45,239. This type car holds its value pretty well but historically depreciates at 12% per year. How much will the car be worth in 3 years?

SOLUTION

$$A_3 = 45,239(1-0.12)^3 = \$30,829.11$$

Exercises 9.2

1. 6.669 versus determine the APR on each of the following investments:
   a. 5% per year compounded monthly
   b. 5% per year compounded quarterly
   c. 5% per year compounded semi-annually

2. Determine the APR on each of the following investments:
   a. 5% per year compounded monthly
   b. 7% per year compounded quarterly
   c. 9% per year compounded semi-annually

3. Determine which of the following is a better investment:
   a. An investment paying 6.5% per year compounded monthly
   b. An investment paying 6.75% per year compounded semi-annually

4. Which is a better option for investment?
   a. 3.5% per year compounded monthly
   b. 3.6% simple interest

5. Which is a better option for investment?
   a. 9% per year compounded monthly
   b. 9.1% simple interest

6. You are offered two credit plans. One has an interest rate of 19.8% annual interest compounded monthly. The other offers 19% compounded quarterly. Which plan should you choose?

7. Find the APR if money is invested at 15% per year with continuous compounding.

8. Find the APR if money is invested at 5.5% per year with continuous compounding.

9. Sam borrows $6450 for 4 years at 6% simple discount. Determine the amount received and the APR for this transaction.

10. Sally borrows $450 for 4 years at 7% simple discount. Determine the amount received and the APR for this transaction.

11. We buy new machinery for our printing business for $12,500. The depreciation schedule is 5.5% per year. What will this machine be worth in 5 years?

12. We buy new computers for our company and spend $20,000. The depreciation is 25% per year. What is the value of the computers after 3 years?

9.3 Present Value

Often we are interested in finding the principal P that must be invested today to obtain some future amount.

9.3.1 Net Present Value and Internal Rate of Return

The following is a cash flow of an investment project:

Year	0	1	2	3	4	5
Cash Flow	−500	120	130	140	150	160

(in 000)

Is the investment worthwhile?

Costs and returns must be brought back to present value.

NPV = present value of cash inflows − present value of cash outflows

Decision Rule:

If net present value (NPV) > 0, then the investment project is worthwhile.

If NPV < 0, then do not invest in the project.

For our example, we are assuming an 8% discount rate ($r = 0.08$) and use the formula:

$$P_0 = \frac{P_t}{(1+r)^t}$$

Year 1	$120,000/(1 + 0.08)^1 = 111,111$
Year 2	$130,000/(1 + 0.08)^2 = 111,454$
Year 3	$140,000/(1 + 0.08)^3 = 111,137$
Year 4	$150,000/(1 + 0.08)^4 = 110,254$
Year 5	$160,000/(1 + 0.08)^5 = 108,893$
Total present value is 552,849.	

$$NPV = 500,000 - 552,849 = 52,849$$

Since 52,849 > 0, then we should go with the investment project.

9.3.2 Internal Rate of Return

The internal rate of return (IRR) is the interest rate for which the NPV is zero. A project is viable if the interest rate is less than the IRR but not profitable if the interest rate is larger than the IRR.

Calculation of the IRR

Example 9.12: *NPV Investment*

Let us return to our investment example.
We computed the NPV for various interest rates (8%, 9%, 10%, 11%, 12%).

Rate	NPV
8	52849.46
9	37868.63
10	23512.43
11	9747.676
12	−3456.7

The IRR is the point where the curve crosses the horizontal axis shown in Figure 9.1. This is between 11% and 12%. We will interpolate as follows:

$$IRR = \frac{(R_1 \cdot NPV_2) - (R_2 \cdot NPV_1)}{NPV_2 - NPV_1} = \frac{(0.11 \cdot (-3456.7)) - (0.12 \cdot 9747.67)}{(-3456.7 - 9747.67)} = 11.73\%$$

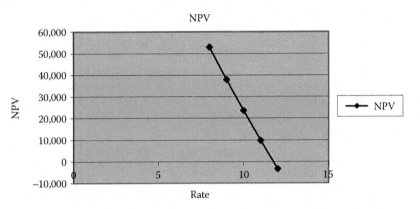

FIGURE 9.1
NPV for various interest rates.

Example 9.13: Alternative Investments

Let us consider a situation where you have two alternatives (Project A and Project B) from which to choose for possible investments.

Year	0	1	2	3	4
Project A	−100,000	−30,000	40,000	60,000	80,000
Project B	−50,000	−20,000	10,000	30,000	50,000

Assume a discount rate of 6% and assume you want to maximize profits. Project A:

$$NPV = -100000 - 28301 + 35599 + 50377 + 63367 = \$21042$$

$$Project\ B\ NPV = -50000 - 18867 + 8899 + 25188 + 39604 = \$4825$$

Project A is more profitable than Project B.

Project A			Project B		
Investment	**(1+r)^t**	**I/(1+r)^t**	**Investment**	**(1+r)^t**	**P/(1+r)^t**
−100,000	1	−100,000	−50,000	1	−50000
−30,000	1.06	−28301.9	−20,000	1.06	−18867.9
40,000	1.1236	35599.86	10,000	1.1236	8899.964
60,000	1.191016	50377.16	30,000	1.191016	25188.58
80,000	1.262477	63367.49	50,000	1.262477	39604.68
		21042.62			4825.302

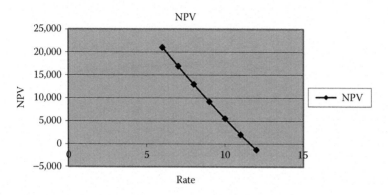

FIGURE 9.2
NPV plot for Example 9.13.

Let us find the IRR for Project A.

Rate	NPV
6	21042
7	16909
8	12948
9	9149
10	5505
11	2007
12	−1349

We approximate the IRR from analysis for Figure 9.2 as $(0.11)(-1349)-(0.12)$
$(2207)/(-1349-2207) = 11.62\%$.

Exercises 9.3

Given the cash-flow for investments in the following, determine the NPV
and determine if the investment is worthwhile.

1. Assume an 8% discount rate

Year	1	2	3	4	5	6	7	8
Cash Flow	−2000	1000	700	550	400	400	350	250

2. Assume an 8% discount rate

Year	1	2	3	4	5	6	7	8
Cash Flow	−1500	700	600	500	300	300	250	250

3. Assume a 5% discount rate

Year	0	1	2	3	4
Cash Flow	−200	120	130	140	150

4. Assume a 5% discount rate

Year	0	1	2	3	4
Cash Flow	−500	220	230	240	250

5. Resolve 1–4 with the following discount rates
   a. 6%
   b. 7%
   c. 9%
   d. 10%
   e. 11%
   f. 12%

In problems 6–10, determine the IRR for the aforementioned problems 1–4.

Projects 9.3

We have an option for alternatives for a major project shown in the following:

Year	0	1	2	3	4	5
Project 1	−20,000	−2,000	2,300	2,400	2,500	3,500
Project 2	−5,000	−1,000	1,100	1,250	1,350	1,500
Project 3	−8,000	0	1,500	3,000	4,500	6,000

Required.
    Determine which project to recommend. Include in your analysis the NPV and the IRR for each project alternative.

9.4 Bonds, Annuities, and Shrinking Funds

9.4.1 Government Bonds

A bond is a cash investment made to the government for an agreed number of years. In return, the government pays the investor a fixed sum at the end of each year: in addition, the government repays the original value of the bond with the final payment.

Treasury notes and bonds* are securities that pay a fixed rate of interest every six months until your security matures, which is when we pay you their par value. The only difference between them is their length until maturity. Treasury notes mature in more than a year, but not more than 10 years from their issue date. Bonds*, on the other hand, mature in more than 10 years from their issue date. You usually can buy notes and bonds* for a price close to their par value.

Treasury sells two kinds of notes: fixed-principal and inflation-indexed. Both pay interest twice a year, but the principal value of inflation-indexed securities is adjusted to reflect inflation as measured by the consumer-priced index—the Bureau of Labor Statistics' Consumer Price Index for All Urban Consumers (CPI-U). With inflation-indexed notes and bonds*, we calculate your semiannual interest payments and maturity payment based on the inflation-adjusted principal value of your security.

Saving bonds are treasury securities that are payable only to the person to whom they are registered. Saving bonds can earn interest for up to 30 years, but you can cash them after 6 months if purchased before February 1, 2003 or 12 months if purchased on or after February 1, 2003.

$$\text{Annual Payment} = r\,x\,\left(\text{price on bond}\right)$$

To determine if the government bond is attractive for investment, we find the NPV of the bond.

Example 9.14: Government Bond

A 5-year government bond values at \$5000 are purchased at a market rate of interest of 10%.

a. Find the annual repayment.
b. Find the NPV if the interest rate were 5%.
c. Calculate the present value of the bond based after 5 years.

SOLUTION

a. 0.20 * 5000 = \$1000 a year
b. NPV = $1000/1.05 + 1000/1.05^2 + 1000/1.05^3 + 1000/1.05^4 + 1000/1.05^5 = \$4329.47$
c. $5000/(1.1)^5 = \$3104.60$, so

Total $NPV = 3104.60 + 4329.47 = \$7434.07$
The investment made \$2434.07 over the 5 years.

9.4.2 Annuities and Sinking Funds

An annuity is a series of regular periodic payments. Examples of annuities include:

- Interest-bearing saving account
- Mortgage payments
- Insurance premium payments
- Retirement benefits
- Social security benefits

9.4.2.1 Ordinary Annuities

A series of equal payments are made at the end of a compounding period. The term is the period from the beginning of the first payment to the end of the last payment.

Example 9.15: Annuity Account

At the end of each month, $100 is invested into an annuity account paying 3.5% per year compounded monthly. What is the value of this account after the fifth payment?

SOLUTION

First payment earns interest for 4 months
Second payment earns interest for 3 months
Third payment earns interest for 2 months
Fourth payment earns interest for 1 month
Fifth payment earns interest for 0 months

Using the compound interest formula

$$\text{Value} = 100\left(1+\frac{0.035}{12}\right)^4 + 100\left(1+\frac{0.035}{12}\right)^3 + 100\left(1+\frac{0.035}{12}\right)^2$$

$$+100\left(1+\frac{0.035}{12}\right)^1 + 100\left(1+\frac{0.035}{12}\right)^0$$

$$= \$503.21$$

To derive a general formula let

A_0 = value of regular payment
i = interest rate per compounding period

n = number of regular payments (number of periods in the terms)
V_n = value of the annuity at the end of n compounding periods

$$V_n = A_0(1+1)^{n-1} + A_0(1+i)^{n-2} + \ldots + A_0(1+i) + A_0$$

$$V_n = A_0\{(1+1)^{n-1} + (1+i)^{n-2} + \ldots + (1+i) + 1\}$$

$$= A_0\left[\frac{(1+i)^n - 1}{(1+i)-1}\right] = A_o\left[\frac{(1+i)^n - 1}{i}\right]$$

This is also called the uniform series amount.

Example 9.16: Company Investments

A small company invests \$3000 at the end of six months in a fund paying 5% compounded semi-annually. Find the value of the investment after 8 years.

SOLUTION

$A_0 = 3000$, $r = 0.08$, so $i = 0.08/2 = 0.04$, $n = (8)(2) = 16$ periods in 8 years

$$V_{18} = 3000\left[\frac{(1+0.04)^{18} - 1}{0.04}\right] = \$76,936.23$$

SINKING FUNDS

A series of regular deposits used to accumulate a sum of money at some future time are called sinking funds.

Example 9.17: New Equipment Sinking Fund

A small firm anticipates capital expenditures of \$120,000 to buy new equipment in 6 years. How much should the firm be depositing in a funding earning 8% per year compounded quarterly to buy the new equipment?

SOLUTION

$A_0 = ?$ $I = 0.08/4 = 0.02$, $n = (6)(4) = 24$ periods, $V_{24} = 120,000$

$$V_{24} = 120,000 = A_0\left[\frac{(1+0.02)^{24} - 1}{0.02}\right]$$

We solve for A_0, $A_0 = \$3944.53$. Thus, the firm must invest \$3499.53 each period.

9.4.3 Present Value of an Annuity

The present value of an annuity is the amount of money today that is equivalent to a series of equal payments at some time in the future.
Let us start with an example.

Example 9.18: Retirement Plans

Retirement plans are on most Americans' minds. Consider a couple who wishes to make a lump sum investment paying 8% per year compounded annually to receive payments of $20,000 per year for 5 years. How much should they invest?

SOLUTION

$A_0 = \$20,000$ $i = 0.08$, $n = 5$
The present value of the annuity is

$$\frac{20000}{(1.08)} + \frac{20000}{(1.082)} + \frac{20000}{(1.083)} + \frac{20000}{(1.08)4} + \frac{20000}{(1.085)} = \$79,854$$

To derive a general formula, let us again look at the geometric series:
Let $1 + i = r$ and let the term be n years. Then,
$A_0[1/r + 1/r^2 + \ldots + 1/r^n]$ represents the present value.
A geometric series $1 + r + r^2 + \ldots + r^n$ can be represented by its sum that makes the sum of the first n terms, $S_n = a_1 (1 - r^n)/(1 - r)$.
The sum $[1/(1 + i) + 1/(1 + i)^2 + \ldots + 1/(1 + i)^n]$ can be represented as $((1 + i)^n - 1)/i$.
So, $A_0\left[\left((1+i)^n - 1\right)/i\right]$ is the value of the series of payments.
If V_0 is sufficient to provide for these payments, then the future value of V_0 at the end of the term n is $V_0(1 + i)^n$ and this value should be equal to the value of the payments. Therefore,

$$V_0(1+i)^n = A_0\left[\frac{(1+i)^n - 1}{i}\right]$$

Solving for V_0, we find

$$V_0 = A_0\left[\frac{1-(1+i)^{-n}}{i}\right]$$

Example 9.19: Annuity Payout Schedule

How much should you pay for annuity of quarterly payments of $15,000 for 5 years assuming an interest rate of 5% per year?

SOLUTION

$V_0 = ?$ $A_0 = 15,000$ $n = (5)(4) = 20$ periods
$i = 0.05/4 = 0.0125$

$$V_0 = 15,000 \left[\frac{1-(1+0.0125)^{-20}}{0.0125} \right] = \$262989.74$$

Example 9.20: Paying off a Loan

A loan of $50,000 is to be repaid in equal monthly payments over 5 years assuming fixed rate of 4.75% per year. Find the amount of each payment and the total interest paid.

SOLUTION

$V_0 = 50,000$ $A_0 = ?$ $n = (5)(12) = 60$ periods
$i = 0.0475/12 = 0.003958$

$$V_0 = A_0 \left[\frac{1-(1+i)^{-n}}{i} \right]$$

$$50,000 = A_0 \left[\frac{1-(1+0.003958)^{-60}}{0.003958} \right]$$

$$A_0 = \$2605.19$$

The total payments are $2605.19 (60) = $156311.82
The interest paid is $156311.82 – 50,000 = $106311.82

Exercises 9.4

1. A 10-year government bond values at $1000 are purchased at a market rate of interest of 8%.
   a. Find the annual repayment.
   b. Find the NPV if the interest rate were 4%.
   c. Calculate the present value of the bond based after 5 years.

2. A $500 saving bonds is bought in 2015. It matures in 2020 (meaning it is worth $500 in 2020). It will continue to earn 2.5% each year after 2020. How much will it be worth in 2030 and 2040?

3. A 15-year government bond values at $50,000 are purchased at a market rate of interest of 4.5%.

   a. Find the annual repayment.

   b. Find the NPV if the interest rate were 2%.

   c. Calculate the present value of the bond based after 20 years.

4. A small company invests $5000 at the end of six months in a fund paying 6% compounded semi-annually. Find the value of the investment after 10 years.

5. A company invests $30,000 at the end of 6 months in a fund paying 2.5% compounded semi-annually. Find the value of the investment after 4 years.

6. Microsoft invests $3 million at the end of nine months in a fund paying 5% compounded monthly. Find the value of the investment after 2 years.

7. A small firm anticipates capital expenditures of $20,000 to buy new equipment in 4 years. How much should the firm be depositing in a funding earning 3.5% per year compounded quarterly in order to buy the new equipment?

8. A university anticipates capital expenditures of $220,000 to buy new equipment within 10 years. How much should the firm be depositing in a funding earning 4% per year compounded quarterly to buy the new equipment?

   a. A loan of $150,000 is to be repaid in equal monthly payments over 15 years assuming a fixed rate of 5.75% per year. Find the amount of each payment and the total interest paid.

   b. A loan of $5000 is to be repaid in equal monthly payments over 4 years assuming fixed rate of 4.25% per year. Find the amount of each payment and the total interest paid.

   c. A car is bought at 3.75% interest per year for 72 monthly payments. Find the amount of each payment and the total interest paid if the car costs:

      a. $20,000

      b. $22,000

      c. $25,000

      d. $28,900

      e. $30,000

      f. $62,000

9.5 Mortgages and Amortization

An interest-bearing debt is said to be amortized if the principal and interest are paid by a sequence of equal payments made over a prescribed time period. What items do we buy over time? Many items such as cars, homes, and boats but also many convenience items when we buy using our credit card. Let us examine this.

Recall the mortgage problem from Chapter 7 where a home was purchased at $80,000 at a monthly interest rate of 1%. We present a concept called conjecturing solutions to nonhomogeneous DDS (adapted from MA 103 Study Guide, West Point, NY, 1990).

Suppose we have a linear, first-order, nonhomogeneous DDS, where the nonhomogeneous part is a constant, $a(n + 1) = r\, a(n) + b$. Let us look at an example:

$$a(n+1) = a(n) + 0.01\, a(n) - b$$

This DDS is composed of a homogeneous part and a nonhomogeneous part.

$$a(n) = 1.01a(n) - b$$

We already know that the solution to the homogeneous part for $a(n + 1) = 1.01$ $a(n)$ is $a(k) = c(1.01)k$. Let us conjecture that for the aforementioned system, the solution to the homogeneous part is not directly affected by the solution to the nonhomogeneous part. For example, the amount of drug in the bloodstream after n hours, $0.75\, a(n)$, is not affected by the 100 mg added after $n + 1$ hours. So, let us say that the solution to the DDS is a linear combination of the solution of the homogeneous part and the solution for the nonhomogeneous part:

$a(k) = c(1.01)k+$ solution for nonhomogeneous part

Since the nonhomogeneous part is a constant, let us conjecture that the solution might be a constant, say d.

$$a(k) = c(1.01)k + d$$

We need to verify the conjectured solution to see if it satisfies the DDS, for some value of d.

$$a(k) = c(1.01)k + d$$

$$a(n) = c(1.01)n + d$$

$$a(n+1) = c(1.01)n^{+1} + d$$

Substituting these into the DDS, we perform the following algebraic manipulation to find d:

$$c(1.01)n^{+1} + d = 1.01 \ [c(1.01)n + d] \ -b$$
$$c(1.01)n(1.01)^{1} + d = c(1.01)(1.01)n + \ 1.01d - b$$
$$c(1.01)n(1.01) + d = c(1.01)n(1.01) \ + \ 1.01d - b$$
$$d = \ 1.01d - b$$
$$-0.01d = \ -b$$
$$d = \frac{b}{0.01} \ = 100b$$

Therefore, d must equal $100b$ for our conjecture to satisfy the DDS and $a(k) = c(1.01)k + 100b$ must be general solution. A linear, first-order, nonhomogeneous DDS, where the nonhomogeneous term is a constant, may have an equilibrium value.

For our problem, $a(n + 1) = 1.01 \ a(n)–b$, we can find the equilibrium value, $ev = 100 \ b$. Thus the ev is the missing part of the general solution.

$$A(k) \ = C(1.01^{k}) + 100b$$

But we know that initially $A(0) = 80{,}000$ and finally after 240 months or payments $A(240) = 0$. Let us substitute these into our general solution to obtain two equations and two unknowns.

$$80{,}000 \ = C(1.01^{0}) + 100 \ b$$
$$0 = C(1.012^{40}) + \ 100 \ b$$
$$80{,}000 = C + 100 \ b$$
$$0 \ = \ 10.89255C \ + \ 100b$$
$$C = -8086.893 \ b = 880.8693$$

Therefore, the amount owed is

$$A(k) = -8086.893(1.01)^{k} + 8800.8693$$

Example 9.21: Credit Card

You charge $1500 on your discover card that charges 19.99% annual interest compounded monthly on the unpaid balance. You can afford $35 a month. How long until you pay off your bill assuming you do not charge anything until it is paid off?

$$A(n+1) = A(n) + \left(\frac{0.1999}{12}\right)A(n) \ \ 35, \ A(0) = 1500$$

It will take you over 75 months to pay this off. You bought $1500 worth of merchandise and paid $35 month for 75 months. You paid (75)(35) = $2625 for the items.

Periodic payment on an amortized loan, Equation 9.8, is

$$A_0 = V_0 \left[\frac{i}{1 - (1+i)^{-n}} \right] \tag{9.8}$$

Example 9.22: Stock Portfolio

You want to invest in a certain stock portfolio. The minimum investment value is $25,000. You currently have $10,000 cash and can set aside $500 per month in other investments until the total reaches $25,000. How long will it take?

An interest-bearing company is said to be amortized if the principle and the interest are paid by a sequence of equal payments over equal time periods. Mortgages and loans (e.g., car loans) are applications that use the formula,

$$V_0 = A_0 \left[\frac{1 - (1+i)^{-n}}{i} \right]$$

Recall our mortgage problem from before. If we have a mortgage of $80,000 for 20 years at 12% per year, find the monthly payments.

$A_0 = ?$ $V_0 = 80,000$

$N = 20(12) = 240$ periods $i = 0.12/12 = 0.01$

$$80,000 = A_0 \left[\frac{1 - (1+0.01)^{-240}}{0.01} \right]$$

$A_0 = \$880.87$

Remark: This now gives us several methods to find these payments.

Example 9.23: Car Buying

You go out car shopping and you budget car payments at most $275 per month. You found several cars that you are interested in buying: a Ford Focus that is now selling for $15,500 at 3.75% over 5 years and a Toyota Corolla selling for $16,500 at 4.25% over 6 years. Which of these cars, if any, can you afford to buy? Assume you can put no money down.

SOLUTION

Car 1 (Ford Focus)

$A_0 = ?$, $V_0 = 15,500$

$N = 5(12) = 60$ periods $i = 0.00375/12 = 0.003125$

$$15,500 = A_0 \left[\frac{1 - (1+0.003125)^{-60}}{0.003125} \right]$$

$$A_0 = \$283.71$$

Car 2 (Toyota Corolla)
$A_0 = ?$, $V_0 = 16{,}500$
$N = 6(12) = 72$ periods $i = 0.0425/12 = 0.003542$

$$16{,}500 = A_0\left[\frac{1-(1+0.003542)^{-72}}{0.003542}\right]$$

$$A_0 = \$260.03$$

Decision: You can afford to buy the Toyota Corolla but not the Ford Focus.

Exercises 9.5

1. You owe \$2000 on your credit card that charges 1.5% interest per month. Determine your monthly payment if you want to pay off the card in

   a. 6 months

   b. 1 year

   c. 2 years

2. You are considering a 30-year \$150,000 mortgage that charges 0.5% each month. Determine the monthly payment schedule and total amount paid.

3. You charge \$2500 on your master card that charges 9.9% annual interest compounded monthly on the unpaid balance. You can afford \$50 a month. How long until you pay off your bill assuming you do not charge anything until it is paid off?

4. You charge \$5000 on your master card that charges 12% annual interest compounded monthly on the unpaid balance. You can afford at most \$100 a month. How long until you pay off your bill assuming you do not charge anything until it is paid off?

5. You go out car shopping and you budget car payments at most \$275 per month. You found several cars that you are interested in buying: a Ford Focus, which is now selling for \$15,500 at 3.75% over 5 years and a Toyota Corolla selling for \$16,500 at 4.25% over 6 years. Which of these cars, if any, can you afford to buy? Assume you can put no money down.

6. You go out car shopping and you budget car payments at most \$475 per month. You found several cars that you are interested in buying: a preowned SUV, which is now selling for \$19,500 at 7.75% over 5 years and a Toyota Highlander selling for \$24,500 at 2.25% over 6 years. Which of these cars, if any, can you afford to buy? Assume you can put no money down.

9.6 Financial Models Using Previous Techniques

Example 9.24: Pensions and Retirement Strategies

Our goal modeling's aim is to create the best portfolio for each client based upon the current state of the stock available and the state of the market. In our problem we were given 5769 possible stocks for a retirement account and 32 criterion measures in eight groups of four. These eight main criteria, as defined by our retirement manager, are: total return (TR), batting average (BA), R-square (RS), sharpe (SH), downside deviation (DD), sortino (SO), standard deviation (SD), and tracking error (TE). There are four separate columns of data for each criterion. We also need to tell the computer program if we desire to maximize or minimize each criterion. This ability is a function that makes technique for order of preference by similarity to ideal solution (TOPSIS) a very useful multiattribute tool. The following criteria are maximized: TR, BA, and RS. The following criteria are minimized: DD, SD, SH, SO, and TE. For simplicity we will use the abbreviations in the tables.

The first step was to get the priorities and the pairwise comparisons for these eight criteria from our pension experts. These weights are biased, so sensitivity analysis is required.

From the priorities and the pairwise comparison, we obtained our pairwise matrix.

	TR	SD	SH	SO	DD	BA	RS	TE
TR	1	1	2	3	5	9	9	9
SD	1	1	2	3	5	7	7	7
SH	½	½	1	3	4	6	6	6
SO	1/3	1/3	1/3	1	3	4	4	5
DD	1/5	1/5	¼	1/3	1	1	3	9
BA	1/9	1/7	1/6	¼	1	1	3	3
RS	1/9	1/7	1/6	¼	1/3	1/3	1	5
TE	1/9	1/7	1/6	1/5	1/9	1/3	1/9	1

We computed the *consistency ratio* (CR) to insure that the pairwise comparison was consistent with a CR <0.1. Our pairwise matrix had a CR value of 0.02116, which is less than 0.1.

Our initial criteria weights were found as:

Total return	0.2917768
Standard deviation	0.2857942
Sharpe	0.1677186
Sortino	0.0955286
Downside deviation	0.0564665
Batting average	0.0374478
R-square	0.0360425
Tracking error	0.0292252

We applied these criterion weights in our TOPSIS analysis using the raw data supplied by our pension company.

Our results ordered the 5679 stocks based upon these weighted criteria. We do not present these here due to the nature of the real pension business. However, we illustrate a sample. We used 45 stocks and we ran the TOPSIS process. We provide only the output for the top 20 coded stocks.

TOPSIS Value	Final Rank	Stock code
0.66502	1	A41
0.6608	2	A40
0.61242	3	A14
0.54392	4	A42
0.53963	5	A43
0.51237	6	A45
0.50885	7	A18
0.50599	8	A21
0.50149	9	A31
0.49432	10	A27
0.49139	11	A11
0.48677	12	A29
0.48571	13	A30
0.48541	14	A9
0.48469	15	A38
0.48168	16	A26
0.48039	17	A35
0.48007	18	A5
0.47957	19	A32
0.47803	20	A17

SENSITIVITY ANALYSIS

Sensitivity analysis in any modeling is critical. This can be applied in two ways: (1) we can alter our decision criterion weights and measure the changes in the outcomes and (2) we can take successive cycles of raw data and compute the differences in the stocks positions (ranking).

We take the decision weights and we alter them using the method described in Chapter 4.

The top 20 are pretty different as seen in the following:

TOPSIS Value	Final Rank	
0.7837	1	A41
0.66443	2	A8
0.62969	3	A35
0.62803	4	A30
0.62737	5	A5
0.62454	6	A22
0.60984	7	A33
0.60978	8	A6
0.60256	9	A25
0.60207	10	A37
0.60072	11	A14
0.60052	12	A42
0.59781	13	A4
0.59735	14	A17
0.59653	15	A21
0.59644	16	A10
0.59516	17	A38
0.59281	18	A1
0.59216	19	A31
0.59216	20	A16

Stock A41 is still first but the remaining are different. This illustrates the importance of good decision rights and shows how the changes in priority and weights *affect the choice of stocks.*

Example 9.25: Bond Portfolio Manager with Linear Programming

A bond portfolio manager has $100,000 to allocate to two different bonds: one corporate bond and one government bond. The corporate bond has a yield of 4%, a maturity of 3 years, and an A rating from a rating agency that is translated into a numerical rating of 2 for computational purposes. In contrast, the government bond has a yield of 3%, a maturity of 4 years, and rating of Aaa with the corresponding numerical rating of 1 (lower numerical ratings correspond to higher quality bonds). The portfolio manager would like to allocate his or her funds so that the average rating for the portfolio is no worse than Aa (numerical equivalent 1.5) and average maturity of the portfolio is at most 3.6 years. Any amount not invested in the two bonds will be kept in a cash account that is assumed to earn no interest for simplicity and does not contribute to the average rating or maturity computations 1. How should the manager allocate his or her funds between these two bonds to achieve his or her objective of maximizing the yield from this investment? By letting variables x_1 and x_2 to denote the allocation of funds to the corporate and government bond, respectively, (in thousands of dollars) we obtain the following formulation for the portfolio manager's problem:

$$\max Z = 4x_1 + 3x_2$$

subject to:

$$x_1 + x_2 \leq 100$$
$$2x_1 + x_2\, 100 \leq 1.5$$
$$3x_1 + 4x_2\, 100 \leq 3.6$$
$$x_1, x_2 \geq 0$$

We solve and find $Z = 3.54$ when $x_1 = 0.48$ and $x_2 = 0.54$. We provide a screenshot of the answer, Figure 9.3.

Sensitivity analysis is also essential. We run the sensitivity report and obtain Figure 9.4. We examine the shadow prices and find that constraints two and three provide good analysis. The shadow price of constraint two is 1.4 and the shadow price of constraint three is 0.4. We know that

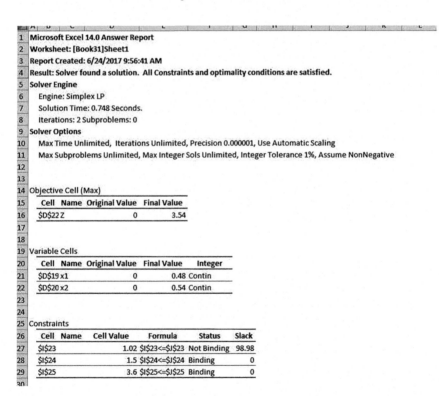

1	Microsoft Excel 14.0 Answer Report
2	Worksheet: [Book31]Sheet1
3	Report Created: 6/24/2017 9:56:41 AM
4	Result: Solver found a solution. All Constraints and optimality conditions are satisfied.
5	Solver Engine
6	Engine: Simplex LP
7	Solution Time: 0.748 Seconds.
8	Iterations: 2 Subproblems: 0
9	Solver Options
10	Max Time Unlimited, Iterations Unlimited, Precision 0.000001, Use Automatic Scaling
11	Max Subproblems Unlimited, Max Integer Sols Unlimited, Integer Tolerance 1%, Assume NonNegative
12	
13	
14	Objective Cell (Max)

Cell	Name	Original Value	Final Value
$D$22	Z	0	3.54

Variable Cells

Cell	Name	Original Value	Final Value	Integer
$D$19	x1	0	0.48	Contin
$D$20	x2	0	0.54	Contin

Constraints

Cell	Name	Cell Value	Formula	Status	Slack
$I$23		1.02	$I$23<=$J$23	Not Binding	98.98
$I$24		1.5	$I$24<=$J$24	Binding	0
$I$25		3.6	$I$25<=$J$25	Binding	0

FIGURE 9.3
Screenshot of answer from portfolio problem.

A	B	C	D	E	F	G	H	I
1	Microsoft Excel 14.0 Sensitivity Report							
2	Worksheet: [Book31]Sheet1							
3	Report Created: 6/24/2017 9:56:42 AM							
4								
5								
6	Variable Cells							
7			Final	Reduced	Objective	Allowable	Allowable	
8	Cell	Name	Value	Cost	Coefficient	Increase	Decrease	
9	$D$19	x1	0.48	0	4	2	1.75	
10	$D$20	x2	0.54	0	3	2.333333333	1	
11								
12	Constraints							
13			Final	Shadow	Constraint	Allowable	Allowable	
14	Cell	Name	Value	Price	R.H. Side	Increase	Decrease	
15	$I$23		1.02	0	100	1E+30	98.98	
16	$I$24		1.5	1.4	1.5	0.9	0.6	
17	$I$25		3.6	0.4	3.6	2.4	1.35	
18								

FIGURE 9.4
Screenshot of sensitivity report.

Z changes for an increase in the available resource for either constraint, by $\Delta\lambda$, where λ is the shadow price. Thus, an increase of 1.4 is better than an increase of 0.4 if all costs for obtaining an additional resource unit are equivalent.

Example 9.26: Cash Flow as a Linear Program (LP)

Consider a short-term financing problem.

Month	Jan	Feb	Mar	Apr	May	Jun
Cash Flow	−150	−100	200	−200	50	300

For example, in January ($i = 1$), there is a cash requirement of $150. To meet this requirement, the company can draw an amount x_1 from its line of credit and issue an amount y_1 of commercial paper. Considering the possibility of excess funds z_1 (possibly 0), the cash flow balance equation is as follows: $x_1 + y_1 - z_1 = 150$. Next, in February ($i = 2$), there is a cash requirement of $100. In addition, principal plus interest of 1.01×1 is due on the line of credit and $1.003z_1$ is received on the invested excess funds. To meet the requirement in February, the company can draw an amount x_2 from its line of credit and issue an amount y_2 of commercial paper. So, the cash flow balance equation for February is as follows:

$$x_2 + y_2 - 1.01x_1 + 1.003z_1 - z_2 = 100$$

Similarly, for March we get the following equation:

$$x_3 + y_3 - 1.01x_2 + 1.003z_2 - z_3 = -200$$

For the months of April, May, and June, issuing a commercial paper is no longer an option, so we will not have variables y_4, y_5, and y_6 in the formulation. Furthermore, any commercial paper issued between January and March requires a payment with 2% interest 3 months later. Thus, we have the following additional equations:

$$x_4 - 1.02y_1 - 1.01x_3 + 1.003z_3 - z_4 = 200$$
$$x_5 - 1.02y_2 - 1.01x_4 + 1.003z_4 - z_5 = -50$$
$$- 1.02y_3 - 1.01x_5 + 1.003z_5 - v = -300$$

Note that x_i is the balance on the credit line in month i, not the incremental borrowing in month i. Similarly, z_i represents the overall excess funds in month i. This choice of variables is quite convenient when it comes to writing down the upper bound and nonnegativity constraints.

$$0 \le x_i \le 100$$

$$y_i \ge 0 z_i \ge 0$$

This gives us the complete model of this problem:
Max v

$$x_1 + y_1 - z_1 = 150$$
$$x_2 + y_2 - 1.01x_1 + 1.003z_1 - z_2 = 100$$
$$x_3 + y_3 - 1.01x_2 + 1.003z_2 - z_3 = -200.$$
$$x_4 - 1.02y_1 - 1.01x_3 + 1.003z_3 - z_4 = 200$$
$$x_5 - 1.02y_2 - 1.01x_4 + 1.003z_4 - z_5 = -50$$
$$- 1.02y_3 - 1.01x_5 + 1.003z_5 - v = -300$$

$$x_1 \le 100$$
$$x_2 \le 100$$
$$x_3 \le 100$$
$$x_4 \le 100$$
$$x_5 \le 100$$

$$x_i, \ y_i, \ z_i \ge 0$$

We solve this LP to obtain $x_1 = x_2 = x_3 = x_4 = 0$, $x_5 = 52$, $z_1 = z_2 = z_4 = z_5 = 0$, $z_3 = 351.9442$, $y_1 = 150$, $y_2 = 100$, $y_3 = 151.9442$, and $v = 92.49695$.

Example 9.27: Bank's Financial Planning

A bank makes four kinds of loans to its personal customers and these loans yield the following annual interest rates to the bank:

- First mortgage 14%
- Second mortgage 20%
- Home improvement 20%
- Personal overdraft 10%

The bank has a maximum foreseeable lending capability of $250 million and is further constrained by the policies:

1. First mortgages must be at least 55% of all mortgages issued and at least 25% of all loans issued (in dollars).
2. Second mortgages cannot exceed 25% of all loans issued (in dollars).
3. To avoid public displeasure and the introduction of a new windfall tax the average interest rate on all loans must not exceed 15%.

Formulate the bank's loan problem as an LP so as to maximize interest income while satisfying the policy limitations.

Note here that these policy conditions, while potentially limiting the profit that the bank can make, also limit its exposure to risk in a particular area. It is a fundamental principle of risk reduction that risk is reduced by spreading money (appropriately) across different areas.

FINANCIAL PLANNING SOLUTION

Note here that as in *all* formulation exercises, we are translating a verbal description of the problem into an *equivalent* mathematical description.

A useful tip when formulating LP's is to express the variables, constraints, and objectives in words before attempting to express them in mathematics.

VARIABLES

Essentially we are interested in the amount (in $) that the bank has loaned to customers in each of the four different areas (not in the actual number of such loans). Thus,

x_i = amount loaned in area i in millions of dollars (where $i = 1$ corresponds to first mortgages, $i = 2$ to second mortgages etc.) and note that $x_i \geq 0$ ($i = 1,2,3,4$).

Note here that it is a necessary and sufficient condition in linear programming problems to have all variables takes on non-negative real values. Any variable (X, say) that can be positive *or* negative can be written as $X_1 - X_2$ (the difference of two new variables) where $X_1 \geq 0$ and $X_2 \geq 0$.

CONSTRAINTS

a. Limit on amount lent

$$x_1 + x_2 + x_3 + x_4 \leq 250$$

b. Policy condition 1

$$x_1 \geq 0.55(x_1 + x_2)$$

That is, first mortgages ≥ 0.55(total mortgage lending) and also

$$x_1 \geq = 0.25(x_1 + x_2 + x_3 + x_4)$$

That is, first mortgages ≥ 0.25(total loans)

c. Policy condition 2

$$x_2 \leq 0.25(x_1 + x_2 + x_3 + x_4)$$

d. Policy condition 3—we know that the total annual interest is $0.14x_1 + 0.20x_2 + 0.20x_3 + 0.10x_4$ on total loans of $(x_1 + x_2 + x_3 + x_4)$. Hence the constraint relating to policy condition (3) is

$$0.14x_1 + 0.20x_2 + 0.20x_3 + 0.10x_4 \leq 0.15\left(x_1 + x_2 + x_3 + x_4\right)$$

OBJECTIVE FUNCTION

To maximize interest income (which is given earlier), that is, maximize $0.14x_1 + 0.20x_2 + 0.20x_3 + 0.10x_4$

COMPLETE FORMULATION

$$\text{Maximize } Z = 0.14x_1 + 0.20x_2 + 0.20x_3 + 0.10x_4$$

Subject to:

$$x_1 + x_2 + x_3 + x_4 \leq 250$$
$$x_1 - 0.55(x_1 + x_2) \geq 0$$
$$x_2 - 0.25(x_1 + x_2 + x_3 + x_4) \leq 0$$
$$0.14x_1 + 0.20x_2 + 0.20x_3 + 0.10x_4 - 0.15(x_1 + x_2 + x_3 + x_4) \leq 0$$

All variables are nonnegative.

SOLUTION

The optimal solution to this LP is $x_1 = 208.33$, $x_2 = 41.67$, and $x_3 = x_4 = 0$. Note here that this optimal solution is not unique. Using the method of find alternate optimal solutions we find the other variable values: $x_1 = 62.50$, $x_2 = 0$, $x_3 = 100$, and $x_4 = 87.50$ also satisfy all the constraints and have exactly the same (maximum) solution value of $Z = 37.5$.

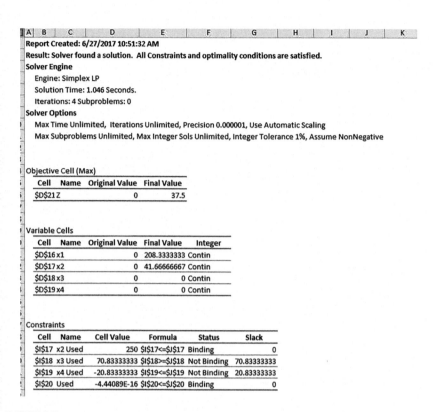

FIGURE 9.5
Screenshot of answer report for Example 9.27.

We also find that an increase in the amount that we can allocate and policy condition (d) are the only two that can be altered to improve the output (Figures 9.5 and 9.6).

Example 9.28: Minimum Variance of Expected Investment Returns (Fox et al., 2013)

A new company has \$5,000 to invest but the company needs to earn about 12% interest. A stock expert has suggested three mutual funds {A, B, and C} in which the company could invest. Based upon previous year's returns, these funds appear relatively stable. The expected return, variance on the return, and covariance between funds are shown as follows:

Expected Value	A	B	C
	0.14	0.11	0.10
Variance	A	B	C
	0.2	0.08	0.18
Covariance	AB	AC	BC
	0.05	0.02	0.03

	A	B	C	D	E	F	G	H	I

Microsoft Excel 14.0 Sensitivity Report
Worksheet: [LP examples chapter 9.xlsx]Sheet3
Report Created: 6/27/2017 10:51:32 AM

Variable Cells

Cell	Name	Final Value	Reduced Cost	Objective Coefficient	Allowable Increase	Allowable Decrease
$D$16	x1	208.3333333	0	0.14	0.06	0
$D$17	x2	41.66666667	0	0.2	0	2.22045E-16
$D$18	x3	0	-2.22045E-16	0.2	2.22045E-16	1E+30
$D$19	x4	0	0	0.1	0	1E+30

Constraints

Cell	Name	Final Value	Shadow Price	Constraint R.H. Side	Allowable Increase	Allowable Decrease
$I$17	x2 Used	250	0.15	250	1E+30	250
$I$18	x3 Used	70.83333333	0	0	70.83333333	1E+30
$I$19	x4 Used	-20.83333333	0	0	1E+30	20.83333333
$I$20	Used	-4.44089E-16	1	0	1.25	2.5

FIGURE 9.6
Screenshot of sensitivity report for Example 9.27.

Formulation:
We use laws of expected value, variance, and covariance in our model.
Let x_j be the number of dollars invested in funds j ($j = 1,2,3$).

Minimize $V_I = var(Ax_1 + Bx_2 + Cx_3)$

$$= x_1{}^2 Var(A) + x_2{}^2 Var(B) + x_3{}^2 Var(C) + 2x_1x_2 Cov(AB)$$

$$+ 2 x_1 x_3 Cov(AC) + 2x_2x_3 Cov(BC)$$

$$= 0.2\, x_1{}^2 + 0.08x_2{}^2 + 0.18\, x_3{}^2 + 0.10\, x_1x_2 + 0.04\, x_1 x_3 + 0.06\, x_2x_3$$

Our constraints include

1. The expectation to achieve at least the expected return of 12% from the sum of all the expected returns:

$$0.14x_1 + 0.11x_2 + 0.10x_3 > (0.12 \times 5000)\ or$$

$$0.14\, x_1 + 0.11x_2 + 0.10\, x_3 > 600$$

2. The sum of all investments must not exceed the $5000 capital:

$$x_1 + x_2 + x_3 \leq \$5000$$

The optimal solution via LINGO is:
$x_1 = 1904.80$, $x_2 = 2381.00$, $x_3 = 714.20$, $z = \$1880942.29$, or a SD of $1371.50.
The expected return is $14(1904.8) + .11(2381) + .1(714.2)/5000 = 12\%$
This example was used as a typical standard for investment strategy.

Projects

1. College

A parent must decide between investment alternatives for their child's college fund. Your child is currently in the fifth grade. You have heard about the South Carolina Tuition Prepayment Program (SCTPP) but you wonder if it is a good deal. It allows you to prepay your child's tuition and guarantee payment to attend any South Carolina public four-year institution for 4 years. The options are as follows:

a. You can pay lump sum right now.

b. You can pay 48 fixed monthly payments starting now.

c. You can pay fixed monthly installments starting now until your child attends college.

The following is a table of in-state tuition and fees for some South Carolina (SC) public colleges:

Public school	93–94	94–95	96–97	97–98	98–99	99–00	00–01	01–02	02–03	03–04
Citadel	3080	3176	3275	3297	3631	3396	3404	3727	4067	4999
Clemson	2954	3036	3112	3112	3252	3344	3470	3590	5090	5834
Coastal	2470	2710	2800	2910	3100	3220	3340	3500	3770	4350
College of Charleston	2950	3060	3090	3190	3290	3390	3520	3630	3760	4858
Francis Marion	2800	2920	3010	3010	3270	3350	3350	3600	3790	4340
Winthrop	3470	3620	3716	3818	3918	4032	4126	4262	4668	5600
Medical University of South Carolina	2560	2819	2910	3202	3648	4034	4626	5180	5824	6230
University of South Carolina (USC) Beaufort										4208

Source: Courtesy of http://www.che.sc.gov/Finance/Abstract/Abstract2003.pdf.

Situation: Current Lump Sum Payment

Tenth grade	3	2006–2007	25,566	NA	954.00	29
Ninth grade	4	2007–2008	25,386	NA	694.00	41
Eighth grade	5	2008–2009	25,207	601.00	552.00	53
Seventh grade	6	2009–2010	25,028	597.00	464.00	65
Sixth grade	7	2010–2011	24,849	592.00	403.00	77
Fifth grade	8	2011–2012	24,671	588.00	358.00	89
Fourth grade	9	2012–2013	24,494	584.00	325.00	101

Your child is currently in the fifth grade and your lump sum payment is $24,671.

Suppose you have the money but decide to put the money into a money market account currently paying 5.84% annual interest compounded monthly. How much would be in the account when your child enters college? Which colleges, if any of those listed, will you be able to afford for your child?

Monthly payments.

Another option is 48 monthly payments of $588 or 89 payments of $358. What is the NPV of each option?

If you could invest that amount in a money market account paying 5.85% interest compounded monthly, how much money would you have available when your child needs to enter college?

2. Car Financing

It is time to purchase the new college graduate a new car. As you look through JD Power and Associates reports on the best cars, you have narrowed your search to the following three cars:

Mazda Miata ($25,000) 4.25% for 60 months, 3.75% for 48 months

Ford Mustang ($32,200) 2.5% for 48 months, 3.7% for 60 months

Mitsubishi Eclipse ($24,600) 3.5% for 60 months

Dodge SRT-4 ($20,400) 4.5% for 48 months plus $1000 cash back

Toyota Celica ($17,700) 4.7% for 48 months

Based on the job that you are planning to get after graduation, you have allocated at most $400 for your car payment. You would also like to have your loan paid off within 48–60 months if at all possible.

Determine which of these cars, you can afford to buy.

You find that Ford has a leasing plan for the Mustang. You pay 1.5% of the purchase price down and then pay $400 a month for the life of the lease (36- or 48-month lease).

Mazda also has a leasing plan. It has no money down but the monthly payments increase by $15 per month with the first month being $50. The life of the lease is 48 months.

Contrast and compare these lease options. Which option do you recommend? Is leasing better or worse than buying in these cases?

References and Suggested Readings

Giordano, F., W. Fox, and S. Horton. 2014. *A First Course in Mathematical Modeling.* Boston, MA: Cengage Publishers.

Newman, D. 1988. *Engineering Economic Analysis,* (3rd ed.). San Jose, CA: Engineering Press.

Answers to Selected Exercises

Chapter 1

Exercises

1. The population of deer in your community.

   Let us assume that we are going to model a population growth of a single species similar to the presentations in Chapter 7. You might want to refer to that section of the text for a more thorough discussion than is given here.

   Let us restrict our attention to a single species and *identify the problem* as follows: For a given species with a known current population, predict the population P at some future time. Lots of modeling assumptions can be made. For a single species, population growth depends on the birth rate, the death rate, and the availability of resources. The birth rate depends on the kind of species, the habitat, food supply, crowding conditions (some species produce fewer off-spring when overcrowding occurs), predators (some species with many predators will produce more off-spring), the general health of the species population, the number of females, and so forth. (If the species is a human population, the birth rate is influenced by such factors as infant mortality rate, attitudes toward and availability of contraceptives, attitudes toward abortion, health care during pregnancy, and so forth.) The death rate depends on the species, the number of predators in the habitat, the health of the species, availability of food and water, environmental factors, and so forth. (For human populations, the death rate is affected by sanitation and public health, war, pollution, diet, medicines, psychological stress and anxiety, age and gender, distribution of the current population, and so forth. Other factors affecting the human population growth in a particular region include immigration and emigration, living space restrictions, and epidemics. We can see that many factors influence the population growth. Initially, we might assume that the rate of change of the population over time is proportional to the present population level. This assumption leads to the differential equation $(dP/dt) = kP$ where k is a proportionality factor or a discrete dynamical system (DDS) model $P(n + 1) = P(n) + k*P(n)$, $P(0) = P_0$. When $k > 0$, the population is growing; for $k < 0$, the population is on the decline. A simple model might assume that k is a constant. It is probably more realistic to assume that k varies with the population P. As the population increases, for example, competition for a limited food supply and living space will cause k to decrease so that k is a decreasing function of P. It may also be true that k is time dependent. The assumptions that $k > 0$ is a constant and that $k = r(M - P)$ for $r > 0$ and $M > 0$ constant, are investigated later in the text.

2. A new outdoor shopping mall is being constructed. How should you design the illumination?

   There are many different types and designs of shopping malls/ centers. You should pick one to consider for your design. The newest design in the contemporary malls is similar to Santana Row in San Jose, CA where we have many outdoor stores linked together by nice streets and area. Parking is usually in two distinct areas: (1) a parking garage and (2) a large outdoor space. After you have picked the type malls then brainstorm the assumptions for lighting. Such assumptions include safety for patrons, safety for the cars, and security issues for both.

   Perhaps start with a schematic, for example:

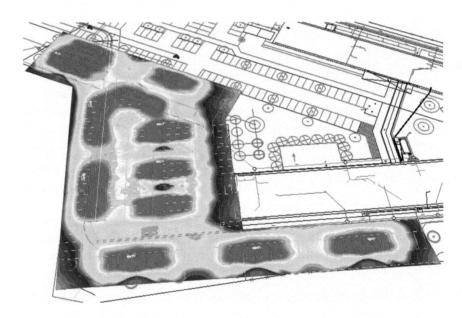

   Consider the type of lights: LED, fluorescent, and so on.

   Consider the height of the lighting for maximum coverage.

   Consider the distribution of the lights for maximum coverage.

We could consider the following discussion:

The light core was at a height of 9 m therefore the LED street light Hawkeye 3.0 with a strong intensity of light was used. An optimum illumination can be ensured with a light flux of 4466 lm and special optic system (see DIALUX).

You might also consider the client's opinion:

We found the optic system of the lights and the expected low-energy consumption quite impressive. Another important reason to decide in favor of the LED lights was that the normal lights installed in the existing parking lot during the first construction phase kept falling off and had to be replaced completely. The new LED lights have a longer life. After the completion of the second phase of construction, we are very satisfied with the light effects at night. We are happy that we opted for a specific brand of LED lights.

We identify the following problem: minimize the cost of procuring, installing, and operating a lighting system while meeting acceptable standards. (Alternatively, the store may wish to maximize the lighting possible at a given cost level, including procurement, installation, operation, and maintenance. The store may wish to take into account traffic patterns in the parking lot so that those areas exhibiting the greatest traffic flow and parking utilization are best illuminated.) We will assume that the size and shape of the parking lot have already been determined and are fixed. A question to answer is during what hours the store desires to operate the lighting system. If the management desires to retain some lighting after closing hours to deter crime, how much lighting is desired and where in the parking lot? Tentatively, we can say the total cost to be minimized is the sum of three costs: (1) procurement and installation, (2) operation, and (3) maintenance of the system. We treat each submodel separately.

Procurement/installation: This cost is a function of the size of the lot, illuminating capacity of the lighting fixtures (which will determine the number needed), construction costs to erect the fixtures, power requirements (special cables, etc.), parking schemes (which will affect the density and placement of fixtures), weather considerations (hurricane country would require stronger supports), vandalism considerations, and so forth. The esthetic qualities may also be a consideration.

Operation: This cost is affected by the type of lighting (neon, sodium vapor, etc.), number of hours the system is operated each day, whether there is a timing device to turn the system on/off when light is needed, number of lighting fixtures, cost per kilowatt-hour of electricity, and so forth.

Maintenance: This cost depends on the number of fixtures, service requirements to replace failed units, life span of illumination device (i.e., bulb or neon tube or whatever), life span of fixtures and support systems, esthetic upkeep of fixtures (such as painting), and so forth. Average values can be assumed for many of these variables based on manufacturer reliability data.

3. A farmer wants a successful season with his crops. He thinks that this is accomplished by growing a lot of anything as long as he uses all his land.

Pick a particular crop or crops that grow and sell well in your area.

Problem identification: *Maximize* the profit from crops sales.

Assumptions: Good weather and that is how the plant grows. We further assume that we can sell all that we grow. This might lead to nice optimization problem; first, what exactly does the farmer means by *yield*? Does he mean long-term or short-term yield? Second, why is he focusing on a *certain* crop? A better long-term yield, for instance, might be obtained by crop rotation. On the other hand, have contracts been signed that commit the farmer to a particular crop (such as in a COOP)? Are there any alternatives for use of the land not planted with the crop?

Alternative objectives:

a. Maximize long-term/short-term productivity.

b. Maximize long-term/short-term profits.

c. Minimize the acreage farmed to produce a specified yield of crop.

d. Produce a specified yield while simultaneously not exceeding a definite level of *damage* to the soil.

e. Ensure *survivability* by planting a low-risk crop that has a high probability of generating revenues next year.

4. Ford Motor Company bought Volvo. Are Volvos still *top quality* cars?

We might begin by defining what we mean by *top quality* cars. Top quality can consider many aspects from reliability of systems and subsystems (such as engine, transmission, and brakes) to impact design in a crash. It might also consider the cost to purchase new versus resale value over time. Defining the metric is our first priority.

Possible problem statement could include the following:

Problem Statement: Maximize resale value of a car

Maximize reliability for 5 years

Maximize safety of passengers and drivers

Maximize sales of Volvo

5. A minor Forbes 500 company wants to go mobile with Internet access and computer upgrades but cost might be a problem. Brainstorm the variables of concern. Costs being a major concern might lead one to formulate a problem such as minimize cost subject to constraints of capital, public reachable through the Internet, and similar approaches.

6. Starbucks has many varieties of coffee available. How can Starbucks make more money? This can be approached from many aspects. Is your issue of scheduling workers in both peak and nonpeak times? May we cut back on workers at certain times on a normal day? How does Starbucks charge for coffee: size, size, and type of coffee? Do they consider premium coffee the same as regular coffee? Is how the coffee distributed to them is different?

Problem Statement: Schedule workers to minimize costs

Coffee available to maximize sales

7. A student does not like math or math-related courses. How can a student maximize their chances for a good grade in a math class to improve their overall GPA?

Lots to be considered here. Does the student ask around for the easy professor? Does the student gather old notes and quizzes on all previous professors? Does the schedule (time of the class) act as a more important feature than the professor? Will any grade really improve their GPA? Is the math course the best option? Maybe a problem statement could be as follows:

Problem Statement: Maximize GPA after next term additions

8. Freshmen think that their first semester courses should be pass–fail for credit or no-credit only. We can look at the impact of the pass–fail versus grades in the freshman's year. We could look at other schools that have freshman pass–fail in lieu of grades. Do we look at graduation rates? Do we look at transfer rate?

9. Alumni are clamoring to fire the college's head football coach.

What is the goal? Is it to have a winning team or is it to win the national championship?

Problem Statement: Maximize profit of having football

Maximize winning percentage

Maximize recruiting efforts

Maximize recruiting results

Minimize cost of the coach

10. Stocking a fish pond with bass and trout.

    Define first the purpose of the pond. Is it a local fishing pond or one person's property?

    How many fishermen will visit the pond? Do we sell fishing licenses or can people fish without licenses? Can the fish survive together? Are there enough food sources for both the fish? What is the level of competition between the fish? We decide to model the species competition.

    Let $B(n)$ is the number of bass in the pond after period n and $T(n)$ is the number of trout in the pond after period n. Define $B(n)*T(N)$ as the interaction terms between the two species. Then a model could be

    $$B(n+1) = \text{rate in } B(n) \quad \text{rate out } B(n)*T(n)$$

    $$T(n+1) = \text{rate in } * T(n) \quad \text{rate out } B(n)*T(n) \text{ with}$$

    $$B(0) = B_0 \text{ and } T(0) = T_0$$

Chapter 2

Exercises 2.2

1. Average is 94
2. Average is $1799/19 = 94.68421053$
3. Mean at ATM is 1.81
4. $875
5. $232 profit
6. 465 lb

Exercises 2.3

1. No, sometimes caution is more important than value. For example, saving money for a college fund might be an example where caution with a lower expected value might be better than high risk with a larger expected value.
2. Bid on high school.

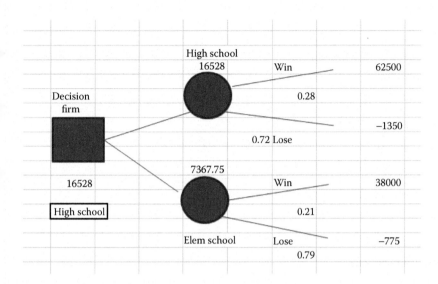

3. $E[A] = 13.24$, $E[B] = 14.72$, $E[C] = 13.96$, decision alternative B is better.
4. Stock 1 unit of strawberry, $E[1] = 0.446 > E[2] = 0.198 > E[0] = 0 > E[3] = -0.249 > E[4] = -3.13$.

Exercises 2.4

1. $E[\text{large}] = \$52{,}250 > E[\text{small}] = \$46{,}250 > E[\text{No warehouse}] = \$0$

Exercises 2.5

1. a. $E[A] = 885$, $E[B] = 105$, $E[C] = 930$, so choose alternative B because $E[B]$ is smallest.
   b. Opportunity loss

Alternatives	$p = 0.30$ #1	$p = 0.35$ #2	$p = 0.20$ #3	$p = 0.15$ #4
A	0	500	500	600
B	300	0	0	500
C	450	300	500	0

$E[\text{loss with } A] = 365$
$E[\text{loss with } B] = 165$, Best option
$E[\text{loss with } C] = 340$

1. (a) Laplace, $E[A] = 14300/3$, $E[B] = 13200/3$, $E[C] = 12500/3$, so option A is better; (b) Maximin $A \to 3500$, $B \to 1500$, $C \to 3500$, both A and C are better; (c) Maximax $A \to 6300$, $B \to 9500$, $C \to 4500$, B is better; (d) coefficient of Optimisim (assume that $x = 0.65$) B is better $E[B] = 5320$.

2. (a) $E[\text{Stocks}] > E[\text{Bonds}]$, $6444 > 5684$; (b) $p = 0.6$ makes the expected values equal with 5180, so if $p > 0.60$ then choose stocks, otherwise choose bonds.

3. (a) Laplace, choose a; (b) Maximin, choose A; (c) Maximax choose A; (d) Choose A.

4. (a) Lapalce hotel best; (b) Maximin, restaurant is best; (c) Maximax, hotel is best; (d) Hotel is best.

Chapter 3

Exercises 3.2

1. Let A = # drink A produced, Let B = # drink B produced

   Max $Z = 15A + 18B$

   Subject to $3A + 2.5B \le 35$

   $$2A + 2.5B \le 40$$
   Nonnegativity

2. Let x = # stream ships produced, y = # sailboats produced, w = # submarines produced

   Max $Z = 6x + 3y + 2w$

   Subject to $x + 3y + 2w \le 45$

   $$2x + 3y + 3w \le 50$$
   $$4x + 2y + w \le 60$$
   Nonnegativity

3. Let S = # small units, M = # medium units, L = # large units produced

   Max $Z = 9S + 12M + 15L$

   Subject to $2S + 2M + L \ge 10$

   $$2S + 3M + L \ge 12$$
   $$S + M + 5L \ge 14$$
   Nonnegativity

4. In thousands,

Let x_1 = # ad units for day time TV, x_2 = # ad units for prime time TV, x_3 = # ad unit for radio, x_4 = # ad units for magazines

Max $Z = 400x_1 + 900x_2 + 500x_3 + 200x_4$

Subject to $40x_1 + 75x_2 + 30x_3 + 15x_4 \leq 800$

$$80x_1 + 40x_2 + 20x_3 + 10x_4 \geq 200$$
$$40x_1 + 75x_2 \leq 500$$
$$x_1 \geq 3$$
$$x_2 \geq 2$$
$$5 \leq x_3 \leq 10$$
$$5 \leq x_4 \leq 10$$

5. Max $Z = 0.08X_{wa} + 0.08X_{wb} + 0.05X_{pa} + 0.05X_{pb}$

Subject to $X_{wa} + X_{pa} \leq 5000$

$$X_{wb} + X_{pb} \leq 10000$$
$$2X_{wa} - 0.8X_{wb} \geq 0$$
$$9X_{pa} - 0.1X_{pb} \geq 0$$
$$\text{Nonnegativity}$$

6. Let x_1 = pound of beef

x_2 = pounds of pork

x_3 = pounds of veal

We know $x_3 = 1000 - x_1 - x_2$.

Maximize Z = Revenue-cost = $400 + 0.1x_1 + 0.2x_2$

Subject to: $400 \leq x_1 \leq 600$

$$200 \leq x_2 \leq 300$$
$$100 \leq x_3 \leq 400$$
$$x_1 + x_2 \geq 600$$
$$x_1 + x_2 \leq 900$$
$$1.5\, x_1 + 2.5x_2 \leq 1500$$
$$\text{Nonnegativity}$$

7. We use the Bonds name {A,B,C,D,E} to represent how much to invest in each type bond.

Maximize $Z = 0.043A + 0.027B + 0.025C + 0.022D + 0.045E$

Subject to:

$$B + C + D \geq 4 \text{ million}$$

$A + B + C + D + E \leq 10$ million

$2A + 2B + C + D + 5E - 1.4*(A + B + C + D + E) \leq 0$

$9A + 15B + 4C + 3D + 2E - 5*(A + B + C + D + E) \leq 0$

Nonnegativity

8. Minimize cost $= (25)(32)/Q_1 + (24)(18)/Q_2 + (20)(20)/Q_3 + Q_1/2 + 1.5Q_2/2 + 2Q_3$ subject to: $4Q_1 + 3Q_2 + 2Q_3 < 200$ sq ft.

9. Maximize $P(L,K) = 1.2L^{0.3} K^{0.6}$ Subject to $10{,}000L + 7{,}000K = C$, where C is either 63,940, 55,060, or 71,510.

10. Maximize profit : $= (339 - 0.01x_1 - 0.03x_2)x_1 + (339 - 0.01x_2 - 0.04x_1)x_2$

$$-400{,}000 - 195x_1 - 225x_2$$

11. Maximize profit: $= (339 - 0.01x_1 - 0.03x_2)x_1 + (339 - 0.01x_2 - 0.04x_1)x_2$

$$- 400{,}000 - 195x_1 - 225x_2$$

Subject to

$$8{,}000x_1 + 5{,}000x_2 \geq 10{,}000$$
$$x_1, x_2 \geq 0$$

Exercises 3.3 and 3.4

1. Max: $Z = 21$, $x = 6$, $y = 3$. Min: $Z = 6$, $x = 3$, $y = 0$
2. Max $Z = 144$, $x = 24$, $y = 0$. Min: $x = y = Z = 0$
3. Max is unbounded. Min: $x = y = 3$, $Z = 33$
4. Max and min are both unbounded
5. Max $Z = 400$, $x = 80$, $y = 0$. Min: $x = y = Z = 0$

Exercises 3.5

1. Objective function is the same. New constraints const: $= 45x_1 + 1.5x_2 \leq 19{,}200$, $15x_1 + 42x_2 \leq 60{,}000$ and const: $= 15x_1 + 42x_2 \leq 60{,}000$, $45x_1 + 1.5x_2 \leq 19{,}200$

   New solution: all in x_2. maximize (obj, const, NONNEGATIVE) and $x_1 = 0$, $x_2 = 1428.571429$

Exercises 3.6

1. Maximize $Z = 17{,}000x_1 + 23{,}000x_2 + 13{,}000x_3 + 9{,}000x_4$
   s.t. $6{,}000x_1 + 8{,}000x_2 + 5{,}000x_3 + 4{,}000x_4 \leq 21{,}000$
   $x_j = 0$ or 1 ($j = 1,2,3,4$)

2. Maximize $Z = 17{,}000x_1 + 23{,}000x_2 + 13{,}000x_3 + 9{,}000x_4$
   s.t. $6{,}000x_1 + 8{,}000x_2 + 5{,}000x_3 + 4{,}000x_4 \leq 21{,}000$
   $x_j = $ integer ($j = 1,2,3,4$)

3. $\text{dist} = \sum_{i=1}^{5} \sqrt{(x - X_i)^2 + (y - Y_i)^2}$

$$= \sqrt{\begin{aligned} &(x-10)^2 + (y-50)^2 + (x-35)^2 + (y-85)^2 + (x-60)^2 + (y-72)^2 + \\ &(x-75)^2 + (y-60)^2 + (x-80)^2 + (y-35)^2 \end{aligned}}$$

Chapter 4

Exercises 4.2

1. Given the input–output table in the following for three hospitals where inputs are number of beds and labor hours in thousands per month and outputs, all measured in hundreds are patient-days for patients under 14, patient-days for patients between 14 and 65, and patient-days for patients over 65. Determine the efficiency of the three hospitals.

Hospital	Inputs		Outputs		
	1	2	1	2	3
1	5	14	9	4	16
2	8	15	5	7	10
3	7	12	4	9	13

Hospital efficiency: Since no units are given and the scales are similar, we decided not to normalize the data. We define the following decision variables:

t_i is the value of a single unit of output of decision-making units (DMU$_i$), for $i = 1,2,3$

w_i is the cost or weights for one unit of inputs of DMU$_i$, for $i = 1,2$

Efficiency$_i$ = DMU$_i$ = (total value of i's outputs)/(total cost of i's inputs), for $i = 1,2,3$

The following modeling assumptions are made:

1. No DMU will have an efficiency of more than 100%.
2. If any efficiency is less than 1, then it is inefficient.
3. We will scale the costs so that the costs of the inputs equal 1 for each linear program. For example, we will use $5w_1 + 14w_2 = 1$ in our program for DMU$_1$.
4. All values and weights must be strictly positive, so we use a constant such as 0.0001 in lieu of 0.

To calculate the efficiency of DMU_1, we define the linear program using Equation 4.2 as

Maximize $DMU_1 = 9t_1 + 4t_2 + 16t_3$

Subject to

$$-9t_1 - 4t_2 - 16t_3 + 5w_1 + 14w_2 \geq 0$$
$$-5t_1 - 7t_2 - 10t_3 + 8w_1 + 15w_2 \geq 0$$
$$-4t_1 - 9t_2 - 13t_3 + 7w_1 + 12w_2 \geq 0$$
$$5w_1 + 14w_2 = 1$$
$$t_i \geq 0.0001, i = 1,2,3$$
$$w_i \geq 0.0001, i = 1,2$$

Nonnegativity

To calculate the efficiency of DMU_2, we define the linear program using Equation 4.2 as

Maximize $DMU_2 = 5t_1 + 7t_2 + 10t_3$

Subject to

$$-9t_1 - 4t_2 - 16t_3 + 5w_1 + 14w_2 \geq 0$$
$$-5t_1 - 7t_2 - 10t_3 + 8w_1 + 15w_2 \geq 0$$
$$-4t_1 - 9t_2 - 13t_3 + 7w_1 + 12w_2 \geq 0$$
$$8w_1 + 15w_2 = 1$$
$$t_i \geq 0.0001, i = 1,2,3$$
$$w_i \geq 0.0001, i = 1,2$$

Nonnegativity

To calculate the efficiency of DMU_3, we define the linear program using Equation 4.2 as

Maximize $DMU_3 = 4t_1 + 9t_2 + 13t_3$

Subject to

$$-9t_1 - 4t_2 - 16t_3 + 5w_1 + 14w_2 \geq 0$$
$$-5t_1 - 7t_2 - 10t_3 + 8w_1 + 15w_2 \geq 0$$
$$-4t_1 - 9t_2 - 13t_3 + 7w_1 + 12w_2 \geq 0$$
$$7w_1 + 12w_2 = 1$$
$$t_i \geq 0.0001, i = 1,2,3$$
$$w_i \geq 0.0001, i = 1,2$$

Nonnegativity

The linear programming (LP) solutions show the efficiencies as $DMU_1 = DMU_3 = 1$, $DMU_2 = 0.77303$.

Interpretation: DMU_2 is operating at 77.303% of the efficiency of DMU_1 and DMU_3. Management could concentrate some improvements or best practices from DMU_1 or DMU_3 for DMU_2. An examination of the dual prices for the linear program of DMU_2 yields $\lambda_1 = 0.261538$, $\lambda_2 = 0$, and $\lambda_3 = 0.661538$. The average output vector for DMU_2 can be written as

$$0.261538 \begin{bmatrix} 9 \\ 4 \\ 16 \end{bmatrix} + 0.661538 \begin{bmatrix} 4 \\ 9 \\ 13 \end{bmatrix} = \begin{bmatrix} 5 \\ 7 \\ 12.785 \end{bmatrix}$$

and the average input vector can be written as

$$0.261538 \begin{bmatrix} 5 \\ 14 \end{bmatrix} + 0.661538 \begin{bmatrix} 7 \\ 12 \end{bmatrix} = \begin{bmatrix} 5.938 \\ 11.6 \end{bmatrix}$$

In our data, output #3 is 10 units. Thus, we may clearly see the inefficiency is in output #3 where 12.785 units are required. We find that they are short 2.785 units (12.785 − 10 = 2.785). This helps focus on treating the inefficiency found for the output #3.

Sensitivity analysis: Sensitivity analysis in a linear program is sometimes referred to as *what if* analysis. Let us assume that without management engaging some additional training for DMU_2 that DMU_2 output #3 dips from 10 to 9 units of output, whereas the input 2 hours increases from 15 to 16 hours. We find that these changes in the *technology coefficients* are easily handled in resolving the LPs. Since DMU_2 is affected, we might only modify and solve the LP concerning DMU_2. We find with these changes that DMU's efficiency is now only 74% as effective as DMU_1 and DMU_3.

2. Resolve problem 1 with the following inputs and outputs.

	Inputs		Outputs		
Hospital	1	2	1	2	3
1	4	16	6	5	15
2	9	13	10	6	9
3	5	11	5	10	12

Decision variables:

t_i is the value of a single unit of output of DMU_i, for $i = 1,2,3$

w_i is the cost or weights for one unit of inputs of DMU_i, for $i = 1,2$

Efficiency$_i$ = DMU_i = (total value of i's outputs)/(total cost of i's inputs), for $i = 1,2,3$

The following modeling assumptions are made:

1. No DMU will have an efficiency of more than 100%.
2. If any efficiency is less than 1, then it is inefficient.
3. We will scale the costs so that the costs of the inputs equal 1 for each linear program. For example, we will use $5w_1 + 14w_2 = 1$ in our program for DMU_1.
4. All values and weights must be strictly positive, so we use a constant such as 0.0001 in lieu of 0.

To calculate the efficiency of DMU_1, we define the linear program using Equation 4.2 as

Maximize $DMU_1 = 6t_1 + 5t_2 + 15t_3$

Subject to

$$-6t_1 - 5t_2 - 15t_3 + 4w_1 + 6w_2 \geq 0$$
$$-10t_1 - 6t_2 - 9t_3 + 9w_1 + 13w_2 \geq 0$$
$$-5t_1 - 10t_2 - 12t_3 + 5w_1 + 11w_2 \geq 0$$
$$4w_1 + 16w_2 = 1$$
$$t_i \geq 0.0001, i = 1,2,3$$
$$w_i \geq 0.0001, i = 1,2$$

Nonnegativity

To calculate the efficiency of DMU_2, we define the linear program using Equation 4.2 as

Maximize $DMU_2 = 10t_1 + 6t_2 + 9t_3$

Subject to

$$-6t_1 - 5t_2 - 15t_3 + 4w_1 + 6w_2 \geq 0$$
$$-10t_1 - 6t_2 - 9t_3 + 9w_1 + 13w_2 \geq 0$$
$$-5t_1 - 10t_2 - 12t_3 + 5w_1 + 11w_2 \geq 0$$
$$9w_1 + 13w_2 = 1$$
$$t_i \geq 0.0001, i = 1,2,3$$
$$w_i \geq 0.0001, i = 1,2$$

Nonnegativity

To calculate the efficiency of DMU_3, we define the linear program using Equation 4.2 as

Maximize $DMU_3 = 5t_1 + 10t_2 + 12t_3$

Subject to

$$-6t_1 - 5t_2 - 15t_3 + 4w_1 + 6w_2 \geq 0$$
$$-10t_1 - 6t_2 - 9t_3 + 9w_1 + 13w_2 \geq 0$$

$$-5t_1 - 10t_2 - 12t_3 + 5w_1 + 11w_2 \geq 0$$
$$5w_1 + 11w_2 = 1$$
$$t_i \geq 0.0001, i = 1,2,3$$
$$w_i \geq 0.0001, i = 1,2$$
Nonnegativity

The LP solutions show the efficiencies as $DMU_1 = DMU_3 = 1$, $DMU_2 = 0.76923$.

Interpretation: DMU_2 is operating at 76.923% of the efficiency of DMU_1 and DMU_3.

Resolving the LPs. Since DMU_2 is affected, we might only modify and solve the LP concerning DMU_2. We find with these changes that DMU's efficiency is now only 74% as effective as DMU_1 and DMU_3.

3. Consider ranking four bank branches in a particular city. The inputs are as follows:

Input 1 = labor hours in hundred per month

Input 2 = space used for tellers in hundreds of square feet

Input 3 = supplies used in dollars per month

Output 1 = loan applications per month

Output 2 = deposits made in thousands of dollars per month

Output 3 = checks processed thousands of dollars per month

The following data table is for the bank branches.

Branches	Input 1	Input 2	Input 3	Output 1	Output 2	Output 3
1	15	20	50	200	15	35
2	14	23	51	220	18	45
3	16	19	51	210	17	20
4	13	18	49	199	21	35

$DMU_1 = 96\%$ efficient and DMU_2, DMU_3, and DMU_4 are 100%.

> $LPSolve(obj1a, \{-200\, t_1 - 15\, t_2 - 35\, t_3 + 15\, w_1 + 20\, w_2 + 50\, w_3$
> $\geq 0, -220\, t_1 - 18\, t_2 - 45\, t_3 + 14\, w_1 + 23\, w_2 + 51\, w_3 \geq 0,$
> $-210\, t_1 - 17\, t_2 - 20\, t_3 + 6\, w_1 + 19\, w_2 + 51\, w_3 \geq 0, -199\, t_1$
> $- 21\, t_2 - 35\, t_3 + 13\, w_1 + 18\, w_2 + 49\, w_3 \geq 0, 15\, w_1 + 20\, w_2$
> $+ 50\, w_3 = 1\}, assume = nonnegative, maximize);$
>
> $[0.960145840238648, [t_1 = 0.00470003314550878, t_2 = 0., t_3$
> $= 0.000575406032482603, w_1 = 0., w_2 = 0.0153463705667882,$
> $w_3 = 0.0138614517732847]]$

> *LPSolve(obj2a, {−200 t_1 − 15 t_2 − 35 t_3 + 15 w_1 + 20 w_2 + 50 w_3*
>   *≥ 0, − 220 t_1 − 18 t_2 − 45 t_3 + 14 w_1 + 23 w_2 + 51 w_3 ≥ 0,*
>   *−210 t_1 − 17 t_2 − 20 t_3 + 6 w_1 + 19 w_2 + 51 w_3 ≥ 0, −199 t_1*
>   *− 21 t_2 − 35 t_3 + 13 w_1 + 18 w_2 + 49 w_3 ≥ 0, 14 w_1 + 23 w_2*
>   *+ 51 w_3 = 1}, assume = nonnegative, maximize);*

[1.00000000000000, [t_1 = 0.00454545454545455, t_2 = 0., t_3= 0., w_1
= 0., w_2 = 0., w_3 = 0.0196078431372549]]

> *LPSolve(obj3a, {−200 t_1 − 15 t_2 − 35 t_3 + 15 w_1 + 20 w_2 + 50 w_3*
>   *≥ 0, − 220 t_1 − 18 t_2 − 45 t_3 + 14 w_1 + 23 w_2 + 51 w_3 ≥ 0,*
>   *−210 t_1 − 17 t_2 − 20 t_3 + 6 w_1 + 19 w_2 + 51 w_3 ≥ 0, −199 t_1*
>   *− 21 t_2 − 35 t_3 + 13 w_1 + 18 w_2 + 49 w_3 ≥ 0, 16 w_1 + 19 w_2*
>   *+ 51 w_3 = 1}, assume = nonnegative, maximize);*

[1.00000000000000, [t_1 = 0.00476190476190476, t_2 = 0., t_3= 0., w_1
= 0., w_2 = 0.0119047619047619, w_3 = 0.0151727357609711]]

> *LPSolve(obj4a, {−200 t_1 − 15 t_2 − 35 t_3 + 15 w_1 + 20 w_2 + 50 w_3*
>   *≥ 0, − 220 t_1 − 18 t_2 − 45 t_3 + 14 w_1 + 23 w_2 + 51 w_3 ≥ 0,*
>   *−210 t_1 − 17 t_2 − 20 t_3 + 6 w_1 + 19 w_2 + 51 w_3 ≥ 0, −199 t_1*
>   *− 21 t_2 − 35 t_3 + 13 w_1 + 18 w_2 + 49 w_3 ≥ 0, 13 w_1 + 18 w_2*
>   *+ 49 w_3 = 1}, assume = nonnegative, maximize);*

[1.00000000000000, [t_1 = 0.00371593724194880, t_2
= 0.0124061185167708, t_3 = 0., w_1 = 0., w_2 = 0., w_3
= 0.0204081632653061]]

Exercises 4.3

In each problem, use SAW to find the ranking under these weighted conditions:

a. All weights are equal.

b. Choose and state your weights.

1. For a given hospital, rank order the procedure using the data in the below-mentioned table.

	Procedure			
	1	2	3	4
Profit	$200	$150	$100	$80
X-Ray times	6	5	4	3
Laboratory Time	5	4	3	2

a. Normalize then add. All weights are equal.

   Procedure 1 best, (1.06) followed by 2 (0.845), 3 (0.625), 4 (0.464).

b. Depends on weights selected.

2. Procedure 2 is best, followed by 4, 1, 3.

3.

								Ranks
Dirty bomb threat	0.096852	0.295804	0.171429	0.249032	0.209302	0.046875	1.069293	3
Anthrax-bio terror threat	0.108959	0.023664	0.011429	0.177089	0.174419	0.1875	0.683059	5
DC-road and bridge network threat	0.084746	0.000148	0.342857	0.047039	0.139535	0.125	0.739325	4
NY subway threat	0.176755	0.354964	0.228571	0.348644	0.162791	0.078125	1.349851	1
DC metro threat	0.16707	0.325384	0.228571	0.138351	0.162791	0.078125	1.100292	2
Major bank robbery	0.196126	5.92E-06	0.011429	0.031544	0.046512	0.25	0.535616	6
FAA threat	0.169492	2.96E-05	0.005714	0.008301	0.104651	0.234375	0.522563	7

4. Consider a scenario where we want to move, rank the cities. Normalize and then use 1/crime so that larger values are better.

City	Affordability of Housing (Average Home Cost in Hundreds of Thousands)	Cultural Opportunities— Events per Month	Crime Rate— Number of Reported Crimes per Month (in Hundreds)	Quality of Schools on Average (Quality Rating between [0,1])
1	250	5	10	0.75
2	325	4	12	0.6
3	676	6	9	0.81
4	1,020	10	6	0.8
5	275	3	11	0.35
6	290	4	13	0.41
7	425	6	12	0.62
8	500	7	10	0.73
9	300	8	9	0.79

City					SUM	Ranks
1	0.061561	0.09434	9.2	0.127986	9.483887	5
2	0.08003	0.075472	7.666667	0.102389	7.924557	8
3	0.166461	0.113208	10.22222	0.138225	10.64012	2
4	0.25117	0.188679	15.33333	0.136519	15.9097	1
5	0.067717	0.056604	8.363636	0.059727	8.547684	6
6	0.071411	0.075472	7.076923	0.069966	7.293772	9
7	0.104654	0.113208	7.666667	0.105802	7.99033	7
8	0.123122	0.132075	9.2	0.124573	9.579771	4
9	0.073873	0.150943	10.22222	0.134812	10.58185	3

Chapter 5

Exercises 5.3

1. Sotto Attack City I, Blotto defend City II, $V = 13$.

2. a. The value of a: $a < b, a > c, a < d$.

b. The value of b: $b < a, b > d, b < c, c < d$.

c. The value of c: $c < d, c > a, b < d$.

d. The value of d: $d > b, d < c, d < a$.

3. $R_1C_2 = 1/2$.

4. $V = 0.250$ when R_2C_2.

5. $R_1C_2, V = 40$.

6. $R_2C_3, V = 55$.

7. $R_1C_3, V = 195$.

8. R_2C_1 (3, −3).

9. $R_2C_2, V = 0.55$.

10. Run right, Nickle, $V = 5$ yards.

11.

	Payoff Tableau	Colin	
		C_1	C_2
Rose	R_1	a	b
	R_2	c	d
	Oddments	$a - c$	$d - b$
	Probability	$(d - b)/((a - c)+(d - b))$	

where $a > d > b > c$. Show that Colin plays C_1 and C_2 with probabilities x and $(1 - x)$ that

$$x = \frac{d-b}{(a-c)+(d-b)}$$

Method 2

$E[R_1] = E[2]$

$ax + b(1 - x) = cx + d(1 - x)$

$ax - bx + b = cx - dx + d$

$a - bx + dx - cx = d - b$

$x = (d - b)/((a - c) + (d - b))$

12. $R_1C_2, v = 3$.

13. Play 5/11 R_1 6/11 R_2 6/11 C_1 5/11 $C_2, V = 8/11$.

14. 0.752 R_1, 0.248 R_2, 0.5179C_1, 0.4821 $C_2, V = 0.296$. Solve the hitter–pitcher duel for the following players:

15. $R_3C_3, V = 4$.

16. 5 R_1, 0 R_2, 5 R_3, 5 C_1, 0 C_2, 5 $C_3, V = 0.7$.

17. Either R_1C_2 or $R_3C_2, V = 5$.

18. $4/7\ R_1$, $3/7\ R_2$, $2/7\ C_1$, $5/7\ C_2$, $V = 3.142857$.

19. $4/7\ R_1$, $3/7\ R_2$, $0\ C_1$, $2/7\ C_2$, $5/7\ C_3$, $0\ C_4$, $V = 1.142857$.

20. $4/5\ R_1$, $1/5\ R_2$, $3/5\ C_1$, $2/5\ C_2$, $V = 0.260$.

21. $0.4\ R_1$, $6\ R_2$, $14/14\ C_1$, $1/15\ C_2$, $v = 1.6$.

22. $0.4\ R_1$, $0.4\ R_2$, $0.2\ R_3$, $0.4\ C_1$, $0.4\ C_2$, $0.2\ C_3$, $v = 1.6$.

Exercises 5.4

1. U.S. Strike Militarily, State Sponsor Terrorism (2,4).

2. U.S. Do not Strike, State: Sponsor Terrorism (4,2.5).

3. Disarm, Disarm (3,3).

4. BC (2,4) and A, D (2,4), trying to get there for each player leaves us at (1,1).

5. Nash equilibrium at (3,4).

6. Nash equilibrium at (0,0).

7. Nash equilibrium at (1,5).

Chapter 6

Exercises 6.2

1. (a) Correlations of x, y is 0.904. p-value $= 0.035$. (b) $y = -0.100 + 0.700x$

```
Predictor     Coef    SE Coef       T        P
Constant    -0.1000    0.6351    -0.16    0.885
x            0.7000    0.1915     3.66    0.035

S = 0.605530   R-Sq = 81.7%   R-Sq(adj) = 75.6%

Analysis of Variance

Source            DF      SS        MS       F       P
Regression         1   4.9000   4.9000   13.36   0.035
Residual Error     3   1.1000   0.3667
Total              4   6.0000
```

(c) The regression equation is $y = 0.158x^2$

```
Predictor     Coef    SE Coef       T        P
Noconstant
x^2        0.15832    0.01931    8.20    0.001

S = 0.604080
```

```
Analysis of Variance

Source            DF       SS       MS       F       P
Regression         1    24.540   24.540   67.25   0.001
Residual Error     4     1.460    0.365
Total              5    26.000

Unusual Observations

Obs    x^2       y      Fit    SE Fit   Residual   St Resid
 5    25.0    4.000   3.958    0.483      0.042       0.12 X
```

2. Correlations 0.999, *p*-value = 0.000

Regression analysis: y (´ 105) versus x (´ 10–3)

The regression equation is $y = 3.70\,x$

```
Predictor       Coef    SE Coef       T       P
Noconstant
x (´ 10-3)    3.70331  0.07435    49.81   0.000

S = 14.5939

Analysis of Variance

Source            DF       SS       MS         F        P
Regression         1   528351   528351   2480.72   0.000
Residual Error    10     2130      213
Total             11   530481
```

The regression equation is
y (´ 105) = - 25.4 + 4.07 x (´ 10-3)

```
Predictor       Coef    SE Coef       T       P
Constant      -25.407    2.653    -9.58   0.000
x (´ 10-3)    4.06933  0.04483    90.77   0.000

S = 4.59866 R-Sq = 99.9% R-Sq(adj) = 99.9%

Analysis of Variance

Source            DF       SS       MS         F        P
Regression         1   174251   174251   8239.69   0.000
Residual Error     9      190       21
Total             10   174441
```

Unusual Observations

x ('Obs 10-3)	y (' 105)	Fit	SE Fit	Residual	St Resid
11 100	390.00	381.53	2.62	8.47	2.24R

The regression equation is
y (' 105) = 0.0449 X_2

Predictor	Coef	SE Coef	T	P
X_2	0.044871	0.002839	15.81	0.000

S = 45.1822

Analysis of Variance

Source	DF	SS	MS	F	P
Regression	1	510067	510067	249.86	0.000
Residual Error	10	20414	2041		
Total	11	530481			

Unusual Observations

Obs	X_2	y (' 105)	Fit	SE Fit	Residual	St Resid
11	10000	390.0	448.7	28.4	-58.7	-1.67 X

X denotes an observation whose X value gives it large
leverage.

The regression equation is
$y = 0.148 \ x^2$

Predictor	Coef	SE Coef	T	P
Noconstant				
xsq	0.148134	0.005111	28.98	0.000

S = 19.1431

Analysis of Variance

Source	DF	SS	MS	F	P
Regression	1	307873	307873	840.13	0.000
Residual Error	13	4764	366		
Total	14	312637			

Unusual Observations

```
Obs   xsq      y_1     Fit   SE Fit   Residual   St Resid
 14  1764   260.00  261.31     9.02      -1.31      -0.08 X
```

The regression equation is
y = 0.00412 x^3

```
Predictor        Coef       SE Coef        T       P
Noconstant
xcube        0.0041205    0.0001441    28.59   0.000
```

S = 19.4025

Analysis of Variance

```
Source            DF       SS        MS        F       P
Regression         1   307743    307743   817.47   0.000
Residual Error    13     4894       376
Total             14   312637
```

Unusual Observations

```
Obs   xcube      y_1      Fit   SE Fit   Residual   St Resid
 10   35937   187.00   148.08     5.18      38.92      2.08R
 14   74088   260.00   305.28    10.68     -45.28     -2.79RX
```

The regression equation is
y = 200 - 0.0136 x^3 + 1.29 x^2 - 28.7 x

```
Predictor     Coef      SE Coef         T        P
Constant     199.6        163.1      1.22    0.249
X³        -0.013617     0.006936    -1.96    0.078
X²           1.2911       0.6177     2.09    0.063
x           -28.71        17.73     -1.62    0.137
```

S = 9.34210 R-Sq = 99.1% R-Sq(adj) = 98.8%

Analysis of Variance

```
Source            DF       SS        MS        F       P
Regression         3    91762     30587   350.47   0.000
Residual Error    10      873        87
Total             13    92635
```

3. Kepler' data model $y = 5.46E - 10*x^{(3/2)}$

Summary Output	
Regression Statistics	
Multiple R	0.999996
R Square	0.999992
Adjusted R Square	0.857135
Standard error	6349934
Observations	8

ANOVA					
	df	**SS**	**MS**	**F**	**Significance F**
Regression	1	3.52E + 19	3.52E + 19	873891.9	1.0114E − 16
Residual	7	2.82E + 14	4.03E + 13		
Total	8	3.52E + 19			

	Coefficients	**Standard Error**	**t Stat**	**P-value**	**Lower 95%**
Intercept	0	#N/A	#N/A	#N/A	#N/A
$x^{(3/2)}$	5.46E − 10	5.84E − 13	934.8219	4.23E − 19	5.44664E − 10

Exercises 6.3

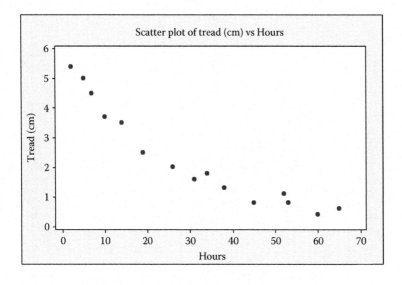

1. Nonlinear regression or polynomial regression

2. Regression Analysis: ln(Z) versus ln(x), ln(y)

```
The regression equation is
ln(Z) = - 0.045 + 2.50 ln(x) - 4.34 ln(y)
which converts to Z = 0.955997 x^2.5/y^4.34

Predictor          Coef    SE Coef         T       P
Constant        -0.0455     0.7534     -0.06   0.952
ln(x)            2.4972     0.1630     15.32   0.000
ln(y)          -4.33833    0.06268    -69.21   0.000


S = 0.196990 R-Sq = 99.6% R-Sq(adj) = 99.5%

Analysis of Variance

Source             DF         SS        MS         F       P
Regression          2    194.121    97.061   2501.24   0.000
Residual Error     22      0.854     0.039
Total              24    194.975
```

3a. Graph

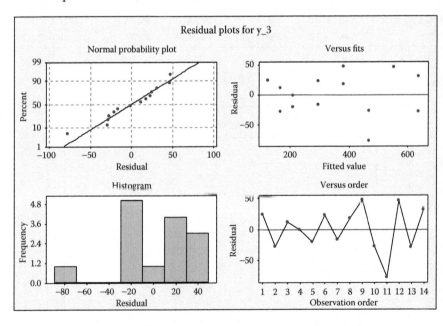

Regression analysis: y versus x

```
The regression equation is
y = - 45.6 + 1.71 x

Predictor      Coef   SE Coef       T      P
Constant     -45.55     25.47   -1.79  0.099
x_4         1.71143   0.09969   17.17  0.000

S = 36.7485  R-Sq = 96.1%  R-Sq(adj) = 95.8%

Analysis of Variance

Source          DF       SS       MS       F      P
Regression       1   398030   398030  294.74  0.000
Residual Error  12    16205     1350
Total           13   414236

Unusual Observations

Obs   x_4      y_3      Fit   SE Fit  Residual  St Resid
 11   300   390.00   467.88    11.73    -77.88    -2.24R

R denotes an observation with a large standardized
residual.
```

3b.

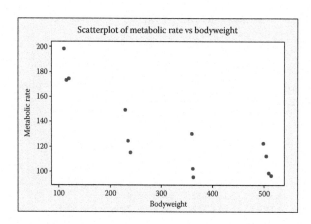

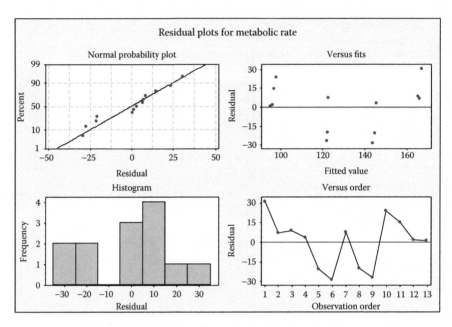

```
metabolic rate = 187 - 0.177 bodyweight

Predictor       Coef    SE Coef      T        P
Constant      186.61      13.04   14.32    0.000
boodyweight  -0.17717    0.03683   -4.81    0.001

S = 19.9657 R-Sq = 67.8% R-Sq(adj) = 64.8%

Analysis of Variance

Source          DF        SS        MS       F        P
Regression       1    9222.8    9222.8   23.14    0.001
Residual Error  11    4384.9     398.6
Total           12   13607.7
```

4. Logistic regression table

```
                                                95% CI
Predictor        Coef     SE Coef      Z       P  Odds Ratio  Lower       Upper
Constant      15.6645     18.7682   0.83   0.404
GRE        -0.0307517   0.0243202  -1.26   0.206        0.97   0.92        1.02
TOP          4.90852      5.02915   0.98   0.329      135.44   0.01  2585632.77
GPA        0.0755136      5.06531   0.01   0.988        1.08   0.00    22100.12


Log-Likelihood = -3.561
Test that all slopes are zero: G = 6.339, DF = 3, P-Value = 0.096
```

```
Goodness-of-Fit Tests

Method              Chi-Square   DF      P
Pearson              5.63758      6   0.465
Deviance             7.12107      6   0.310
Hosmer-Lemeshow      5.63758      8   0.688

Table of Observed and Expected Frequencies:
(See Hosmer-Lemeshow Test for the Pearson Chi-Square
Statistic)

                              Group
Value   1     2     3     4     5     6     7     8     9    10   Total
1
  Obs   0     0     0     0     0     1     1     0     1     1       4
  Exp  0.0   0.0   0.0   0.2   0.3   0.3   0.5   0.6   1.0   1.0
0
  Obs   1     1     1     1     1     0     0     1     0     0       6
  Exp  1.0   1.0   1.0   0.8   0.7   0.7   0.5   0.4   0.0   0.0
Total   1     1     1     1     1     1     1     1     1     1      10

Measures of Association:
(Between the Response Variable and Predicted Probabilities)

Pairs        Number  Percent  Summary Measures
Concordant      22     91.7   Somers' D               0.83
Discordant       2      8.3   Goodman-Kruskal Gamma   0.83
Ties             0      0.0   Kendall's Tau-a         0.44
Total           24    100.0
```

Chapter 7

Exercises 7.2

1. The closed form solution is given followed by the numerical list and its graph.

$$-\frac{1}{10}\left(\frac{1}{2}\right)^{n}+\frac{1}{5}$$

0.1, 0.15, 0.175, 0.1875, 0.19375, 0.196875, 0.1984375,
0.19921875, 0.199609375, 0.1998046875, 0.1999023438,
0.1999511719, 0.1999755860, 0.1999877930, 0.1999938965,
0.1999969482, 0.1999984741, 0.1999992370, 0.1999996185,
0.1999998092, 0.1999999046, 0.1999999523, 0.1999999762,
0.1999999881, 0.1999999940, 0.1999999970, 0.1999999985,
0.1999999992, 0.1999999996, 0.1999999998, 0.1999999999,
0.2000000000, 0.2000000000, 0.2000000000, 0.2000000000,
0.2000000000, 0.2000000000, 0.2000000000, 0.2000000000,
0.2000000000, 0.2000000000, 0.2000000000, 0.2000000000,
0.2000000000, 0.2000000000, 0.2000000000, 0.2000000000,
0.2000000000, 0.2000000000, 0.2000000000

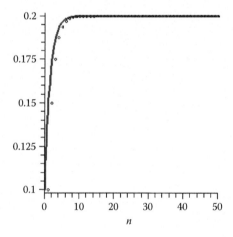

2. The closed form solution is given followed by the numerical list and graph.

$$\frac{1}{5}$$

0.2, 0.20, 0.200, 0.2000, 0.20000, 0.200000, 0.2000000,
0.20000000, 0.200000000, 0.2000000000, 0.2000000000,
0.2000000000, 0.2000000000, 0.2000000000, 0.2000000000,
0.2000000000, 0.2000000000, 0.2000000000, 0.2000000000,
0.2000000000, 0.2000000000, 0.2000000000, 0.2000000000,
0.2000000000, 0.2000000000, 0.2000000000, 0.2000000000,
0.2000000000, 0.2000000000, 0.2000000000, 0.2000000000,
0.2000000000, 0.2000000000, 0.2000000000, 0.2000000000,
0.2000000000, 0.2000000000, 0.2000000000, 0.2000000000,
0.2000000000, 0.2000000000, 0.2000000000, 0.2000000000,
0.2000000000, 0.2000000000, 0.2000000000, 0.2000000000,
0.2000000000, 0.2000000000, 0.2000000000

```
0        10       20       30       40       50
                          n
```

3. The closed form solution is given followed by the numerical list and graph.

$$\frac{1}{10}\left(\frac{1}{2}\right)^{n}+\frac{1}{5}$$

0.3, 0.25, 0.225, 0.2125, 0.20625, 0.2031255, 0.2015625,
0.20078125, 0.200390625, 0.2001953125, 0.2000976562,
0.2000488281, 0.2000244140, 0.2000122070, 0.2000061035,
0.2000030518, 0.2000015259, 0.2000007630, 0.2000003815,
0.2000001908, 0.2000000954, 0.2000000477, 0.2000000238,
0.2000000119, 0.2000000060, 0.2000000030, 0.2000000015,
0.2000000008, 0.2000000004, 0.2000000002, 0.2000000001,
0.2000000000, 0.2000000000, 0.2000000000, 0.2000000000,
0.2000000000, 0.2000000000, 0.2000000000, 0.2000000000,
0.2000000000, 0.2000000000, 0.2000000000, 0.2000000000,
0.2000000000, 0.2000000000, 0.2000000000, 0.2000000000,
0.2000000000, 0.2000000000, 0.2000000000

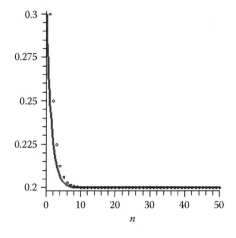

4. The closed form solution is given followed by the numerical list and graph.

$$-10000 \left(\frac{101}{100} \right)^n + 100000$$

90000, 89900.00, 89799.0000, 89696.99000, 89593.95990,
89489.89950, 89384.79850, 89278.64648, 89171.43294,
89063.14727, 88953.77874, 88843.31653, 88731.74970,
88619.06720, 88505.25787, 88390.31045, 88274.21355,
88156.95569, 88038.52525, 87918.91050, 87798.09960,
87676.08060, 87552.84141, 87428.36982, 87302.65352,
87175.68006, 87047.43686, 86917.91123, 86787.09034,
86654.96124, 86521.51085, 86386.72596, 86250.59322,
86113.09915, 85974.23014, 85833.97244, 85692.31216,
85549.23528, 85404.72763, 85258.77491, 85111.36266,
84962.47629, 84812.10105, 84660.22206, 84506.82428,
84351.89252, 84195.41145, 84037.36556, 83877.73922,
83716.51661

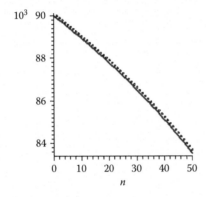

5. The closed form solution is given followed by the numerical list and graph.

100000

100000, $1.0000000 \ 10^5$, $1.000000000 \ 10^5$, $1.000000000 \ 10^5$,
$1.000000000 \ 10^5$, $1.000000000 \ 10^5$, $1.000000000 \ 10^5$,
$1.000000000 \ 10^5$, $1.000000000 \ 10^5$, $1.000000000 \ 10^5$,
$1.000000000 \ 10^5$, $1.000000000 \ 10^5$, $1.000000000 \ 10^5$,
$1.000000000 \ 10^5$, $1.000000000 \ 10^5$, $1.000000000 \ 10^5$,
$1.000000000 \ 10^5$, $1.000000000 \ 10^5$, $1.000000000 \ 10^5$,
$1.000000000 \ 10^5$, $1.000000000 \ 10^5$, $1.000000000 \ 10^5$,
$1.000000000 \ 10^5$, $1.000000000 \ 10^5$, $1.000000000 \ 10^5$,
$1.000000000 \ 10^5$, $1.000000000 \ 10^5$, $1.000000000 \ 10^5$,
$1.000000000 \ 10^5$, $1.000000000 \ 10^5$, $1.000000000 \ 10^5$,
$1.000000000 \ 10^5$, $1.000000000 \ 10^5$, $1.000000000 \ 10^5$,
$1.000000000 \ 10^5$, $1.000000000 \ 10^5$, $1.000000000 \ 10^5$,
$1.000000000 \ 10^5$, $1.000000000 \ 10^5$, $1.000000000 \ 10^5$,
$1.000000000 \ 10^5$, $1.000000000 \ 10^5$, $1.000000000 \ 10^5$,
$1.000000000 \ 10^5$, $1.000000000 \ 10^5$, $1.000000000 \ 10^5$,
$1.000000000 \ 10^5$, $1.000000000 \ 10^5$, $1.000000000 \ 10^5$,
$1.000000000 \ 10^5$

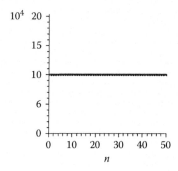

6. The closed form solution is given followed by the numerical list and graph.

$$10000 \left(\frac{101}{100} \right)^n + 100000$$

110000, $1.1010000 \ 10^5$, $1.102010000 \ 10^5$, $1.103030100 \ 10^5$,
$1.104060401 \ 10^5$, $1.105101005 \ 10^5$, $1.106152015 \ 10^5$,
$1.107213535 \ 10^5$, $1.1008285670 \ 10^5$, $1.109368527 \ 10^5$,
$1.110462212 \ 10^5$, $1.111566834 \ 10^5$, $1.112682502 \ 10^5$,
$1.113809327 \ 10^5$, $1.114947420 \ 10^5$, $1.116096894 \ 10^5$,
$1.117257863 \ 10^5$, $1.118430442 \ 10^5$, $1.119614746 \ 10^5$,
$1.120810893 \ 10^5$, $1.122019002 \ 10^5$, $1.123239192 \ 10^5$,
$1.124471584 \ 10^5$, $1.125716300 \ 10^5$, $1.126973463 \ 10^5$,
$1.128243198 \ 10^5$, $1.129525630 \ 10^5$, $1.130820886 \ 10^5$,
$1.132129095 \ 10^5$, $1.133450386 \ 10^5$, $1.134784890 \ 10^5$,
$1.136132739 \ 10^5$, $1.1374494066 \ 10^5$, $1.138869007 \ 10^5$,
$1.140257697 \ 10^5$, $1.141660274 \ 10^5$, $1.143076877 \ 10^5$,
$1.144507646 \ 10^5$, $1.145952722 \ 10^5$, $1.147412249 \ 10^5$,
$1.148886371 \ 10^5$, $1.150375235 \ 10^5$, $1.151878987 \ 10^5$,
$1.153397777 \ 10^5$, $1.154931755 \ 10^5$, $1.156481073 \ 10^5$,
$1.158045884 \ 10^5$, $1.159626343 \ 10^5$, $1.161222606 \ 10^5$,
$1.162834832 \ 10^5$

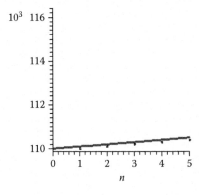

7. The closed form solution is given followed by the numerical list and graph.

$$\frac{7}{23}\left(-\frac{13}{10}\right)^{n} + \frac{200}{23}$$

9, 8.3, 9.21, 8.027, 9.5649, 7.56563, 10.164681, 6.7859147, 11.17831089, 5.46819584, 12.89134541, 3.24125097, 15.78637374, −.52228586, 20.67897162, −6.88266311, 28.94746204, −17.63170065, 42.92121084, −35.79757409, 66.53684632, −66.49790022, 106.4472703, −118.3814514, 173.8958868, −206.0646528, 287.8840486, −354.2492632, 480.5240422, −604.5812549, 806.0856314, −1027.911321, 1356.284717, −1743.170132, 2286.121172, −2951.957521, 3857.544781, −4994.808215, 6513.250680, −8447.225884, 11001.39365, −14281.81174, 18586.35526, −24142.26184, 31404.94039, −40806.42251, 53068.34926, −68968.85404,, 89679.51025, −1.165633633 10^{5}

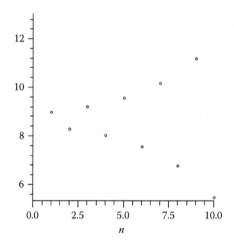

8. The closed form solution is given followed by the numerical list and graph.

$$-(-4)^n + 10$$

9, 14, −6, 74, −246, 1034, −4086, 16394, −65526, 262154,
−1048566, 4194314, −16777206, 67108874, −268435446,
1073741834, −4294967286, 17179869194, −68719476726,
274877906954, −1099511627766, 4398046511114,
−17592186044406, 70368744177674, −281474976710646,
1125899906842634, −4503599627370486,
18014398509481994, −72057594037927926,
288230376151711754, −1152921504606846966,
4611686018427387914, −18446744073709551606,
73786976294838206474, −295147905179352825846,
1180591620717411303434, −4722366482869645213686,
18889465931478580824794, −75557863725914323419126,
302231454903657293676554,
−1208925819614629174706166,
4835703278458516698824714,
−19342813113834066795298806,
77371252455336267181195274,
−309485009821345068724781046,
1237940039285380274899124234,
−4951760157141521099596496886,
19807040628566084398385987594,
−79228162514264997593543950326,
316912650057057350374175801354

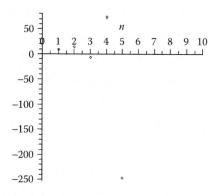

9. The closed form solution is given followed by the numerical list and graph.

$$-\frac{9919}{9}\left(\frac{1}{10}\right)^{n} + \frac{10000}{9}$$

9, 1000.9, 1100.09, 1110.009, 1111.0009, 1111.10009,
   1111.110009, 1111.111001, 1111.111100, 1111.111110,
   1111.111111, 1111.111111, 1111.111111, 1111.111111,
   1111.111111, 1111.111111, 1111.111111, 1111.111111,
   1111.111111, 1111.111111, 1111.111111, 1111.111111,
   1111.111111, 1111.111111, 1111.111111, 1111.111111,
   1111.111111, 1111.111111, 1111.111111, 1111.111111,
   1111.111111, 1111.111111, 1111.111111, 1111.111111,
   1111.111111, 1111.111111, 1111.111111, 1111.111111,
   1111.111111, 1111.111111, 1111.111111, 1111.111111,
   1111.111111, 1111.111111, 1111.111111, 1111.111111,
   1111.111111, 1111.111111, 1111.111111, 1111.111111,

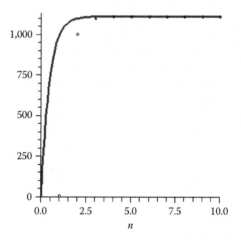

10. The closed form solution is given followed by the numerical list and graph.

$$p1 := -\frac{2414}{13}\left(\frac{9987}{10000}\right)^{n} + \frac{3350}{13}$$

prescribed := 72, 72.2414, 72.48248618, 72.72325895,
72.96371871, 73.20386588, 73.44370085, 73.68322404,
73.92243585, 74.16133668, 74.39992694, 74.63820703,
74.87617736, 75.11383833, 75.35119034, 75.58823379,
75.82496909, 76.06139663, 76.29751681, 76.53333004,
76.76883671, 77.00403722, 77.23893197, 77.47352136,
77.70780578, 77.94178563, 78.17546131, 78.40883321,
78.64190173, 78.87466726, 79.10713019, 79.33929092,
79.57114984, 79.80270735, 80.03396383, 80.26491968,
80.49557528, 80.72593103, 80.95598732, 81.18574454,
81.41520307, 81.64436331, 81.87322564, 81.18574454,
82.33005812, 82.55802904, 82.78570360, 83.01308219,
83.24016518, 83.46695297

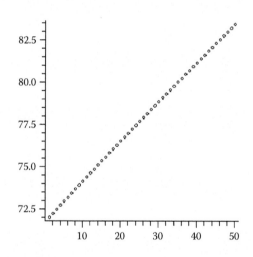

Exercises 7.3

1. a. 0, unstable
   b. 0, stable
   c. 0, stable
   d. No equilibrium value
   e. 84, stable
   f. 500, stable
   g. 500, stable
   h. 55.55, stable

2. a.   Not stable
   b.   Stable
   c.   Stable
   d.   Not stable

Exercises 7.4

1. *prescribed* := 0.2, 0.32, 0.4350, 0.49160192, 0.4998589446,
   0.4999999602, 0.5000000000, 0.5000000000, 0.5000000000,
   0.5000000000, 0.5000000000, 0.5000000000, 0.5000000000,
   0.5000000000, 0.5000000000, 0.5000000000, 0.5000000000,
   0.5000000000, 0.5000000000, 0.5000000000

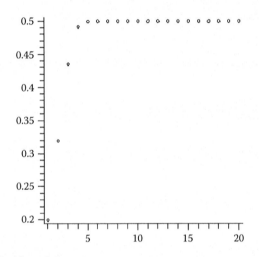

Stable at 0.5

2. *prescribed* := 0.2, 0.48, 0.7488, 0.56429568, 0.7375981965,
 0.5806412910, 0.7304909466, 0.5906217705, 0.7253630841,
 0.5976344409, 0.7214025480, 0.6029427351, 0.7182083799,
 0.6071553087, 0.7155532194, 0.6106104288, 0.7132959990,
 0.6135144504, 0.7113434088, 0.6160018908

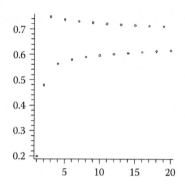

3. *prescribed* := 0.2, 0.576, 0.8792064, 0.3823290223,
 0.8501527475, 0.4586149923, 0.8938342119, 0.3416206088,
 0.8096974866, 0.5547148805, 0.8892226148, 0.3546207220,
 0.8239135158, 0.5222881234, 0.8982116623, 0.3291388992,
 0.7949033432, 0.5869152655, 0.8728046519, 0.3996600895

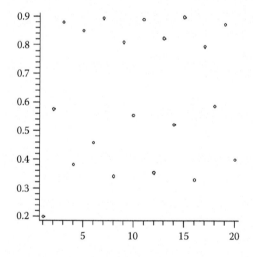

4. *prescribed* := 0.2, 0.592, 0.8936832, 0.3515500907,
    0.8434617107, 0.4885259972, 0.9245128850, 0.2582185987,
    0.7087044898, 0.7638370129, 0.6674431134, 0.8212623741,
    0.5431248018, 0.9181189306, 0.2781532715, 0.7429009078,
    0.7066968513, 0.7669227232, 0.6613833614, 0.8286350194

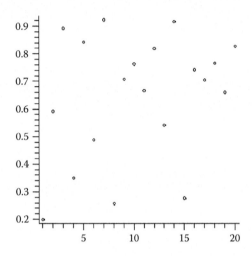

This is an example of chaos.

5. Equilibrium values are 0, 5 and stable at 5.

6. Equilibrium values are 2, 0 and stable at 0.

7. Real equilibrium value at 0 and unstable.

8. Equilibrium values at −1, 2 and −1 is stable.

9. About 15 days.

Exercises 7.5

1. The system goes to about 107 and 142, respectively.

2. (0,0) and (1500,2000). Only (0,0) is achievable if we start at (0,0).

Chapter 8
Exercises 8.2

1. Possible random numbers are

> $random := \text{proc}(n)$ option *remember* for i from 1 to n do $x(i) :=$
> $evalf\left(\dfrac{rand()}{999999999999}\right)$ end do end;

> $seq(x(i), i = 1\ ..20);$

0.2240171515, 0.2008401063, 0.8685719066, 0.5704134665,
0.9920881460, 0.4437752746, 0.4780291372,
0.01294305199, 0.6408831565, 0.2487102377,
0.9700117830, 0.04705144220, 0.5745759906,
0.1717448313, 0.6488303816, 0.4114303688,
0.09704265386, 0.7719464925, 0.2477991384, 0.8950382012

2. Random numbers between −10 and 10 are as follows:
for i **from** 1 **to** 20 **do** $y(i) := evalf(-10 + (20) \cdot x(i))$ **end do;**

$$y(1) := -5.519656970$$

$$y(2) := -5.983197874$$

$$y(3) := 7.37143813$$

$$y(4) := 1.40826933$$

$$y(5) := 9.84176292$$

$$y(6) := -9.112449451$$

$$y(7) := -.439417256$$

$$y(8) := -9.741138960$$

$$y(9) := 2.81766313$$

$$y(10) := -5.025795246$$

$$y(11) := 9.40023566$$

$$y(12) := -9.058971156$$

$$y(13) := 1.49151981$$

$$y(14) := -6.565103374$$

$$y(15) := 2.97660763$$

$$y(16) := -1.771392624$$

$$y(17) := -8.059146923$$

$$y(18) := 5.43892985$$

$$y(19) := -5.044017232$$

$$y(20) := 7.90076402$$

3. Random exponential variables are

> for *i* from 1 to 20 do *expX* (*i*) := −0.5·ln(*x*(*i*)); end do;

expX(1) := 0.7480163305

expX(2) := 0.8026230895

expX(3) := 0.07045245135

expX(4) := 0.2806969006

expX(5) := 0.003971659395

expX(6) := 1.557511038

expX(7) := 0.3690417960

expX(8) := 2.173598080

expX(9) := 0.2224540609

expX(10) := 0.6957333820

expX(11) := 0.01522353007

expX(12) := 1.528256880

expX(13) := 0.2770614590

expX(14) := 0.8808727215

expX(15) := 0.2162919750

expX(16) := 0.4440577432

expX(17) := 1.166302334

expX(18) := 0.1294200208

expX(19) := 0.6975683935

expX(20) := 0.05544443935

4. Normal 0,1 variables are

> seq(X1(i),i=1..20);

1.521070484, −.5768899232, −1.042158196, −.1882228151,
0.8014649537, 1.578949306, −.9882787283, −.9383631568,
0.7018566620, 1.956211964, −.9212502179, 0.1331107801,
−1.379775780, 0.03929782162, −1.090981430,
0.1086640465, 1.269057050, 0.02160515710, −.6052083964,
−1.145969309

> seq(X2(i),i=1..20);

−.4092496291, 0.09498239192, 0.4725129005, −.8563409978,
−.7432717739, 1.365049016, −.4002408685, −2.159175315,
0.8682968698, −1.378212065, 0.4190439735, 2.668672851,
1.175383344, −0.09933300662, −.4249319179,
0.7566029424, −.4004936206, 0.1509887558,
−0.02918414136, −1.073491864

5. Normal 5,0.5 found by using the formula $z = (x - \mu)/\sigma$

seq(.5*x_1 (*i*) +5, *i*=1 .. 20);

    5.760535242, 4.711555038, 4.478920902, 4.905888542,
    5.400732477, 5.789474653, 4.505860636, 4.530818422,
    5.350928331, 5.978105982, 4.539374891, 5.066555390,
    4.310112110, 5.019648911, 4.454509285, 5.054332023,
    5.634528525, 5.010802576, 4.697395802, 4.427015346

Exercises 8.3

1. The approximate area with n = 2000 is 3.12116. The exact area is 3.14.

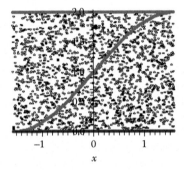

2. Area is approximately with n = 2000 trials is 1.708 sq units. The exact area is 1.89583333 sq units.

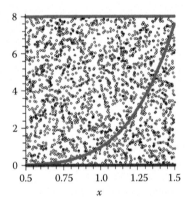

3. Approximate area with $n = 2000$ is 0.7885.

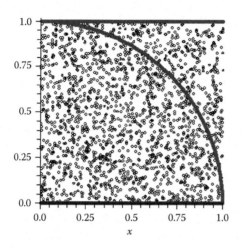

4. The real area is 0.7853981634.
5. This is a quarter region in problem 3, so multiply by 4. Approximation for (is 4*0.7853981634 = 3.14159.
6. Approximate volume is 0.734 cubic units.

Exercises 8.4

1. HHTHTTTHTHHTHTHHHHTT, $P(H) = 0.55$, $P(T) = 0.45$
2. Rolls of a die: 3,5,6,2,3,1,2,5,4,3,2,1,6,6,5,4,2,3,1,4

Exercises 8.5

2. $n = 6, p = 0.9871120530$

3. $>$ *bombsaway* $(1, 10, .45, .4)$;

0.002720977920

$>$ for i from 1 to 10 do *bombsaway* $(i, 10, .45, .3)$ end do;

0.01271138620

0.1041425713

0.3682693412

0.6469328405

0.8483092369

0.9518961010

0.9889778713

0.9981730859

0.9997262239

0.9999225455

$>$

We will have to increase the number of F-15 to 8.

4. $>$ inventorygas $(11450, 7, 500, 0.05, 20, xdat, qdat)$;

5697.695140

$1.979095492 \ 10^5, 5697.695140$

Chapter 9

Exercises 9.1

1. (a) $3710 (b) $4130 (c) $3552.50 (d) $3571.72
2. (a) $550 (b) $575 (c) $625
3. $2500/(1 + 0.065 * 2) = $2212.39
4. (a) $5857.5 or total interest of $357.50 interest (b) $5857.5 or total interest of $357.50 interest (c) $5857.5 or total interest of $357.50 interest (d) $5857.5 or total interest of $357.50 interest
7. (a) $5500*e^{(0.05)} = $5781.99 (c) $5500*e^{(0.05/12)(6)} = $5639.23

Exercises 9.2

1. (a) 5.116% (b) 5.09% (c) 5.06%
3. 6.5% per year compounded monthly (APR 6.69% < 6.86%)

9. $P = S(1 - dt) = 6450(1 - 0.07 * 4) = \$4644.00$, $APR = 9.722\%$

12. $8437.50

Exercises 9.3

	Flow	Discount rate	NPV	
1	−2000	1.08	−1851.85	
2	1000	1.1664	857.3388	
3	700	1.259712	555.6826	
4	550	1.36048896	404.2664	
5	400	1.469328077	272.2333	
6	400	1.586874323	252.0679	
7	350	1.713824269	204.2216	
8	250	1.85093021	135.0672	
			829.0259	Total

1. NPV = 829.02 and since NPV > 0, it is worthwhile.
2. NPV = 729.22 and since NPV > 0, it is worthwhile.
3. NPV = 276.54 and since NPV > 0, it is worthwhile.

Exercises 9.4

1. (a) 16(1000) = 160 (b) NPV = $1297.00 (c) $1469.32

4. $A_0 = 5000$, $r = 0.06$ so $i = 0.06/2 = 0.03$, $n = (10)(2) = 20$ periods in 10 years

$$V_{20} = 5000\left[\frac{(1+0.03)^{20} - 1}{0.03}\right] = \$134351.87$$

Exercises 9.5

1. (a) $351.05 (b) $183.61 (c) $99.84.

2. $899.32 per month and $323,757.30 total paid.

Index

Note: Page numbers followed by f refer to figures respectively.